Solution of
Superlarge Problems
in Computational Mechanics

Solution of
Superlarge Problems
in Computational Mechanics

Edited by

James H. Kane

*Clarkson University
Potsdam, New York*

Arthur D. Carlson

*Naval Underwater Systems Center
New London, Connecticut*

Donald L. Cox

*Naval Underwater Systems Center
New London, Connecticut*

PLENUM PRESS • NEW YORK AND LONDON

Library of Congress Cataloging in Publication Data

Solution of superlarge problems in computational mechanics / edited by James H. Kane,
Arthur D. Carlson, and Donald L. Cox.
    p.   cm.
"Proceedings of a Symposium on Solution of Superlarge Problems in Computational
Mechanics, held October 18–19, 1988, in Mystic, Connecticut."
Includes bibliographical references.

ISBN 978-1-4612-7854-2   e-ISBN-13: 978-1-4613-0535-4
DOI: 10.1007/978-1-4613-0535-4

1. Engineering mathematics—Congresses. 2. Mechanics, Applied—Congresses. I. Kane,
James H. II. Carlson, Arthur D. III. Cox, Donald L. IV. Symposium on Solution of
Superlarge Problems in Computational Mechanics (1988: Mystic, Conn.)
TA329.S65 1989   89-22932
620.1—dc20   CIP

Proceedings of a Symposium on Solution of Superlarge Problems
in Computational Mechanics, held October 18–19, 1988,
in Mystic, Connecticut

© 1989 Plenum Press, New York
Softcover reprint of the hardcover 1st edititon 1989

A Division of Plenum Publishing Corporation
233 Spring Street, New York, N.Y. 10013

PREFACE

There is a need to solve problems in solid and fluid mechanics that currently exceed the resources of current and foreseeable supercomputers. The issue revolves around the number of degrees of freedom of simultaneous equations that one needs to accurately describe the problem, and the computer storage and speed limitations which prohibit such solutions. The goals of this symposium were to explore some of the latest work being done in both industry and academia to solve such extremely large problems, and to provide a forum for the discussion and prognostication of necessary future directions of both man and machine. As evidenced in this proceedings we believe these goals were met.

Contained in this volume are discussions of: iterative solvers, and their application to a variety of problems, e.g. structures, fluid dynamics, and structural acoustics; iterative dynamic substructuring and its use in structural acoustics; the use of the boundary element method both alone and in conjunction with the finite element method; the application of finite difference methods to problems of incompressible, turbulent flow; and algorithms amenable to concurrent computations and their applications. Furthermore, discussions of existing computational shortcomings from the big picture point of view are presented that include recommendations for future work.

Although it is difficult to draw conclusions from the results of two days discussions focused on contemporary work from various areas with differing motivations, one conclusion is clear; the most effective solution of super large problems will require the interdisciplinary involvement of the mathematician, engineer, and computer scientist. This symposium offered such a forum, and for those involved provided a genuine learning experience. It is the hope of the editors that this volume will extend the experience.

In order to put on a symposium requires a great deal of work and cooperation. To this end the Naval Underwater System Center, Worcester Polytechnic Institute and the Office of Naval Research were very successful. In addition to all of our contributors, we extend our special thanks to Spiros Laquidos of ONR and Richard Gallagher now President of Clarkson University who supported the symposium while provost at WPI.

J.H. Kane
A.D. Carlson
D.L. Cox

CONTENTS

ELEMENT-BY-ELEMENT ALGORITHMS FOR
NONSYMMETRIC MATRIX PROBLEMS ARISING IN FLUIDS

Farzin Shakib[1], Thomas J. R. Hughes[2], and Zdeněk Johan[1]

Division of Applied Mechanics
Stanford University
Stanford, California

Abstract

A multi-element group, domain decomposition algorithm is presented for solving linear nonsymmetric systems arising in finite element analysis. The iterative strategy employed is based on the generalized minimum residual (GMRES) procedure originally proposed by Saad and Shultz. Two levels of preconditioning are investigated. Applications to problems of high-speed compressible flow illustrate the effectiveness of the scheme.

1. Introduction

Space-time finite element discretization of the compressible Euler and Navier-Stokes equations leads to a nonlinear time-marching problem, see Shakib [1988]. Linearization of this problem at a given time step results in a linear system of equations,

$$A x = b \tag{1}$$

where $x^T = \{x_1^T, x_2^T, \cdots, x_{n_{np}}^T\}$ is the vector of unknown nodal increments, in which n_{np} is the number of nodes and n_{dof} is the number of degrees of freedom at each node; A is the $n_{eq} \times n_{eq}$ nonsymmetric left-hand-side effective "mass" matrix, where $n_{eq} = n_{np} \times n_{dof}$; and b is the n_{eq}-dimensional right-hand-side out-of-balance "force", or residual, vector. Both A and b are constructed from the assembly of element contributions

$$A = \overset{n_{el}}{\underset{e=1}{\mathbf{A}}} A^e \tag{2}$$

$$b = \overset{n_{el}}{\underset{e=1}{\mathbf{A}}} b^e \tag{3}$$

---

[1] Graduate Research Assistant.
[2] Professor and Chairman of Mechanical Engineering.

1

where A^e and b^e are the element matrices, $\mathbf{A}$ is the assembly operator, and n_{el} is the number of elements. See Hughes [1987] for background.

Direct methods based on Gaussian elimination can be used to solve (1). However, these methods quickly become too memory and CPU intensive to be of practical interest, especially in three-dimensions. Consequently, the finite element community has turned towards iterative solvers based on the generation of Krylov spaces. Most iterative algorithms studied in detail so far are applicable to symmetric systems emanating from finite element discretizations of structures and solids. For example, Hughes, Ferencz and Hallquist [1987] studied and implemented the preconditioned conjugate gradients algorithm on a Cray X-MP, and Nour-Omid, Parlett and Raefsky [1988] studied the performance of different preconditioners for conjugate gradients and the Lanczos algorithm. These methods are based on minimization of energy norms, which are ill-defined for general nonsymmetric systems. Hence, extension of these methods to nonsymmetric systems may not result in robust algorithms. Saad and Schultz [1983] have proposed the Generalized Minimal Residual (GMRES) algorithm, which minimizes the (well-defined) residual norm. Mallet, Periaux and Stoufflet [1987] have employed the GMRES procedure with success to solve compressible flow problems. We also adopt GMRES herein and implement it in the context of the finite element method.

To accelerate convergence of the GMRES algorithm, we have developed a two level preconditioning strategy. In the first level, referred to as a "pre-preconditioning," the system is transformed by way of a diagonal, or nodal block-diagonal, scaling matrix. Within the GMRES iteration phase an element-by-element preconditioning is accommodated. The algorithm allows for a domain decomposition into explicit element groups and implicit element groups with either direct or iterative solvers.

An outline of the remainder of the paper follows: In Section 2 we describe the multi-element group partitioning into subdomains. The structure of the nonsymmetric linear system, which we eventually solve by iteration, is delineated. The pre-preconditioning techniques are described in Section 3 and the preconditioned GMRES algorithm is presented in Section 4. The numerical results of Section 5 speak to the efficiency of the proposed procedures. Particularly striking is the effectiveness of nodal block-diagonal pre-preconditioning. Conclusions are drawn in Section 6.

2. Multi-element Group Partitioning

2.1 General structure of the equation system

To simplify the presentation, we consider a problem consisting of three element groups: an explicit group; an implicit group with direct solver; and an implicit group with iterative solver (see Figure 1). Our computer program accommodates an arbitrary number of element groups of each type. It is important to distinguish between two sets of nodes: *group-interior nodes* and *group-boundary nodes*. Group-boundary nodes are defined to be those nodes that are shared by more than one element group. All remaining nodes are group-interior nodes. Arrays associated with group-interior nodes and group-boundary nodes are denoted by subscripts 1 and 2, respectively.

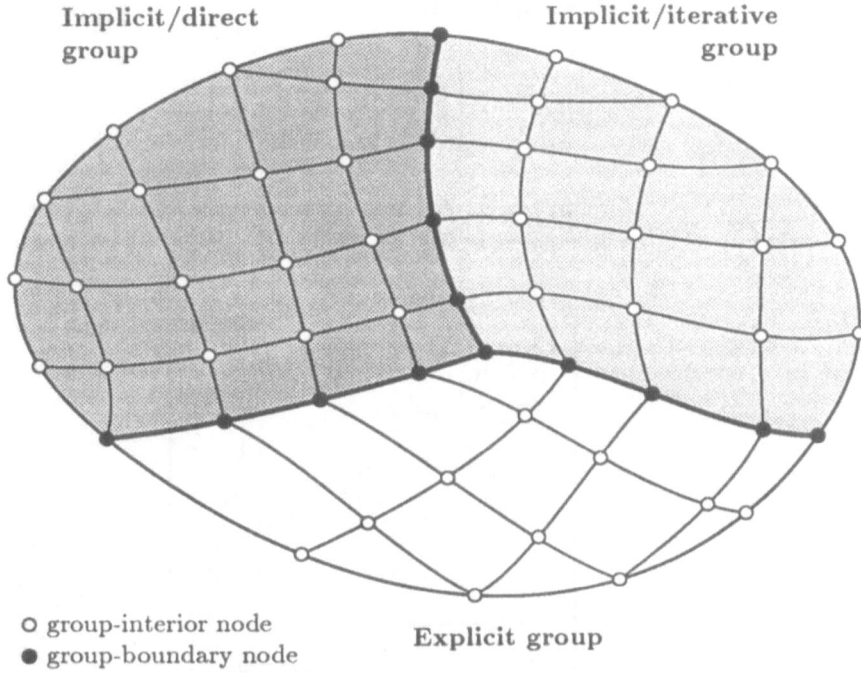

Implicit/direct group

Implicit/iterative group

○ group-interior node
● group-boundary node

Explicit group

Figure 1. Group partitioning.

The left-hand-side matrix consists of contributions from each element group. We shall examine in turn the submatrices associated with each group. For the explicit group we have

$$A^{\mathrm{exp}} = \begin{bmatrix} A_{11}^{\mathrm{exp}} & 0 \\ 0 & A_{22}^{\mathrm{exp}} \end{bmatrix} \tag{4}$$

where A^{exp} is a block-diagonal matrix consisting of $m \times m$ symmetric blocks, m being the number of degrees of freedom per node ($m = 4$ for the compressible Navier-Stokes equations in two dimensions, and $m = 5$ in three dimensions). See Figure 2.

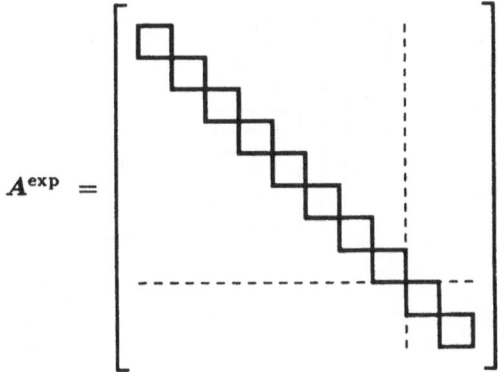

$$A^{\mathrm{exp}} =$$

Figure 2. Explicit element group matrix structure.

For the implicit/direct group we have

$$A^{\text{dir}} = \begin{bmatrix} A^{\text{dir}}_{11} & A^{\text{dir}}_{12} \\ A^{\text{dir}}_{21} & A^{\text{dir}}_{22} \end{bmatrix} \tag{5}$$

where A^{dir} is an assembled nonsymmetric matrix possessing a symmetric profile. It is stored in skyline form as illustrated in Figure 3.

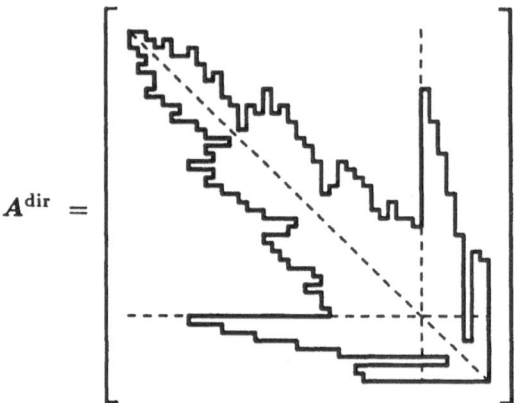

Figure 3. Implicit/direct element group matrix structure.

For the implicit/iterative group we have

$$A^{\text{iter}} = \begin{bmatrix} A^{\text{iter}}_{11} & A^{\text{iter}}_{12} \\ A^{\text{iter}}_{21} & A^{\text{iter}}_{22} \end{bmatrix} \tag{6}$$

where A^{iter} is a nonsymmetric matrix possessing a symmetric profile. However, this is only a formal representation as A^{iter} is never actually assembled, but rather stored in the form of an unassembled element file (see Figure 4).

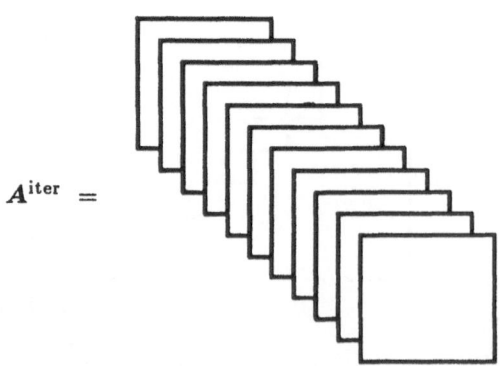

Figure 4. Implicit/iterative element group matrix structure.

With these notations, the system matrix equation can be represented as

$$
\begin{bmatrix}
A_{11}^{\mathrm{exp}} & 0 & 0 & 0 \\
0 & A_{11}^{\mathrm{dir}} & 0 & A_{12}^{\mathrm{dir}} \\
0 & 0 & A_{11}^{\mathrm{iter}} & A_{12}^{\mathrm{iter}} \\
0 & A_{21}^{\mathrm{dir}} & A_{21}^{\mathrm{iter}} & (A_{22}^{\mathrm{exp}} + A_{22}^{\mathrm{dir}} + A_{22}^{\mathrm{iter}})
\end{bmatrix}
\begin{Bmatrix}
\boldsymbol{x}^{\mathrm{exp}} \\
\boldsymbol{x}^{\mathrm{dir}} \\
\boldsymbol{x}^{\mathrm{iter}} \\
\boldsymbol{x}^{\mathrm{bndy}}
\end{Bmatrix}
$$

$$
=
\begin{Bmatrix}
\boldsymbol{b}_1^{\mathrm{exp}} \\
\boldsymbol{b}_1^{\mathrm{dir}} \\
\boldsymbol{b}_1^{\mathrm{iter}} \\
(\boldsymbol{b}_2^{\mathrm{exp}} + \boldsymbol{b}_2^{\mathrm{dir}} + \boldsymbol{b}_2^{\mathrm{iter}})
\end{Bmatrix}
\tag{7}
$$

where $\boldsymbol{x}^{\mathrm{bndy}}$ is the solution subvector for the group-boundary nodes. The right-hand-side vectors $\boldsymbol{b}_1^{\mathrm{exp}}$, $\boldsymbol{b}_1^{\mathrm{dir}}$ and $\boldsymbol{b}_1^{\mathrm{iter}}$ are stored with their respective element group data, whereas $(\boldsymbol{b}_2^{\mathrm{exp}} + \boldsymbol{b}_2^{\mathrm{dir}} + \boldsymbol{b}_2^{\mathrm{iter}})$ is stored with the global data. Prescribed degrees of freedom are retained as active degrees of freedom in order to simplify the data structure so as to facilitate optimization and vectorization.

2.2 Partial reduction of the equation system

Direct solution techniques are used to eliminate $\boldsymbol{x}^{\mathrm{exp}}$ and $\boldsymbol{x}^{\mathrm{dir}}$. The solution of

$$
A_{11}^{\mathrm{exp}} \, \boldsymbol{x}^{\mathrm{exp}} = \boldsymbol{b}_1^{\mathrm{exp}}
\tag{8}
$$

involves the triangularization of uncoupled, $m \times m$, symmetric, positive-definite matrices. A static condensation is used to eliminate $\boldsymbol{x}^{\mathrm{dir}}$. The reduced system has the form

$$
\begin{bmatrix}
A_{11}^{\mathrm{iter}} & A_{12}^{\mathrm{iter}} \\
A_{21}^{\mathrm{iter}} & (A_{22}^{\mathrm{exp}} + \widehat{A}_{22}^{\mathrm{dir}} + A_{22}^{\mathrm{iter}})
\end{bmatrix}
\begin{Bmatrix}
\boldsymbol{x}^{\mathrm{iter}} \\
\boldsymbol{x}^{\mathrm{bndy}}
\end{Bmatrix}
=
\begin{Bmatrix}
\boldsymbol{b}_1^{\mathrm{iter}} \\
(\boldsymbol{b}_2^{\mathrm{exp}} + \widehat{\boldsymbol{b}}_2^{\mathrm{dir}} + \boldsymbol{b}_2^{\mathrm{iter}})
\end{Bmatrix}
\tag{9}
$$

where

$$
\widehat{A}_{22}^{\mathrm{dir}} \overset{\mathrm{def}}{=} A_{22}^{\mathrm{dir}} - A_{21}^{\mathrm{dir}}(A_{11}^{\mathrm{dir}})^{-1} A_{12}^{\mathrm{dir}}
\tag{10}
$$

and

$$
\widehat{\boldsymbol{b}}_2^{\mathrm{dir}} \overset{\mathrm{def}}{=} \boldsymbol{b}_2^{\mathrm{dir}} - A_{21}^{\mathrm{dir}}(A_{11}^{\mathrm{dir}})^{-1} \boldsymbol{b}_1^{\mathrm{dir}}
\tag{11}
$$

Once (9) is solved, $\boldsymbol{x}^{\mathrm{dir}}$ is determined from

$$
\boldsymbol{x}^{\mathrm{dir}} = (A_{11}^{\mathrm{dir}})^{-1}(\boldsymbol{b}_1^{\mathrm{dir}} - A_{12}^{\mathrm{dir}} \boldsymbol{x}^{\mathrm{bndy}})
\tag{12}
$$

For future reference we introduce the following concise notation for the reduced system:

$$
\widetilde{A}\widetilde{\boldsymbol{x}} = \widetilde{\boldsymbol{b}}
\tag{13}
$$

5

where

$$\widetilde{A} = \begin{bmatrix} A_{11}^{\text{iter}} & A_{12}^{\text{iter}} \\ A_{21}^{\text{iter}} & (A_{22}^{\text{exp}} + \widehat{A}_{22}^{\text{dir}} + A_{22}^{\text{iter}}) \end{bmatrix} \tag{14}$$

$$\widetilde{x} = \left\{ \begin{array}{c} x^{\text{iter}} \\ x^{\text{bndy}} \end{array} \right\} \tag{15}$$

$$\widetilde{b} = \left\{ \begin{array}{c} b_1^{\text{iter}} \\ (b_2^{\text{exp}} + \widehat{b}_2^{\text{dir}} + b_2^{\text{iter}}) \end{array} \right\} \tag{16}$$

Note that $\widetilde{A}$ is a nonsymmetric matrix.

2.3 Implementation of static condensation

The computational details are summarized in the following steps:

$$A_{11}^{\text{dir}} = L_{11}^{\text{dir}} U_{11}^{\text{dir}} \qquad \text{(L-U factorization)} \tag{17}$$

$$\widehat{A}_{12}^{\text{dir}} = (L_{11}^{\text{dir}})^{-1} A_{12}^{\text{dir}} \tag{18}$$

$$\widehat{A}_{21}^{\text{dir}} = \left((U_{11}^{\text{dir}})^{-T} (A_{21}^{\text{dir}})^T \right)^T \tag{19}$$

$$\widehat{A}_{22}^{\text{dir}} = A_{22}^{\text{dir}} - \widehat{A}_{21}^{\text{dir}} \widehat{A}_{12}^{\text{dir}} \tag{20}$$

$$\widehat{b}_1^{\text{dir}} = (L_{11}^{\text{dir}})^{-1} b_1^{\text{dir}} \tag{21}$$

$$\widehat{b}_2^{\text{dir}} = b_2^{\text{dir}} - \widehat{A}_{21}^{\text{dir}} \widehat{b}_1^{\text{dir}} \tag{22}$$

$$x^{\text{dir}} = (U_{11}^{\text{dir}})^{-1} (\widehat{b}_1^{\text{dir}} - \widehat{A}_{12}^{\text{dir}} x^{\text{bndy}}) \tag{23}$$

The factors L_{11}^{dir} and U_{11}^{dir} overwrite A_{11}^{dir}; $\widehat{A}_{12}^{\text{dir}}$, $\widehat{A}_{21}^{\text{dir}}$, $\widehat{A}_{22}^{\text{dir}}$, $\widehat{b}_1^{\text{dir}}$ and $\widehat{b}_2^{\text{dir}}$ overwrite A_{12}^{dir}, A_{21}^{dir}, A_{22}^{dir}, b_1^{dir} and b_2^{dir}, respectively.

3. Pre-preconditioning

The matrix system $\widetilde{A}\widetilde{x} = \widetilde{b}$ is not dimensionally homogeneous. To nondimensionalize the system and, at the same time, enhance the performance of the iterative solver, we introduce a scaling transformation which may be viewed as an initial level of preconditioning. We refer to this transformation as a "pre-preconditioning." (A subsequent level of preconditioning will be introduced within the iterative solver in the following section.)

Let $\widetilde{W}$ denote a symmetric, positive-definite, narrowly banded matrix. Two candidates for $\widetilde{W}$ are considered herein:

1) Diagonal pre-preconditioning:

$$\widetilde{W} \stackrel{\text{def}}{=} \text{diag}(\widetilde{A}) \tag{24}$$

The transformed system takes the form:

$$Ax = b \tag{25}$$

where

$$A = \widetilde{W}^{-\frac{1}{2}} \widetilde{A} \widetilde{W}^{-\frac{1}{2}} \tag{26}$$

$$x = \widetilde{W}^{\frac{1}{2}} \widetilde{x} \tag{27}$$

$$b = \widetilde{W}^{-\frac{1}{2}} \widetilde{b} \tag{28}$$

The transformation is well-defined as the diagonal entries of $\widetilde{A}$ are positive for the systems under consideration.

2) Nodal block-diagonal pre-preconditioning:

$$\widetilde{W} \stackrel{\text{def}}{=} \text{block}(\widetilde{A}) \tag{29}$$

The operator "block" assigns to $\widetilde{A}$ the block-diagonal matrix consisting of the $m \times m$ nodal diagonal blocks of $\widetilde{A}$ (see Figure 5). For the systems under consideration, $\widetilde{W}$ turns out to be symmetric and positive-definite. Thus $\widetilde{W}$ possesses a well-defined Cholesky factorization:

$$\widetilde{W} = \widetilde{U}^T \widetilde{U} \tag{30}$$

The transformed system, (25), is defined in this case by

$$A = \widetilde{U}^{-T} \widetilde{A} \widetilde{U}^{-1} \tag{31}$$

$$x = \widetilde{U} \widetilde{x} \tag{32}$$

$$b = \widetilde{U}^{-T} \widetilde{b} \tag{33}$$

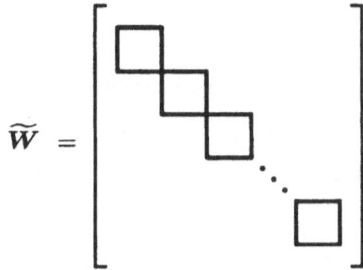

$$\widetilde{W} =$$

Figure 5. Nodal block-diagonal pre-preconditioner.

Remarks:

1. The individual partitions of $\widetilde{A}$, namely, A^{iter}, A_{22}^{exp}, and $\widehat{A}_{22}^{\text{dir}}$, are transformed separately. Both pre-preconditioners maintain the band-profile structure of the individual partitions of $\widetilde{A}$. Hence, $\widetilde{A}$ can be overwritten by A.

2. For diagonal pre-preconditioning, $\text{diag}(A) = I$. For nodal block-diagonal pre-conditioning, $\text{block}(A) = I$.

3. As observed by Ferencz [1988], for the block-diagonal pre-preconditioning, A transforms tensorialy under rotations of the global reference frame, and the iterative solver is unaffected by the particular choice of reference frame. This is not the case for the diagonal pre-preconditioner.

4. The cost of diagonal pre-preconditioning is equal to one Ax product, whereas block-diagonal pre-preconditioning is equal to m Ax products.

4. Preconditioned GMRES Algorithm

The objective is to solve the system $Ax = b$. We shall first describe the main tools employed in the construction of the preconditioned GMRES algorithm and then close this section with a detailed flowchart of the procedure employed in our computer program. An excellent general reference for the numerical algebraic procedures under study herein is Golub and Van Loan [1983].

4.1 GMRES algorithm

Let $\|\cdot\|$ and $(\cdot, \cdot)$ denote the 2-norm and inner product. Consider an approximate solution of the form $x_0 + z$, where x_0 is an initial guess and z is a member of the Krylov space $K = \text{span}\{r_0, Ar_0, A^2r_0, \ldots, A^{k-1}r_0\}$, in which $r_0 = b - Ax_0$, and k is the dimension of K. The GMRES algorithm (without preconditioning) determines z such that the 2-norm of the residual (i.e., $\|b - A(x_0 + z)\|$) is minimized. The algorithm employs an orthonormal basis of K which may be obtained by the *modified Gram-Schmidt procedure*:

$$u_1 = \frac{r_0}{\|r_0\|}$$

For $i = 1, \ldots, k$

$\quad \widetilde{u}_{i+1} = Au_i$

$\quad$ For $j = 1, \ldots, i$

$\quad\quad \beta_{i+1,j} = (\widetilde{u}_{i+1}, u_j)$

$\quad\quad \widetilde{u}_{i+1} \leftarrow \widetilde{u}_{i+1} - \beta_{i+1,j}u_j$

$\quad u_{i+1} = \dfrac{\widetilde{u}_{i+1}}{\|\widetilde{u}_{i+1}\|}$

Let $U_k = [u_1, u_2, \ldots, u_k]$. Then it can be shown that

$$AU_k = U_{k+1}H_k \tag{34}$$

where H_k is the following $(k+1) \times k$ upper Hessenberg matrix:

$$H_k = \begin{bmatrix} \beta_{2,1} & \beta_{3,1} & \cdots & \beta_{k,1} & \beta_{k+1,1} \\ \|\widetilde{u}_2\| & \beta_{3,2} & \cdots & \beta_{k,2} & \beta_{k+1,2} \\ 0 & \|\widetilde{u}_3\| & \ddots & \vdots & \vdots \\ \vdots & 0 & \ddots & \beta_{k,k-1} & \vdots \\ \vdots & \vdots & \ddots & \|\widetilde{u}_k\| & \beta_{k+1,k} \\ 0 & 0 & \cdots & 0 & \|\widetilde{u}_{k+1}\| \end{bmatrix} \tag{35}$$

Let $z = \displaystyle\sum_{j=1}^{k} y_j u_j$ and $e = \{\|r_0\|, 0, \ldots, 0\}^T$, where e has $k+1$ entries. Note that $r_0 = U_{k+1}e$. With these notations we have

$$\|b - A(x_0 + z)\| = \|r_0 - A\big(\sum_{j=1}^{k} y_j u_j\big)\|$$

$$= \|r_0 - AU_ky\|$$

$$= \|U_{k+1}(e - H_ky)\|$$

$$= \|e - H_ky\| \tag{36}$$

8

Thus the minimization problem can be written as

$$\min_{z \in K} \|b - A(x_0 + z)\| = \min_{y \in \mathcal{R}^k} \|e - H_k y\| \tag{37}$$

4.2 Q-R algorithm

The minimization problem is simplified by the "almost" triangular structure of H_k. As suggested by Saad and Schultz [1983], we employ the Q-R algorithm to obtain the minimizer. Let

$$\bar{H} = R H_k \tag{38}$$

where

$$R = R_k R_{k-1} \cdots R_1 \tag{39}$$

and each R_j, $j = 1, 2, \ldots, k$, is a $(k+1) \times (k+1)$ *Givens rotation* of the form

$$R_j \overset{\text{def}}{=} \begin{bmatrix} I_{j-1} & & \\ & \begin{bmatrix} c_j & s_j \\ -s_j & c_j \end{bmatrix} & \\ & & I_{k-j} \end{bmatrix} \tag{40}$$

in which I_i is the identity matrix of dimension i, and c_j and s_j $(c_j^2 + s_j^2 = 1)$ are coefficients. Clearly R is orthogonal. The transformation is constructed so that $\bar{H}$ is upper triangular. Let $\bar{e} = Re$. Then

$$\|e - H_k y\| = \|R^T(\bar{e} - \bar{H}y)\|$$
$$= \|\bar{e} - \bar{H}y\| \tag{41}$$

Consequently,

$$\min_{y \in \mathcal{R}^k} \|\bar{e} - \bar{H}y\| = |\bar{e}_{k+1}| \tag{42}$$

and y satisfies

$$\begin{bmatrix} \bar{H}_{1,1} & \cdots & \bar{H}_{1,k-1} & \bar{H}_{1,k} \\ 0 & \ddots & \vdots & \vdots \\ \vdots & \ddots & \bar{H}_{k-1,k-1} & \bar{H}_{k-1,k} \\ 0 & \cdots & 0 & \bar{H}_{k,k} \end{bmatrix} \begin{Bmatrix} y_1 \\ \vdots \\ y_{k-1} \\ y_k \end{Bmatrix} = \begin{Bmatrix} \bar{e}_1 \\ \vdots \\ \bar{e}_{k-1} \\ \bar{e}_k \end{Bmatrix} \tag{43}$$

which is solved by back-substitution.

The coefficients c_j and s_j are determined in the course of the reduction of the Hessenberg matrix to upper triangular form. The columns of $\bar{H}$ can be formed in sequence from the columns of H_k. The first column of H_k, denoted by $h^{(1)}$, is given by

$$h^{(1)} = \begin{Bmatrix} h_1^{(1)} \\ h_2^{(1)} \\ 0 \\ \vdots \\ 0 \end{Bmatrix} \overset{\text{def}}{=} \begin{Bmatrix} \beta_{2,1} \\ \|\widetilde{u}_2\| \\ 0 \\ \vdots \\ 0 \end{Bmatrix} \tag{44}$$

By choosing

$$r_1 = \sqrt{\left(h_1^{(1)}\right)^2 + \left(h_2^{(1)}\right)^2} \tag{45}$$

$$c_1 = \frac{h_1^{(1)}}{r_1} \tag{46}$$

$$s_1 = \frac{h_2^{(1)}}{r_1} \tag{47}$$

the first column of $\bar{H}$ becomes

$$\bar{h}^{(1)} = R_1 h^{(1)} = \left\{ \begin{array}{c} r_1 \\ 0 \\ \vdots \\ 0 \end{array} \right\} \tag{48}$$

Since all entries of $\bar{h}^{(1)}$ are zero except the first, subsequent Givens rotations have no effect upon $\bar{h}^{(1)}$, viz.

$$\bar{h}^{(1)} = R_k R_{k-1} \cdots R_2 \, \bar{h}^{(1)} \tag{49}$$

Let us assume that we have already obtained the first $j-1$ columns of $\bar{H}$. Thus we have

$$[\bar{h}^{(1)}, \bar{h}^{(2)}, \ldots, \bar{h}^{(j-1)}] = R_{j-1} R_{j-2} \cdots R_1 \, [h^{(1)}, h^{(2)}, \ldots, h^{(j-1)}]$$

$$= \begin{bmatrix} \times & \cdots & \times \\ 0 & \ddots & \vdots \\ \vdots & \ddots & \times \\ \vdots & & 0 \\ \vdots & & \vdots \\ 0 & \cdots & 0 \end{bmatrix} \tag{50}$$

The next column of $\bar{H}$ is formed by transforming $h^{(j)}$ with the first j Givens rotations:

$$\bar{h}^{(j)} = R_j R_{j-1} \cdots R_1 \, h^{(j)}$$

$$= R_j R_{j-1} \cdots R_1 \left\{ \begin{array}{c} \beta_{j+1,1} \\ \vdots \\ \beta_{j+1,j} \\ \|\tilde{u}_{j+1}\| \\ 0 \\ \vdots \\ 0 \end{array} \right\}$$

$$= R_j \left\{ \begin{array}{c} \bar{H}_{1,j} \\ \vdots \\ \bar{H}_{j-1,j} \\ \widehat{h}_1 \\ \widehat{h}_2 \\ 0 \\ \vdots \\ 0 \end{array} \right\} \tag{51}$$

where $\widehat{h}_1$ and $\widehat{h}_2$ denote intermediate results. Choosing

$$r_j = \sqrt{\left(\widehat{h}_1\right)^2 + \left(\widehat{h}_2\right)^2} \tag{52}$$

$$c_j = \frac{\widehat{h}_1}{r_j} \tag{53}$$

$$s_j = \frac{\widehat{h}_2}{r_j} \tag{54}$$

we obtain

$$\bar{h}^{(j)} = \left\{ \begin{array}{c} \bar{H}_{1,j} \\ \vdots \\ \bar{H}_{j-1,j} \\ \bar{H}_{j,j} \\ 0 \\ \vdots \\ 0 \end{array} \right\} = \left\{ \begin{array}{c} \bar{H}_{1,j} \\ \vdots \\ \bar{H}_{j-1,j} \\ r_j \\ 0 \\ \vdots \\ 0 \end{array} \right\} \tag{55}$$

Subsequent Givens rotations leave $\bar{h}^{(j)}$ unaltered.

Remark:

As may be seen from (42), the value of the residual norm is easily computed.

4.3 Preconditioning

The preconditioned system is assumed to take the form

$$(L^{-1}AU^{-1})(Ux) = (L^{-1}b) \tag{56}$$

where L and U are referred to as the left and right preconditioning matrices, respectively. This form of the equation replaces $Ax = b$ in the GMRES algorithm.

The matrices L and U are defined in terms of contributions from the element groups:

$$L = \prod_{g=1}^{n_{eg}} L^{(g)} \tag{57}$$

$$U = \prod_{g=n_{eg}}^{1} U^{(g)} \tag{58}$$

11

where n_{eg} is the number of groups. For implicit/iterative groups, element-by-element preconditioners are employed:

$$L^{\mathrm{iter}} = \prod_{e=1}^{n_{el}} \bar{L}^e \qquad (59)$$

$$U^{\mathrm{iter}} = \prod_{e=n_{el}}^{1} \bar{U}^e \qquad (60)$$

where n_{el} is the number of elements in the group under consideration. The element matrices $\bar{L}^e$ and $\bar{U}^e$ are determined from the regularized element array

$$\bar{A}^e = A^e - \mathrm{diag}(A^e) + I \qquad (61)$$

(Note that A^e is the pre-preconditioned element array.)

Two element-by-element preconditioners have been investigated:

1) Nonsymmetric Cholesky element-by-element preconditioner (NC-EBE):

$$\bar{L}^e \bar{U}^e = \bar{A}^e \qquad (62)$$

In the nonsymmetric Cholesky factorization, the diagonals of the lower and upper triangular factors are the same.

2) Gauss-Siedel element-by-element preconditioner (GS-EBE):

$$\bar{L}^e + \bar{U}^e = \bar{A}^e + I \qquad (63)$$

The off-diagonal components of the Gauss-Seidel factors are identical to the off-diagonal components of A^e. Consequently, no additional storage is entailed in this case. In practice, element factors are calculated with respect to local nodal orderings rather than global (see Hughes, Ferencz and Hallquist [1987]). Further details of element-by-element implementations may be found in Ferencz [1988], Hughes and Ferencz [1987], Nour-Omid, Parlett and Raefsky [1988], and references therein.

Implicit/direct groups may be viewed as "one-element groups" and preconditioners analogous to (62) and (63) may be defined for the implicit 22-partition of A viewed as an "element array." When no preconditioning is employed for a group g, we have

$$L^{(g)} = U^{(g)} = I \qquad (64)$$

4.4 Flowchart of the precondition GMRES algorithm

The flowchart is presented in Box 1.

Remarks:

1. $\varepsilon_{\mathrm{tol}}$ is a user-specified tolerance.

2. The GMRES iteration loop consists of formation of an orthogonal basis for the Krylov space by the modified Gram-Schmidt procedure; triangularization of the Hessenberg matrix by the Q-R algorithm; and back-substitution.

3. If convergence is not attained within the GMRES iteration loop, the latest x is used to initiate the next GMRES iteration loop. The maximum number of GMRES loops allowed is $l_{\max}$, a user-specified value.

Box 1 - Preconditioned GMRES Algorithm.

Given A, b, L, U, k, ε_{tol} and $l_{\max}$, proceed as follows:

(Initialization)

$b \leftarrow L^{-1}b$

$\varepsilon = \varepsilon_{\text{tol}} \|b\|$

$x = 0$

(GMRES cycles)

For $l = 1, \ldots, l_{\max}$

$\qquad u_1 = b - L^{-1}AU^{-1}x$

$\qquad \bar{e}_1 = \|u_1\|$

$\qquad u_1 \leftarrow \dfrac{u_1}{\|u_1\|}$

$\qquad$ (GMRES iteration)

$\qquad$ **For** $i = 1, \ldots, k$

$\qquad\qquad u_{i+1} = L^{-1}AU^{-1}u_i$

$\qquad\qquad$ (Modified Gram-Schmidt orthogonalization)

$\qquad\qquad$ **For** $j = 1, \ldots, i$

$\qquad\qquad\qquad \beta_{i+1,j} = (u_{i+1}, u_j)$

$\qquad\qquad\qquad u_{i+1} \leftarrow u_{i+1} - \beta_{i+1,j}u_j$

$\qquad\qquad$ (End modified Gram-Schmidt orthogonalization)

$\qquad\qquad \bar{h}^{(i)} = \{\beta_{i+1,1}, \ldots, \beta_{i+1,i}, \|u_{i+1}\|\}^T$

$\qquad\qquad u_{i+1} \leftarrow \dfrac{u_{i+1}}{\|u_{i+1}\|}$

$\qquad\qquad$ (Q-R algorithm)

$\qquad\qquad$ **For** $j = 1, \ldots, i-1$

$$\begin{Bmatrix} \bar{h}_j^{(i)} \\ \bar{h}_{j+1}^{(i)} \end{Bmatrix} \leftarrow \begin{bmatrix} c_j & s_j \\ -s_j & c_j \end{bmatrix} \begin{Bmatrix} \bar{h}_j^{(i)} \\ \bar{h}_{j+1}^{(i)} \end{Bmatrix}$$

$\qquad\qquad$ (End j loop)

$\qquad\qquad r = \sqrt{\left(\bar{h}_i^{(i)}\right)^2 + \left(\bar{h}_{i+1}^{(i)}\right)^2}$

$\qquad\qquad c_i = \dfrac{\bar{h}_i^{(i)}}{r}$

$\qquad\qquad s_i = \dfrac{\bar{h}_{i+1}^{(i)}}{r}$

$\qquad\qquad \bar{h}_i^{(i)} \leftarrow r$

$\qquad\qquad \bar{h}_{i+1}^{(i)} \leftarrow 0$

$\qquad\qquad \bar{e}_{i+1} = -s_i\bar{e}_i$

$\qquad\qquad \bar{e}_i \leftarrow c_i\bar{e}_i$

$\qquad\qquad$ (End Q-R algorithm)

$\qquad\qquad$ Convergence check: **If** $|\bar{e}_{i+1}| \leq \varepsilon$, **Exit** i loop

$\qquad$ (End GMRES iteration)

$\qquad$ Solve for y:

$$
\begin{bmatrix}
\bar{h}_1^{(1)} & \cdots & \bar{h}_1^{(i-1)} & \bar{h}_1^{(i)} \\
0 & \ddots & \vdots & \vdots \\
\vdots & \ddots & \bar{h}_{i-1}^{(i-1)} & \bar{h}_{i-1}^{(i)} \\
0 & \cdots & 0 & \bar{h}_i^{(i)}
\end{bmatrix}
\left\{
\begin{array}{c}
y_1 \\
\vdots \\
y_{i-1} \\
y_i
\end{array}
\right\}
=
\left\{
\begin{array}{c}
\bar{e}_1 \\
\vdots \\
\bar{e}_{i-1} \\
\bar{e}_i
\end{array}
\right\}
$$

Update solution:

$$
\boldsymbol{x} \leftarrow \boldsymbol{x} + \sum_{j=1}^{i} y_j \boldsymbol{u}_j
$$

Convergence check: **If** $|\bar{e}_{i+1}| \leq \varepsilon$, **Exit** l loop

(End GMRES cycles)

$\boldsymbol{x} \leftarrow \boldsymbol{U}^{-1} \boldsymbol{x}$

Return

5. Numerical Results

In this section we present numerical results to demonstrate the effectiveness of the proposed procedure as applied to the compressible Euler and Navier-Stokes equations.

All computations were done on a CONVEX-C1 in double precision (64 bits per floating point word), using 2×2 Gaussian quadrature for bilinear elements and 3-point quadrature for linear triangular elements.

5.1 Flow over a flat plate

Problem statement and numerical solution

The problem statement of this two-dimensional Navier-Stokes flow is shown in Figure 6.

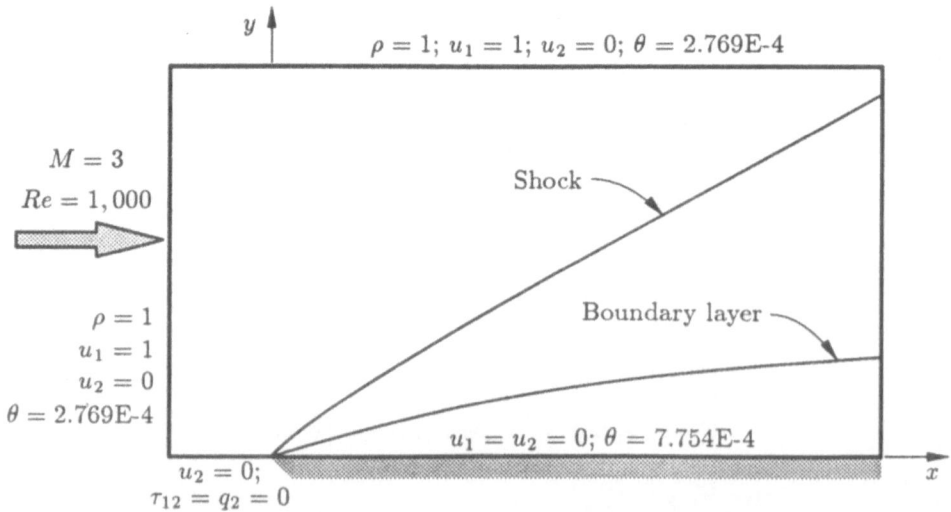

Figure 6. Flow over a flat plate. Problem description.

This problem, known as Carter's problem, consists of a Mach three flow passing over a plate at zero angle of attack, causing a curved shock and boundary layer to develop aft of the leading edge of the plate. The Reynolds number, based on free stream values and distance from the leading edge of the plate, is 10^3. The Sutherland viscosity law, $\mu = 0.0906\theta^{1.5}/(\theta + 0.0001406)$, is employed. This corresponds to a free stream temperature of 216.7°K.

The computational domain covers the area $-0.2 \leq x \leq 1.2$, $0 \leq y \leq 0.8$, with the leading edge of the plate placed at $x = 0$. On the inflow boundary ($x = -0.2$) and top boundary ($y = 0.8$), four quantities, ρ, u_1, u_2 and θ, are prescribed; on the line of symmetry ($y = 0$ and $x < 0$), the symmetry conditions, i.e., $u_2 = \tau_{12} = q_2 = 0$, are imposed; on the plate ($y = 0$ and $x \geq 0$), the no-slip condition ($u_1 = u_2 = 0$) and stagnation temperature ($\theta_{stag} = \theta_{inflow}(1 + \frac{\gamma-1}{2}M_{inflow}^2)$) are prescribed; and finally, on the outflow boundary ($x = 1.2$), no quantity is prescribed.

To solve this problem, we employed five uniform nested meshes, consisting of 14×8, 28×16, 56×32, 112×64, and 224×128 square elements. These correspond to mesh sizes of $h = 0.1$, 0.05, 0.025, 0.0125, and 0.00625, respectively. The solution obtained using the fourth mesh ($h = 0.0125$) is shown in Figure 7.

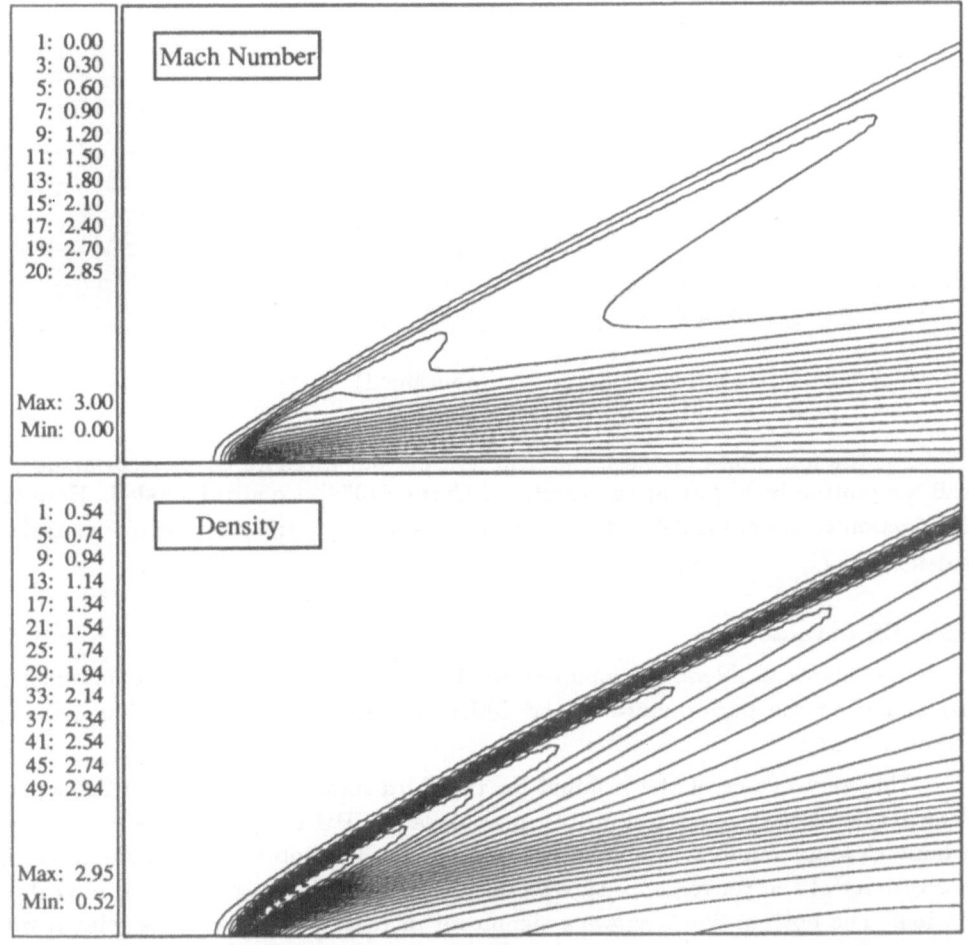

Figure 7. Solution of the flow over a flat plate problem (7,168 elements).

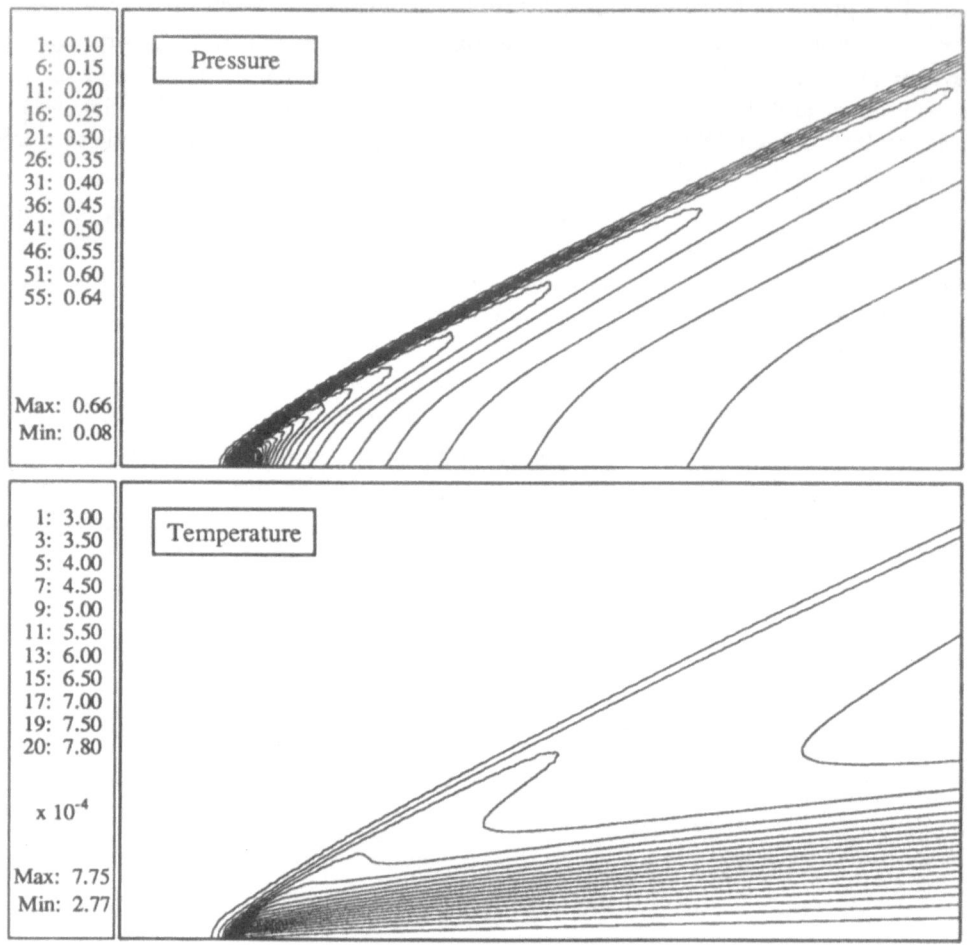

1: 0.10	
6: 0.15	Pressure
11: 0.20	
16: 0.25	
21: 0.30	
26: 0.35	
31: 0.40	
36: 0.45	
41: 0.50	
46: 0.55	
51: 0.60	
55: 0.64	
Max: 0.66	
Min: 0.08	

1: 3.00	Temperature
3: 3.50	
5: 4.00	
7: 4.50	
9: 5.00	
11: 5.50	
13: 6.00	
15: 6.50	
17: 7.00	
19: 7.50	
20: 7.80	
x 10⁻⁴	
Max: 7.75	
Min: 2.77	

Figure 7. (continued)

For the four finest meshes, the skin friction and pressure coefficients along the wall are plotted in Figure 8; the results of Carter [1972] are also included. Here the skin friction coefficient is defined as $C_f = \tau_{wall}/(\frac{1}{2}\,\rho_\infty u_\infty^2)$, and the pressure coefficient is defined as $C_p = (p - p_\infty)/(\frac{1}{2}\,\rho_\infty u_\infty^2)$.

GMRES parameters

The above problem is employed to determine the effects of the parameters ε_{tol} and k on the performance of the GMRES algorithm with block-diagonal pre-preconditioning.

The convergence of the residual for the third mesh, consisting of 1,792 square elements, is plotted as a function of time step and CPU-time in Figure 9. An algorithmic Courant number, C_r, of 25 was employed. The top figure demonstrates the insensitivity of the convergence-rate to the accuracy of the linear solution in each iteration. The bottom figure shows a significant increase in CPU cost associated with decreasing values of ε_{tol}. These results suggest the use of fairly large values for ε_{tol}.

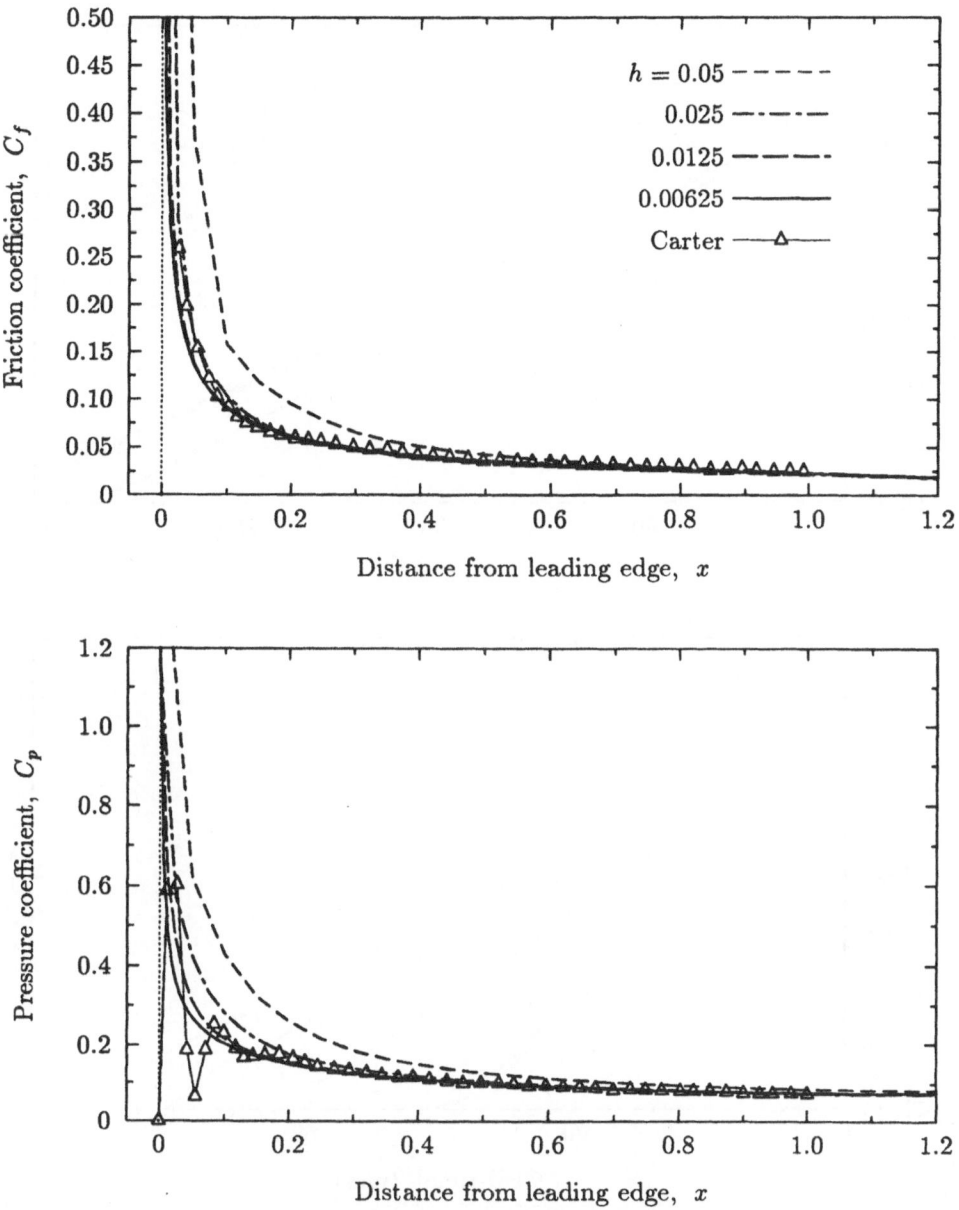

Figure 8. Flow over a flat plate. Wall quantities.

The convergence of the GMRES procedure as a function of CPU-time is plotted in Figure 10, for several values of k. The convergence, which is measured by $||\boldsymbol{b} - \boldsymbol{A}(\boldsymbol{x}_0 + \boldsymbol{z})||$ $(= |\bar{e}_{i+1}|/\varepsilon$ in Box 1), was computed for the first time step. Based on these results, $k = 5$ may be recommended. In the case of diagonal pre-preconditioning, numerical experiments have shown the need of larger Krylov spaces to achieve convergence.

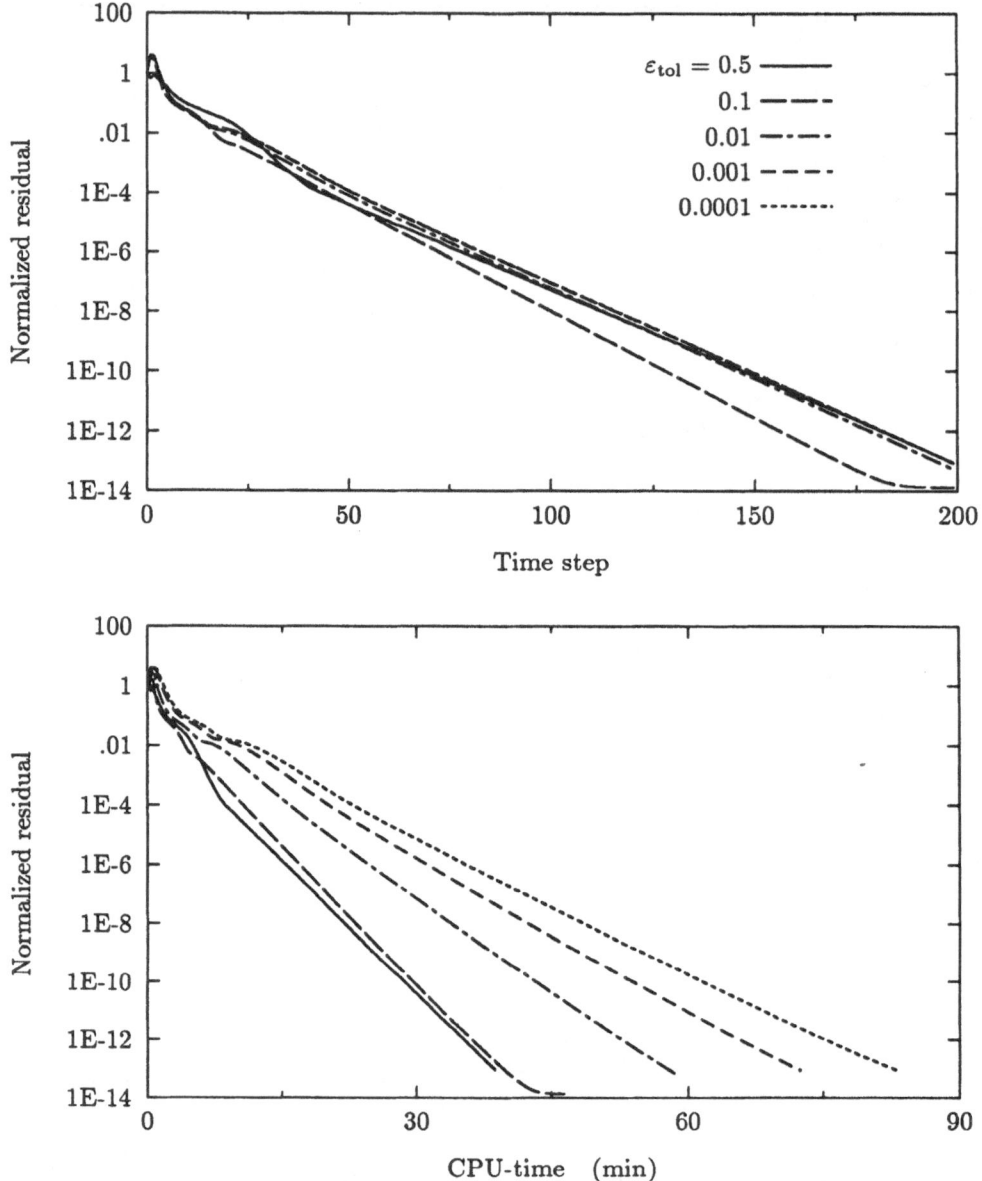

Figure 9. Flow over a flat plate ($C_r = 25$; 1,792 elements). Effect of linear solution accuracy on nonlinear convergence. GMRES procedure with block-diagonal pre-preconditioner.

Performance of implicit solvers

The performance of the GMRES algorithm is compared with an active column direct solver. The results presented in Figure 11 and Tables 1–3 are average values for the first ten time steps; $\varepsilon_{tol} = 0.1$, $k = 5$ was used with block-diagonal pre-

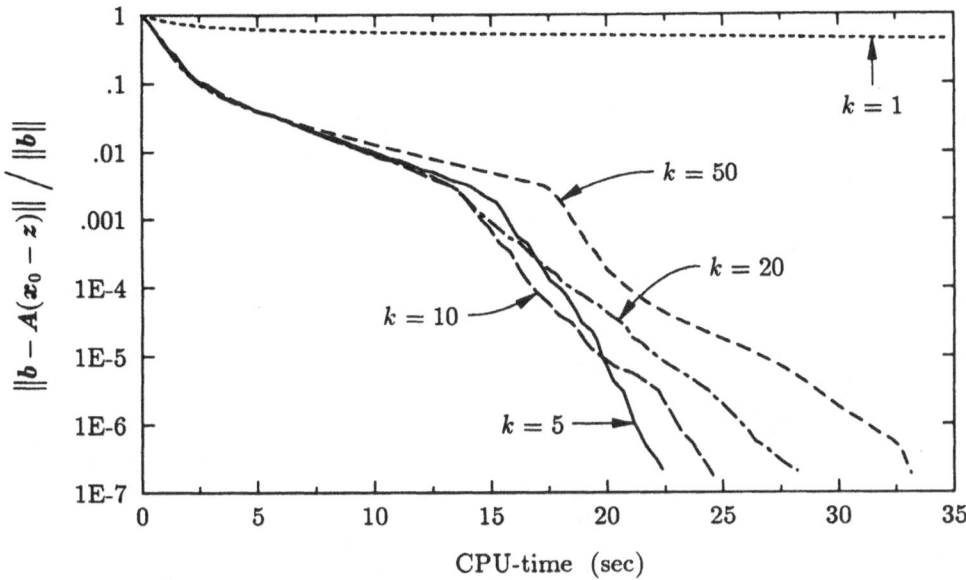

Figure 10. Flow over a flat plate ($C_r = 25$; 1,792 elements). Effect of Krylov space dimension on the convergence of the GMRES algorithm with block-diagonal pre-preconditioning.

preconditioning, and $k = 25$ was used with diagonal pre-preconditioning. The memory required by the direct solver for the 7,168-element mesh and the GMRES solver with NC-EBE for the 28,672-element mesh exceeded the memory available on our machine. Consequently, demand paging (i.e., swapping memory and disk content) was invoked in both cases. Note that for the timing results presented in Table 1 and Figure 11 the time-increment was automatically scaled with the mesh size parameter. Thus, one should be careful in extrapolating the cost of the overall solution of the problem via these results. Finally, note that the CPU-time of the direct solver is proportional to $(n_{el})^{1.5}$ rather than the theoretically predicted $(n_{el})^2$. This discrepancy can be attributed to vectorization. The flop rate of the direct solver increases with vector length, which is proportional to the average bandwidth of the left-hand-side matrix. For this problem, the optimal vector length of 128 had not been achieved until the 1,792-element mesh. In contrast, in the iterative procedures almost optimal vector length was obtained by the 448-element mesh (i.e., vector length = element block size = 112).

5.2 NACA-0012 airfoil

Problem statement and numerical solution

This two-dimensional Navier-Stokes problem consists of a Mach 0.85 flow about a NACA-0012 airfoil at zero angle of attack. The Reynolds number based on the airfoil cord is 500. A constant viscosity is employed. The problem statement and the boundary conditions are shown in Figure 12.

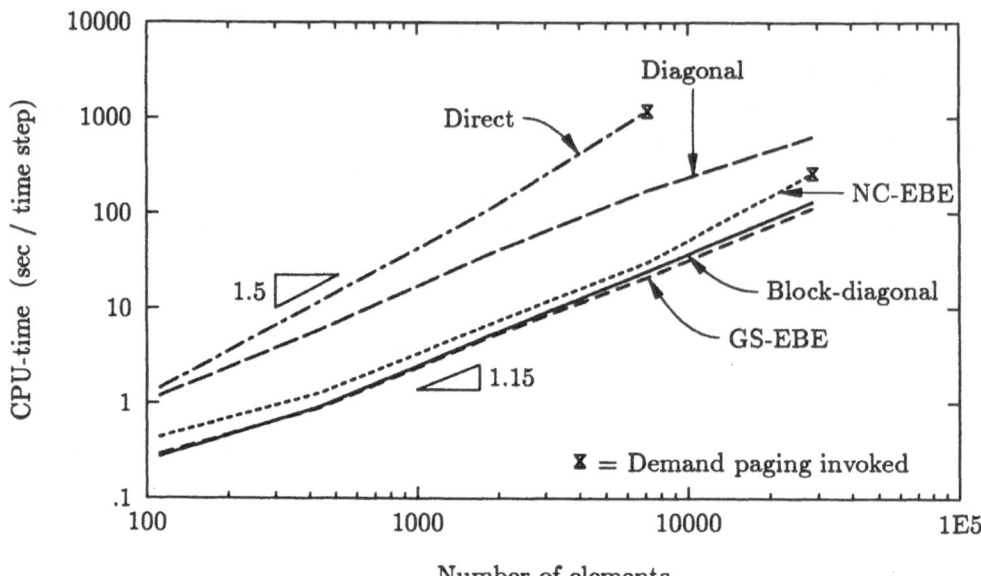

Figure 11. Flow over a flat plate ($C_r = 25$). Comparison of solution times for the direct and iterative procedures.

Table 1. Flow over a flat plate ($C_r = 25$). Comparison of solution times, in seconds per time step, for the direct and iterative procedures.

Elements	Direct	Diagonal no EBE	Block-Diagonal		
			no EBE	NC-EBE	GS-EBE
112	1.5	1.2	0.3	0.4	0.3
448	12.2	6.1	0.9	1.3	0.9
1,792	102.7	35.6	5.0	6.5	4.8
7,168	1,174.4*	175.3	24.5	30.4	21.2
28,672	—	628.3	130.9	263.0*	112.0

* Demand paging was invoked.

Table 2. Flow over a flat plate ($C_r = 25$). Average number of iterations per step required by the GMRES procedure.

Elements	Diagonal no EBE	Block-Diagonal		
		no EBE	NC-EBE	GS-EBE
112	40.6	8.6	3.7	3.8
448	85.1	10.8	4.2	4.3
1,792	127.2	15.6	5.9	5.8
7,168	159.0	19.7	6.7	6.6
28,672	144.8	21.3	6.5	6.9

Table 9. Flow over a flat plate. Storage requirement for the solution procedures, in MBytes.

Elements	Direct	Diagonal no EBE	Block-Diagonal		
			no EBE	NC-EBE	GS-EBE
112	0.35	0.36	0.28	0.51	0.28
448	2.30	1.40	1.10	2.02	1.10
1,792	16.49	5.49	4.38	8.05	4.38
7,168	124.58	21.79	17.44	32.12	17.44
28,672	996.00*	86.82	69.63	128.35	69.63

* Estimated.

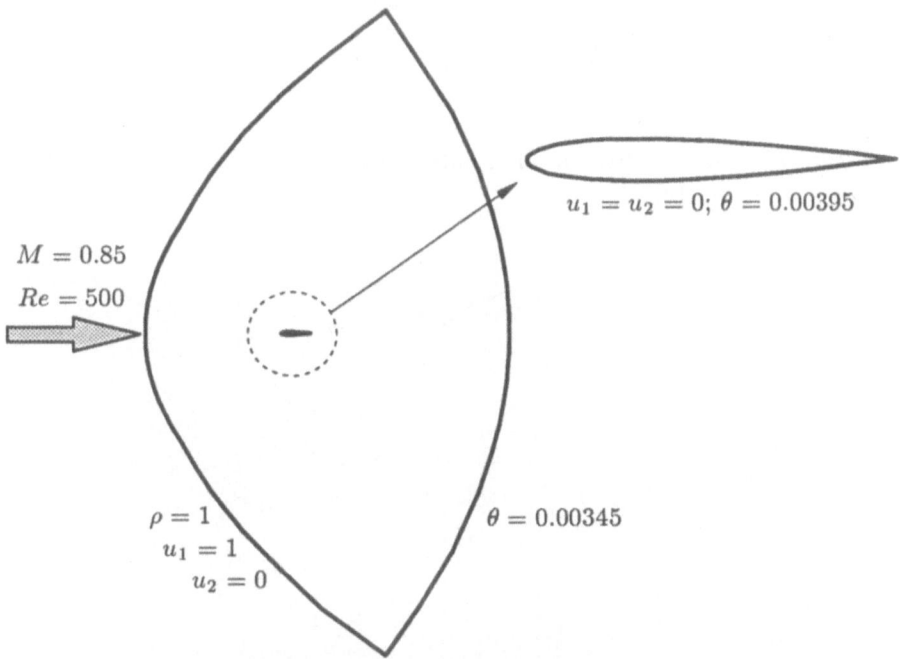

$M = 0.85$

$Re = 500$

$u_1 = u_2 = 0; \ \theta = 0.00395$

$\theta = 0.00345$

$\rho = 1$
$u_1 = 1$
$u_2 = 0$

Figure 12. NACA-0012 airfoil. Problem description.

To solve this problem, we employed three unstructured nested meshes, consisting of 1,516, 6,064 and 24,256 triangular elements, and 801, 3,118 and 12,300 nodes, respectively. The 1,516-element mesh is shown in Figure 13. The refined meshes were constructed in turn by subdividing each triangular element into four triangles. A detail of the 24,256-element mesh is shown in Figure 14.

The Mach number, density, pressure and temperature contours computed using the finest mesh are presented in Figure 15. This figure shows a symmetric pair of detached supersonic pockets, with a maximum Mach number of 1.13, followed by a thick wake behind the airfoil.

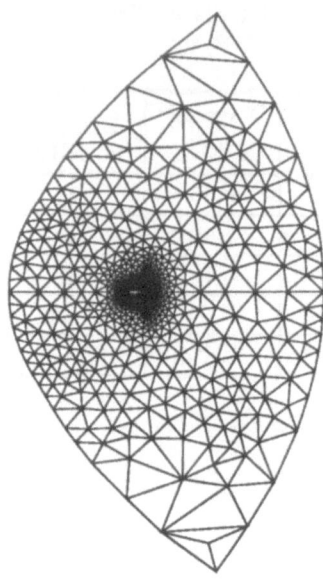

Figure 13. NACA-0012 airfoil. Finite element mesh (801 nodes; 1,516 elements).

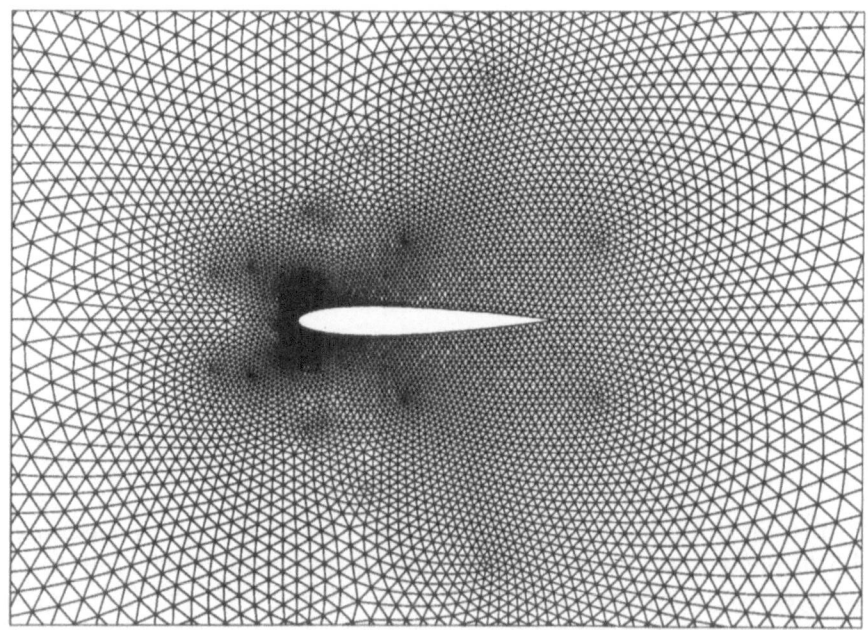

Figure 14. NACA-0012 airfoil. Finite element mesh (12,300 nodes; 24,256 elements).

The pressure and skin friction coefficients are plotted in Figure 16. In the coarse, medium and fine meshes there are only one, two and four mesh points contained in the boundary layer. Thus, the finest mesh is still not sufficiently fine to accurately

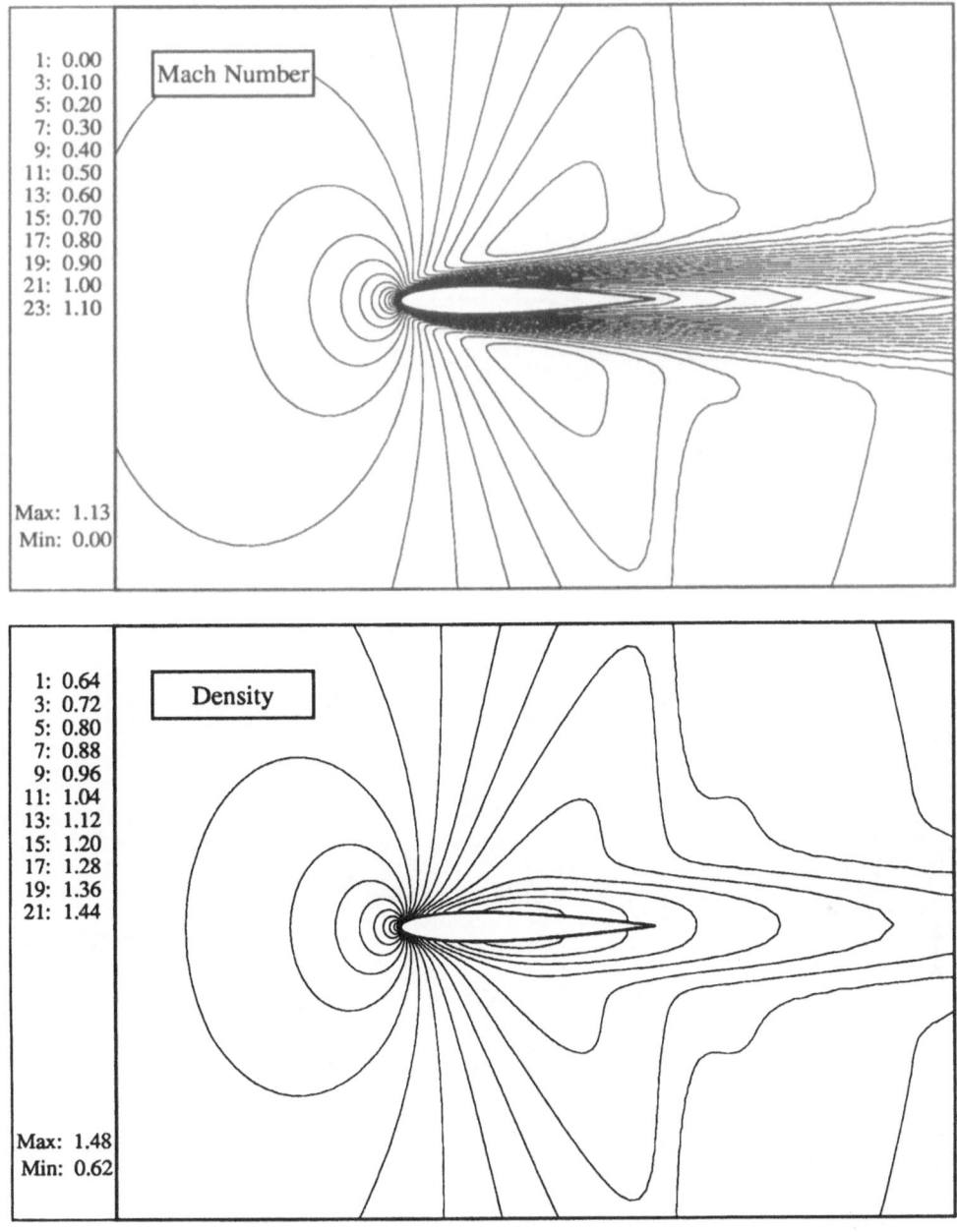

Figure 15. Solution of the NACA-0012 airfoil problem (24,256 elements).

represent all details of the solution. A better solution with fewer elements can be obtained by employing a refinement strategy which results in finer elements only near the airfoil, as opposed to the employed strategy which resulted in uniform refinement over the entire mesh.

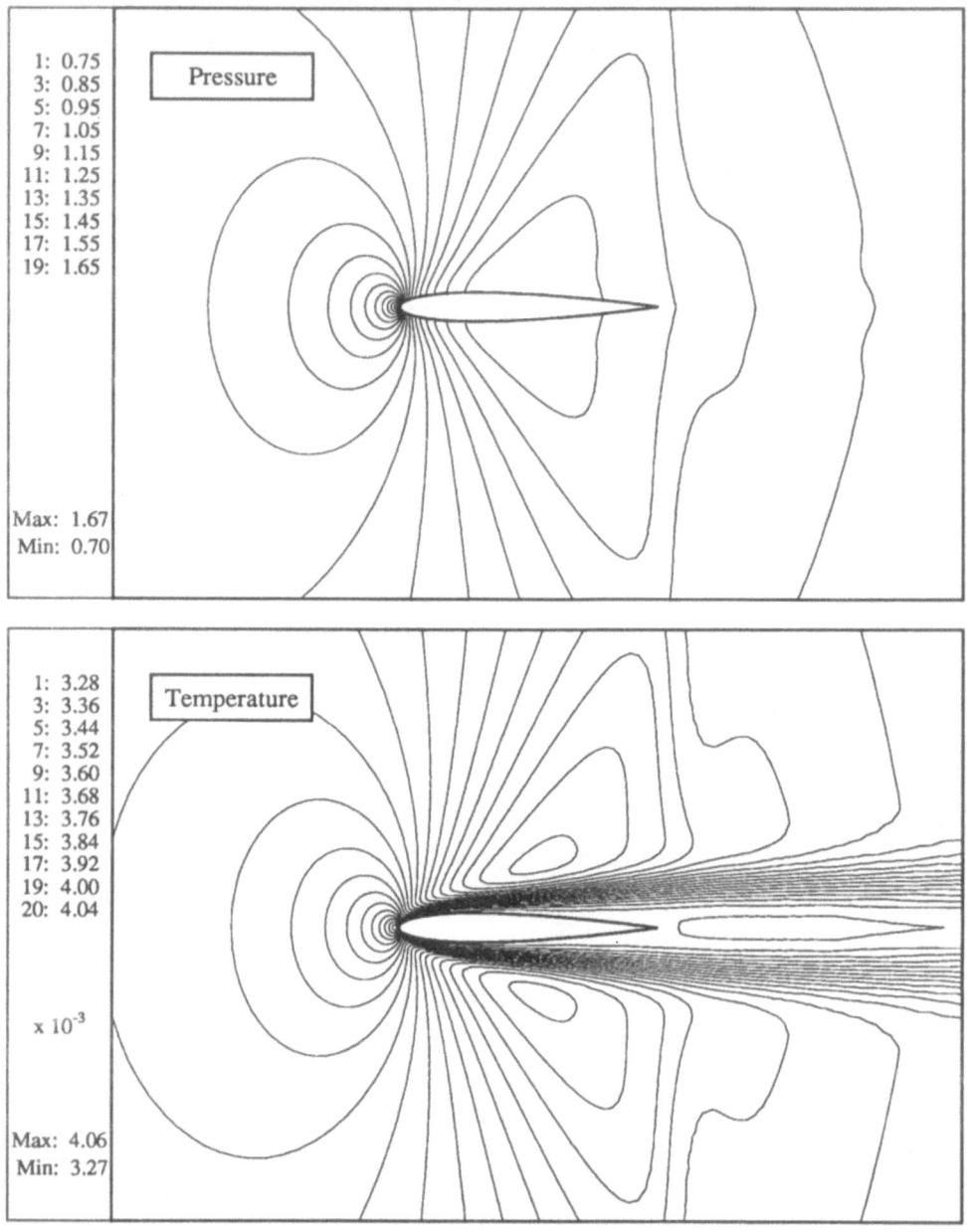

Figure 15. (continued)

Local time-stepping strategy

Convergence results for the 1,516 element mesh using the global and local time-stepping strategies with $C_r = 50$ are presented in Figure 17. The global time-stepping strategy employs one time step for all the nodal points of the domain. This time step is computed such that the maximum Courant number does not exceed a predetermined

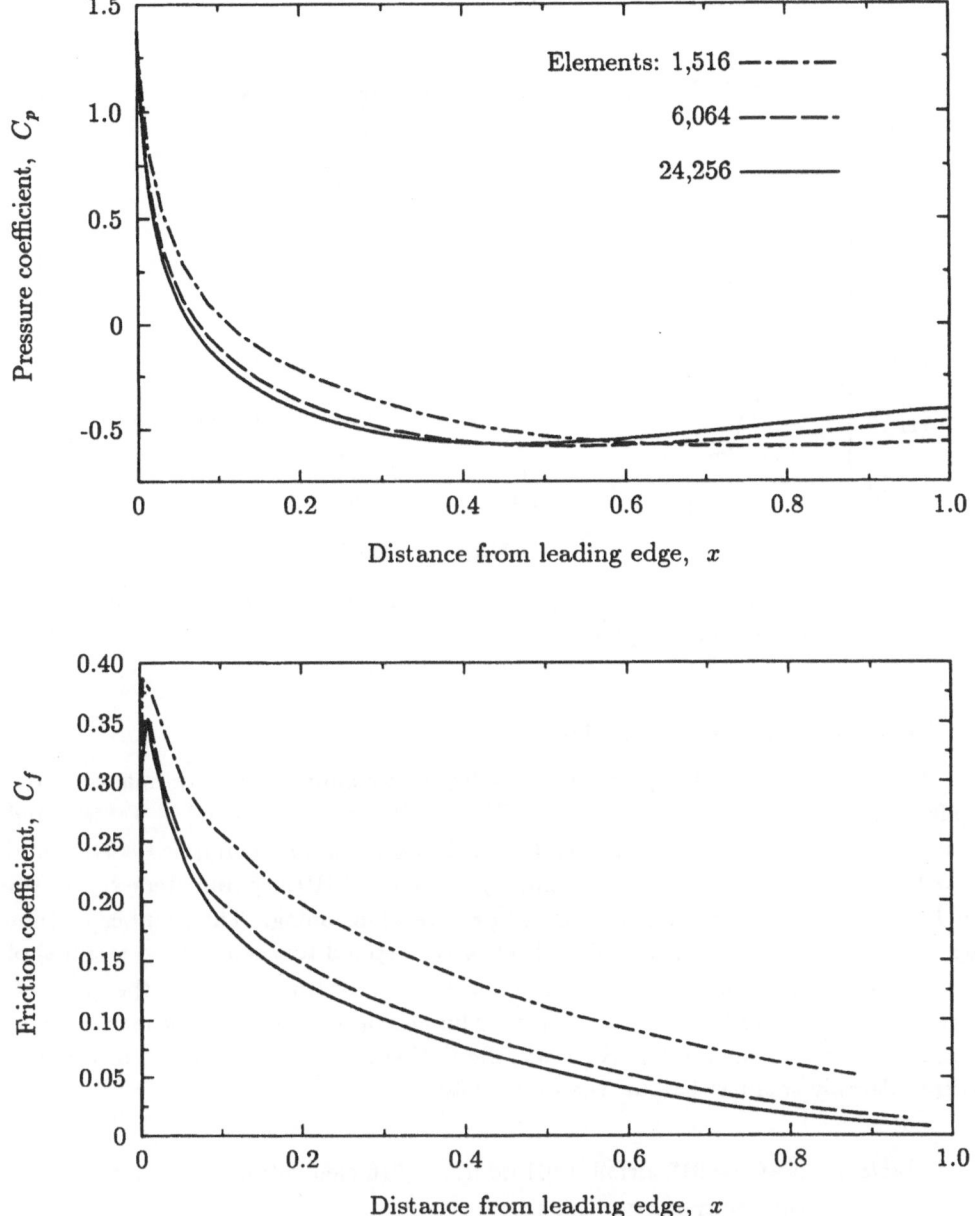

Figure 16. NACA-0012 airfoil. Pressure and skin-friction coefficients.

value set by the user. The local time-stepping strategy employs a different time step for every nodal point of the domain. The nodal time steps are computed such that the Courant number at each node equals the predetermined value. See Shakib [1988] for further details. Note that, in Figure 17, after 150 time steps the solution residual for the local time-stepping strategy has reached machine precision, whereas the solution residual for the global time-stepping strategy has decreased only by two orders of magnitude in the first 1000 time steps.

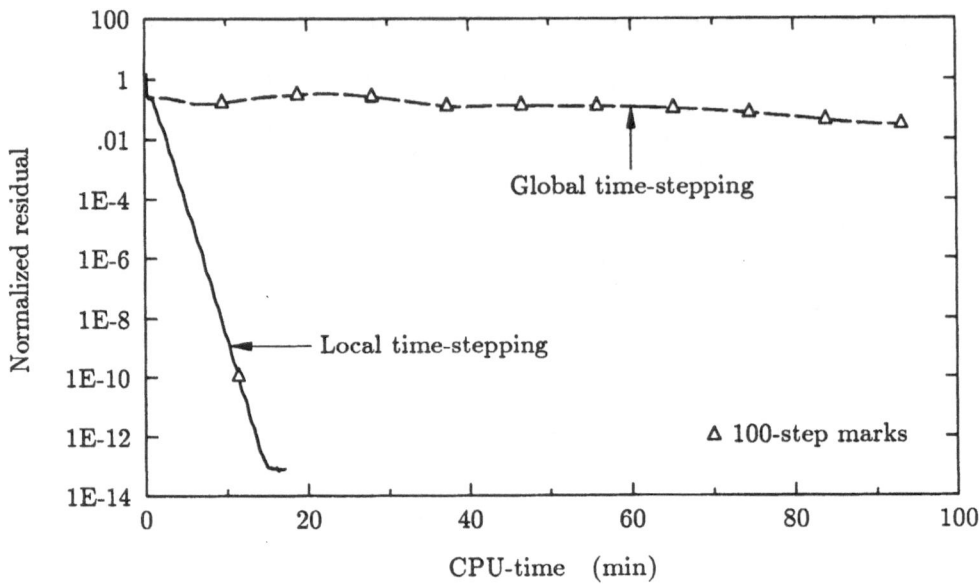

Figure 17. NACA-0012 airfoil (1,516 elements). Convergence of global and local time-stepping strategies.

Performance of solution algorithms

The NACA-0012 airfoil problem is employed to compare the performance of the implicit and explicit solution algorithms. The results are summarized in Table 4 and Figure 18. The local time-stepping strategy with $C_r = 50$ for the implicit solvers and $C_r = 0.5$ for the explicit solver was employed. The GMRES parameters $\varepsilon_{tol} = 0.1$ and $k = 5$ were used with block-diagonal pre-preconditioning. The number of time steps and the CPU-times presented in Table 4 correspond to a normalized residual of 10^{-13}. These results show an increase in the number of time steps for the GMRES procedures as compared to the direct solver. This can be attributed to the lack of linear solution accuracy of the former. However, the CPU-times for the GMRES procedures are considerably smaller than for the direct solver.

Table 4. NACA-0012 airfoil (801 nodes; 1,516 elements). Performance of the solution algorithms.

Procedure	Number of time steps	CPU-time (min)	Ave. iterations per step	Storage (Mbytes)
Explicit	13,977	356.33	—	0.830
Direct	104	118.05	—	9.752
Block-diagonal	147	17.42	16.86	4.249
Block-diag./GS-EBE	132	14.86	6.45	4.249
Block-diag./NC-EBE	132	16.53	6.56	7.354

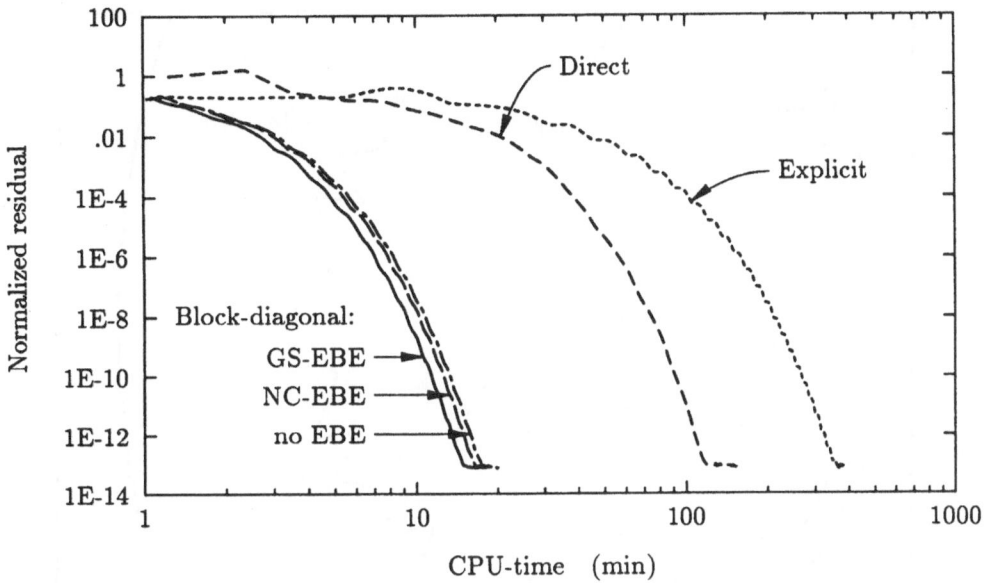

Figure 18. NACA-0012 airfoil (801 Nodes; 1,516 Elements). Comparison of CPU-times of the solution algorithms.

Element group partitioning

The 1,516-element mesh is decomposed into a series of two-element group partitions, with one partition in the inner region (near the airfoil, containing the small elements) and the other partition in the outer region. The boundaries of these partitions are shown in Figure 19.

Two studies were conducted. In the first study, an implicit/iterative group with block-diagonal pre-preconditioning and $C_r = 100$ was employed for the inside partition, and an explicit group with $C_r = 0.5$ was employed for the outside partition. In the second study, the partitions were interchanged. In both cases the local time-stepping strategy was employed. CPU-times, normalized by the fully implicit result, are presented in Figure 20. Contrary to what was anticipated, the CPU-time for the implicit group on the inside increased with increasing implicit group fraction, except for the fully implicit calculation. These results suggest that it is propitious to use implicit elements on the outside. Note that the local time-stepping strategy is an essential ingredient for the effective use of implicit/explicit partitions.

5.3 Bi-NACA-0012 airfoil

A parallel pair of NACA-0012 airfoils, with leading edges at $x = 0$ and $y = \pm 0.24$, are placed in an Euler flow at Mach 0.55 and 6° angle of attack; see Figure 21. The computational domain consists of two connected semicircles of radius 7.5 and centers at $x = 0.5$, $y = \pm 0.24$. On the inflow boundary ($x \leq 0.5$), ρ, u_1 and u_2 are prescribed; on the outflow boundary ($x > 0.5$), temperature is prescribed; and on the airfoils, the slip and adiabatic conditions, $u_n = 0$ and $q_n = 0$, are imposed. At the two trailing-edge vertices, no velocity is prescribed.

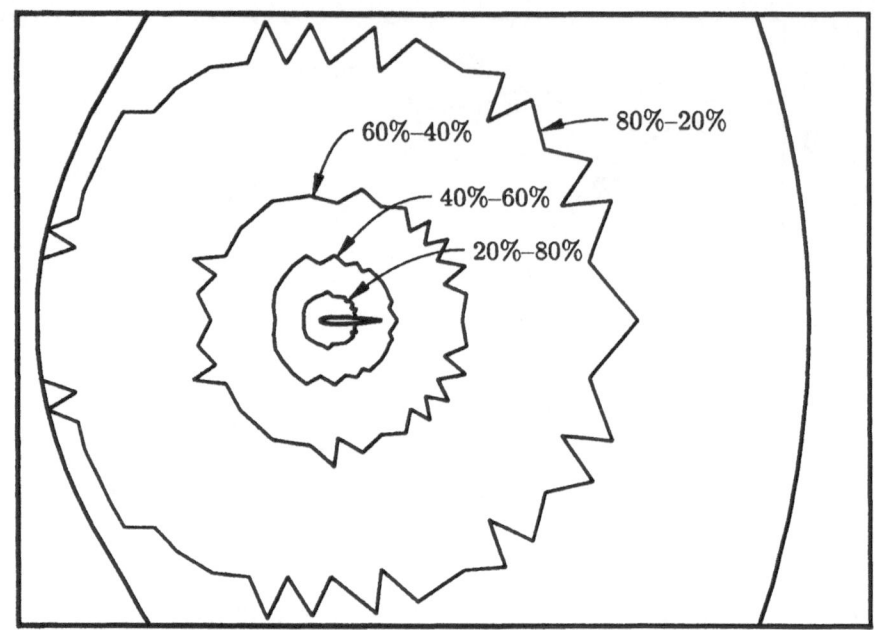

Figure 19. NACA-0012 airfoil (1,516 elements). Two-element group partitions.

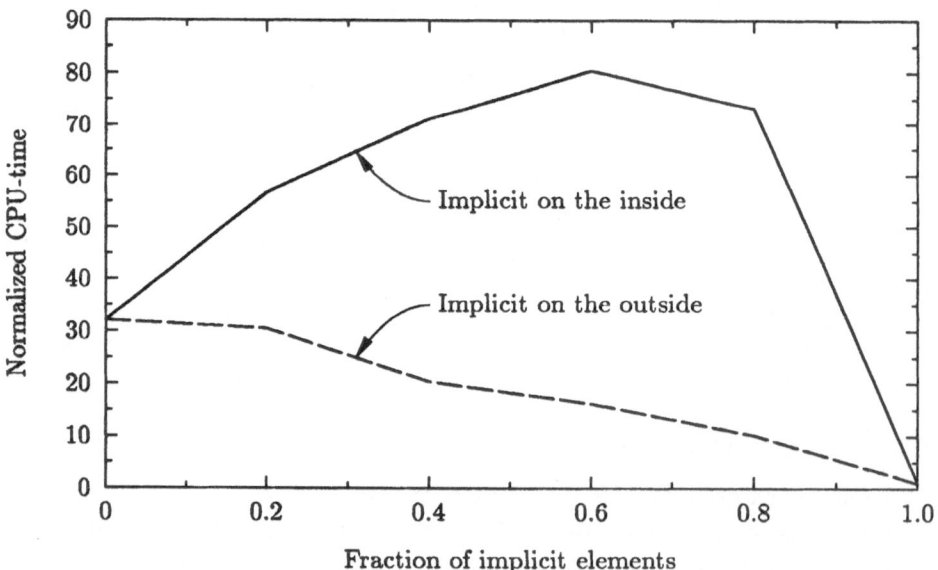

Figure 20. NACA-0012 airfoil (1,516 elements). Performance of two-element group partitions.

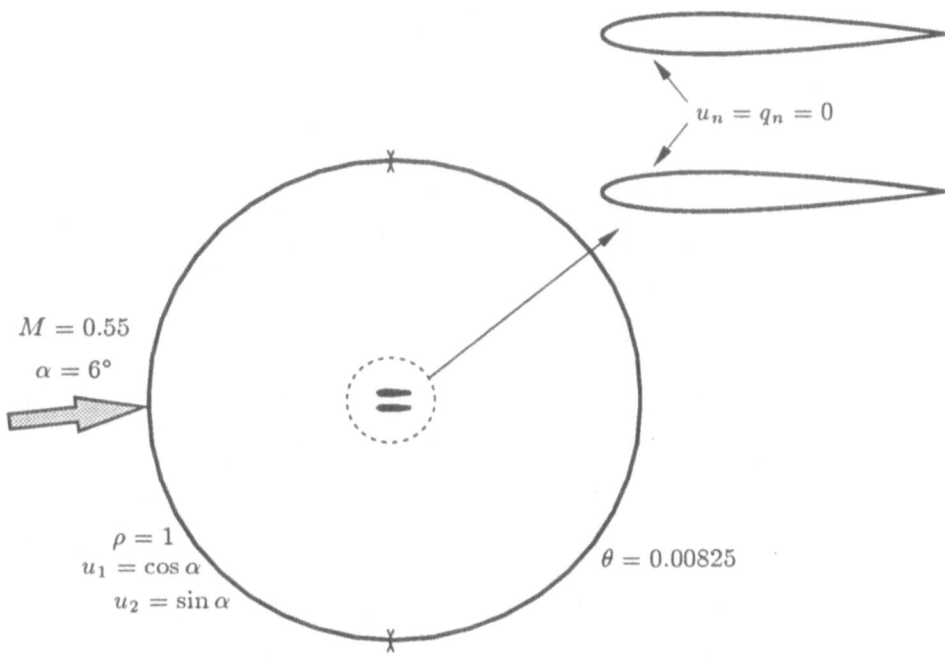

Figure 21. Bi-NACA-0012 airfoil. Problem description.

To solve this problem, we employed an unstructured mesh having 3,572 nodes and 6,871 triangular elements; see Figure 22. The local time-stepping strategy was employed. An implicit/direct calculation was not undertaken because of excessive storage requirements ($\sim$ one Gbyte). An explicit calculation exhibited no residual reduction after 30,000 time steps and was therefore terminated. Contour plots indicated that convergence was nevertheless taking place. However, the cost was becoming prohibitively expensive necessitating abandoning the calculation. The implicit/iterative procedure, on the other hand, converged to machine precision after 5,000 steps (residual reduction of order 10^{-13}). From all appearances, a practically converged solution was obtained much sooner ($< 1,000$ steps). The parameters used in the implicit/iterative calculation were $k = 5$, $\varepsilon_{tol} = 0.1$, and $C_r = 10$. Block-diagonal pre-preconditioning and GS-EBE preconditioning were employed. The Mach number and pressure contours are presented in Figure 23. Note the sharp shock between the two airfoils and the two weak supersonic pockets on the upper surfaces of the two airfoils immediately after the leading edges.

The pressure coefficients on the upper and lower surfaces of the airfoils are plotted in Figure 24. Note that the shocks are captured in two elements.

The conclusion drawn from the preceding studies is that the implicit/iterative solver is superior to both implicit/direct and explicit strategies.

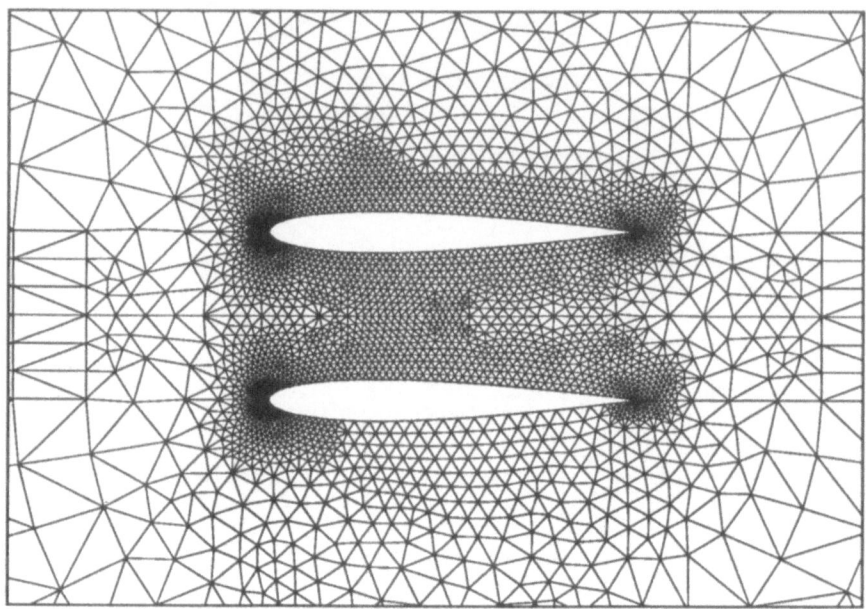

Figure 22. Bi-NACA-0012 airfoil. Finite element mesh (3,572 nodes; 6,871 elements).

6. Conclusions

We have presented a multi-element group preconditioned GMRES iterative algorithm for solving nonsymmetric linear systems arising in finite element analysis. The algorithm is fully vectorizable, employing element-by-element implementational concepts. Applications to high-speed compressible flows governed by the Euler and Navier-Stokes equations indicate the effectiveness of the proposed procedure. Numerical results clearly demonstrate the superiority of the iterative algorithm over implicit/direct and explicit strategies. The iterative algorithm, in its present form, is not applicable to completely general nonsymmetric systems, as we assume either the diagonal entries are positive, or the nodal block-diagonal submatrices admit Cholesky factorization. However, we conjecture these properties are possessed by many finite element discretizations of well-posed physical problems of interest. Current work is focused on reducing the data base of the unassembled element file. This appears to be a crucial step in extending the present work to large three-dimensional problems.

Acknowledgments

The authors would like to express their appreciation to Robert Ferencz, Greg Hulbert, Michel Mallet and Arthur Raefsky for helpful comments. This research was supported by the NASA Langley Research Center under Grant NASA-NAG-1-361, the IBM Almaden Research Center under Grant No. 604912 and Avions Marcel Dassault–Breguet Aviation, St. Cloud, France.

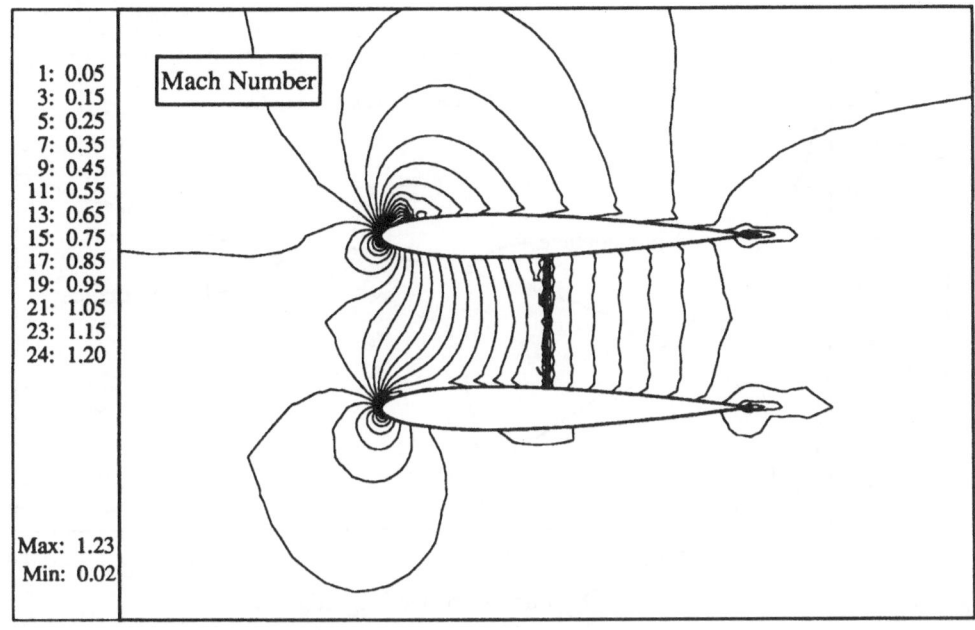

Mach Number

1: 0.05
3: 0.15
5: 0.25
7: 0.35
9: 0.45
11: 0.55
13: 0.65
15: 0.75
17: 0.85
19: 0.95
21: 1.05
23: 1.15
24: 1.20

Max: 1.23
Min: 0.02

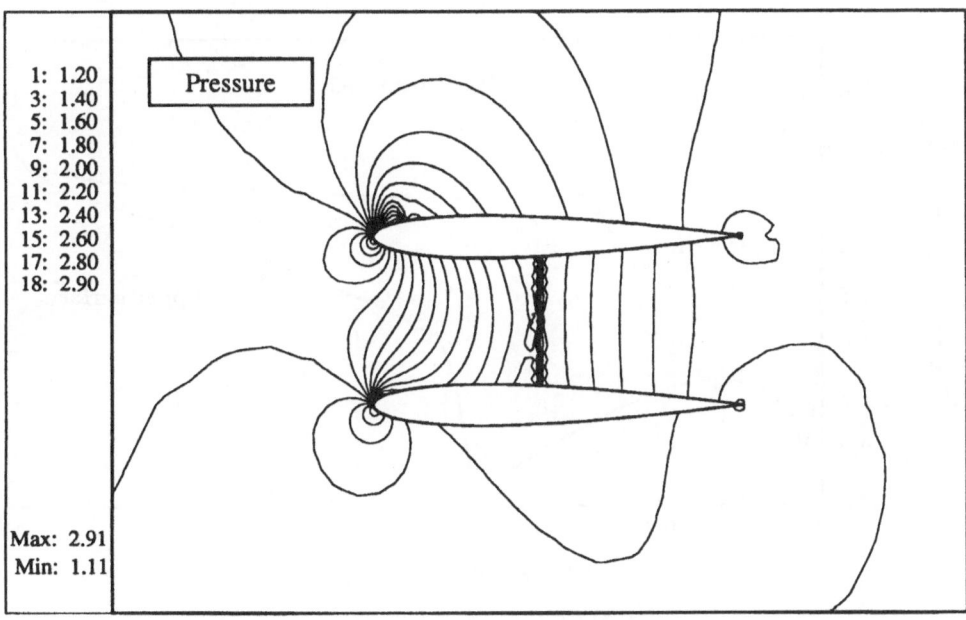

Pressure

1: 1.20
3: 1.40
5: 1.60
7: 1.80
9: 2.00
11: 2.20
13: 2.40
15: 2.60
17: 2.80
18: 2.90

Max: 2.91
Min: 1.11

Figure 23. Solution of the Bi-NACA-0012 airfoil problem.

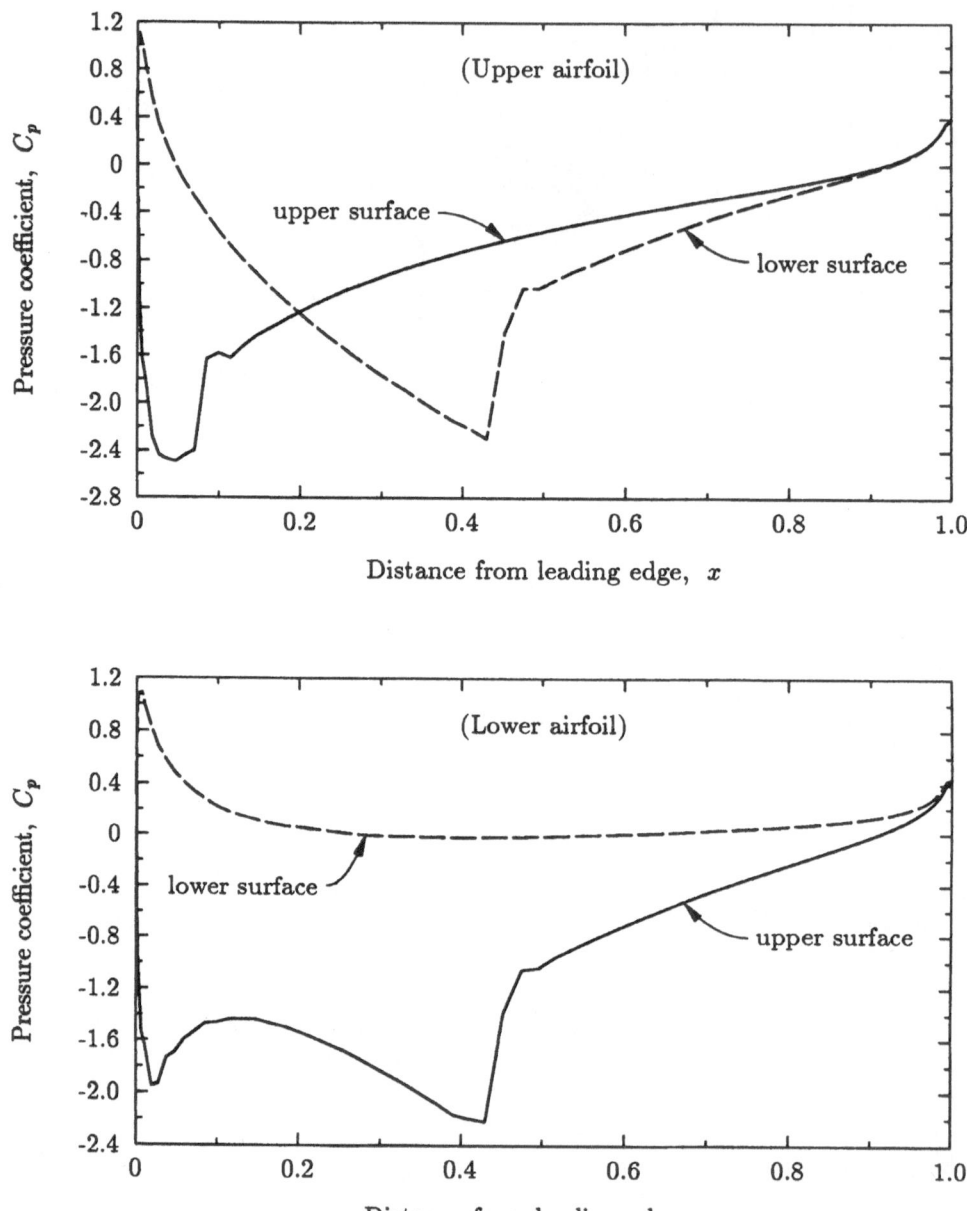

Figure 24. Bi-NACA-0012 airfoil. Pressure coefficients.

References

J.E. Carter [1972], "Numerical Solutions of the Navier-Stokes Equations for the Super-sonic Laminar Flow over a Two-dimensional Compression Corner", *NASA Technical Report*, NASA TR R-385.

R.M. Ferencz [1988], "Element-by-element Preconditioning Techniques for Large-scale, Vectorized Finite Element Analysis in Nonlinear Solid and Structural Mechanics", *Ph.D. Thesis*, Division of Applied Mechanics, Stanford University, in preparation.

G.H. Golub and C.F. Van Loan [1983], *Matrix Computations*, The Johns Hopkins University Press, Baltimore.

T.J.R. Hughes [1987], *The Finite Element Method: Linear Static and Dynamic Finite Element Analysis*, Prentice-Hall, Englewood Cliffs, New Jersey.

T.J.R. Hughes and R.M. Ferencz [1987], "Implicit Solution of Large-scale Contact and Impact Problems Employing an EBE Preconditioned Iterative Solver", Preprint of contribution to *IMPACT 87 International Conference on Effects of Fast Transient Loading in the Context of Structural Mechanics*, Lausanne, Switzerland, August 26-27.

T.J.R. Hughes, R.M. Ferencz and J.O. Hallquist [1987], "Large-scale Vectorized Implicit Calculations in Solid Mechanics on a Cray X-MP/48 Utilizing EBE Preconditioned Conjugate Gradients", *Computer Methods in Applied Mechanics and Engineering*, **61**, 215-248.

M. Mallet, J. Periaux and B. Stoufflet [1987], "Convergence Acceleration of Finite Element Methods for the Solution of the Euler and Navier-Stokes Equations of Compressible Flow", *Proceedings of the 7th GAMM Conference on Numerical Methods in Fluid Dynamics*, to appear.

B. Nour-Omid, B.N. Parlett and A. Raefsky [1988], "Comparison of Lanczos with Conjugate Gradient Using Element Preconditioning", pp. 250-260 in *Proceedings of the First International Symposium on Domain Decomposition Methods for Partial Differential Equations*, (eds. R. Glowinski, G.H. Golub, G.A. Meurant, and J. Periaux), SIAM, Philadelphia.

Y. Saad and M.H. Schultz [1983], "GMRES: a Generalized Minimal Residual Algorithm for Solving Nonsymmetric Linear Systems", *Research Report YALEU/DCS/RR-254*, Department of Computer Science, Yale University.

F. Shakib [1988], "Finite Element Analysis of the Compressible Euler and Navier-Stokes Equations", *Ph.D. Thesis*, Division of Applied Mechanics, Stanford University.

A SUBSTRUCTURE-BASED PARALLEL ALGORITHM

FOR LARGE STRUCTURE EIGENVALUE PROBLEMS

Jeffrey K. Bennighof and Michael C. Sciascia

Dept. of Aerospace Engineering and Engineering Mechanics
The University of Texas at Austin
Austin, Texas 78712

ABSTRACT

In this paper, a parallel algorithm is presented for solving large structure eigenvalue problems. The computation is parallelized on the basis of a division of the structure into substructures. The algorithm resembles the basic subspace iteration algorithm, but the reduced subspace is augmented to include previous iterates for eigenvectors that have not yet converged, and substructure eigenvectors, which are computed by substructure processors when they might otherwise be idle. Including previous iterates results in convergence approximating that of the Lanczos method, and including substructure eigenvectors enhances convergence further. An advantage of the present approach over the Lanczos method is that the reduced subspace does not grow as large, because basis vectors that do not enhance convergence significantly are automatically removed from the subspace, decreasing storage requirements. A numerical example demonstrates the method and shows the results that can be expected.

INTRODUCTION

It is not unusual for finite element models of complex structures to possess tens or even hundreds of thousands of degrees of freedom in current engineering practice. Solving the associated algebraic eigenvalue problems models for approximate natural frequencies and mode shapes is of fundamental importance in the dynamic analysis and design of structures, but it poses significant computational challenges. The importance of developing algorithms for solving these problems on multi-processor computers is obvious. However, because the sparsity structure of the matrices involved is frequently very irregular, dividing the computational task among different processors in such a way that all processors are used efficiently can be difficult.

In this paper, the computation is assigned to processors on the basis of a division of the structure into substructures. This approach exemplifies the domain decomposition approach to parallelization, in which each processor is assigned to a different portion of the problem domain[1]. An advantage of this approach is that communication requirements may be decreased, as the processors only need to have access to the portion of the overall model that is contained within their own domains. Effective parallel implementation requires partitioning of the structure into substructures of approximately equal dimension, which may be difficult for certain types of structures. However, when the bandwidth of system matrices is relatively small, as is the case for many structures, partitioning the model into substructures of approximately equal size should not be difficult. Another approach to the parallel solution of large structure eigenvalue problems was taken by Storaasli, et al. [2], in which different processors were assigned the computation associated with different rows of the matrices involved. This approach requires all processors to have access to all of the model data, and lends itself more readily to implementation on a shared memory machine, whereas the approach presented in this paper would be more appropriate when each processor is assigned its own local memory.

The algorithm presented in this paper is related to a number of substructuring methods that have appeared in the literature. Early references on the component mode synthesis method for dynamic modeling of structures include those by Hurty[3, 4] and by Craig and Bampton[5]. Arora and Nguyen[6] present a method for implementing the basic subspace iteration algorithm using standard substructuring techniques for static analysis of structures. Hale and Meirovitch[7] present a method for iterative refinement of the component modes used to represent substructures, for solving eigenvalue problems. The refinement is done on the substructure level in the context of substructure eigenvalue problems, and coupling between substructures is done approximately by using Lagrange multipliers. Bennighof[8] presents a simpler method for iterative refinement of component modes for eigenvalue problems, in which the coupling between substructures is done exactly. The refinement is based on approximating the dynamic response of substructures to motion at the interfaces in the lowest modes, and the interface motion is approximated by solving reduced eigenvalue problems for the structure. This method exhibits rapid convergence to the lowest modes of the structure and yields a set of optimal component modes for representing substructure motion in those modes.

The algorithm presented in this paper carries out an enhanced version of the subspace iteration algorithm[9] on the substructure level. It assumes that there is a processor available for each substructure and one for structure-level computations. The algorithm makes use of idle time for the substructure processors by setting them to work obtaining substructure eigenpairs, which can then be used very economically to enhance convergence to structure eigenpairs. Additionally, the reduced subspace in a given iteration is augmented by including iterates from the previous iteration for eigenpairs that have not converged. This feature results in a convergence rate approximating that of the Lanczos method[10], but also keeps the reduced subspace small by removing basis vectors that do not enhance convergence significantly. Using the techniques presented in this paper, augmentation of the reduced subspace to include previous iterates and substructure eigenvectors is done on the structure level, without demanding extra effort from substructure processors. This streamlines the computation on the substructure level, so that computation there is the same as if the basic subspace iteration algorithm were being implemented, except for the use of idle time to compute substructure eigenpairs.

A SUBSTRUCTURE-BASED SUBSPACE ITERATION ALGORITHM

The generalized algebraic eigenvalue problem associated with structures in free vibration has the form

$$K\phi_i = \lambda_i M\phi_i, \tag{1}$$

where K and M are symmetric structure stiffness and mass matrices, respectively, here assumed positive definite, and the eigenvectors ϕ_i are associated with mode shapes, and the eigenvalues λ_i are squares of natural frequencies. It is assumed that the eigenvalues are ordered so that $\lambda_1 \leq \lambda_2 \leq \ldots \leq \lambda_n$, and that the structure model has n degrees of freedom. The vector iteration

$$u_{p+1} = K^{-1}Mu_p, \tag{2}$$

where u_p and u_{p+1} are successive trial vectors, serves as the basis for the subspace iteration and

Lanczos algorithms for obtaining lower eigenpairs for complex structures. This vector iteration can be interpreted as a static response of the structure to a force of the form $F = M u_p$. As pointed out in [6], standard substructuring techniques can be used to obtain this static response. If a structure can be partitioned into two substructures, its stiffness matrix can be written in the form

$$K = \begin{bmatrix} K_{LL}^{(1)} & K_{LS}^{(1)} & 0 \\ K_{SL}^{(1)} & K_{SS}^{(1)} + K_{SS}^{(2)} & K_{SL}^{(2)} \\ 0 & K_{LS}^{(2)} & K_{LL}^{(2)} \end{bmatrix}, \tag{3}$$

where superscripts in parentheses tell which substructure a submatrix of K is associated with, and the subscripts L and S refer to degrees of freedom that are *local* to a given substructure, and *shared* with a neighboring substructure, respectively. The vector of displacements can be written in a similar fashion:

$$u = \begin{Bmatrix} u_L^{(1)} \\ u_S \\ u_L^{(2)} \end{Bmatrix}. \tag{4}$$

A matrix which contains a set of displacement vectors, each of which corresponds to a static response of the structure to a unit displacement in one shared degree of freedom, with all other shared degrees of freedom constrained to have zero displacement, is given by

$$\Psi = \begin{bmatrix} -(K_{LL}^{(1)})^{-1} K_{LS}^{(1)} \\ I \\ -(K_{LL}^{(2)})^{-1} K_{LS}^{(2)} \end{bmatrix} = \begin{bmatrix} \Psi_L^{(1)} \\ I \\ \Psi_L^{(2)} \end{bmatrix}. \tag{5}$$

The vectors in this matrix are the "constraint modes" referred to in the component mode synthesis literature[5]. An arbitrary displacement vector u can be represented as $u = \Psi u_S + v$, where u_S contains displacements shared between substructures and v has the form

$$v = \begin{Bmatrix} v^{(1)} \\ 0 \\ v^{(2)} \end{Bmatrix}, \tag{6}$$

i.e., the entries in v are nonzero only in local substructure degrees of freedom. This is true because the constraint mode matrix Ψ allows arbitrary variation in shared degrees of freedom, and the vector v allows arbitrary variation in all other degrees of freedom. With this representation, u_S and v must be determined in order to solve a static response problem of the form $K u = F$. Substituting for u yields

$$K [\Psi u_S + v] = \begin{bmatrix} 0 \\ \hat{K} \\ 0 \end{bmatrix} u_S + \begin{bmatrix} K_{LL}^{(1)} v_1 \\ K_{SL}^{(1)} v_1 + K_{SL}^{(2)} v_2 \\ K_{LL}^{(2)} v_2 \end{bmatrix} = \begin{Bmatrix} F_L^{(1)} \\ F_S \\ F_L^{(2)} \end{Bmatrix}, \tag{7}$$

where

$$\hat{K} = [\, K_{SL}^{(1)} \quad K_{SS}^{(1)} + K_{SS}^{(2)} \quad K_{SL}^{(2)} \,] \, \Psi = \Psi^T K \Psi. \tag{8}$$

The vectors v_1 and v_2 can be obtained without consideration of the global behavior of the structure using the equations

$$v_1 = \left(K_{LL}^{(1)} \right)^{-1} F_L^{(1)}, \quad v_2 = \left(K_{LL}^{(2)} \right)^{-1} F_L^{(2)}, \tag{9}$$

and u_S can then be obtained from the middle partition of Eq. (7) as

$$u_S = \hat{K}^{-1}[F_S - K_{SL}^{(1)} v_1 - K_{SL}^{(2)} v_2]. \tag{10}$$

Substituting for v_1 and v_2 results in

$$\begin{aligned} u_S &= \hat{K}^{-1}[F_S - K_{SL}^{(1)}(K_{LL}^{(1)})^{-1} F_L^{(1)} - K_{SL}^{(2)}(K_{LL}^{(2)})^{-1} F_L^{(2)}] \\ &= \hat{K}^{-1}[F_S + (\Psi_L^{(1)})^T F_L^{(1)} + (\Psi_L^{(2)})^T F_L^{(2)}] \\ &= \hat{K}^{-1} \Psi^T F, \end{aligned} \tag{11}$$

which shows that u_S and the vectors v_1 and v_2 can be obtained simultaneously. In the vector iteration of Eq. (2), F is given by $F = Mu_p$.

The preceding development serves as a basis for a simple substructure-based algorithm for subspace iteration, which will be described in the following. This algorithm is based on the assumption that there are n_{ss} substructures and there is a processor assigned to the computation associated with each substructure, and that another processor is available for structure-level computation. It is further assumed that stiffness and mass matrices are available for each of the substructures, and are stored in such a way that substructure-level processors have access to them. The substructure stiffness and mass matrices can be written in the form

$$K^{(k)} = \begin{bmatrix} K_{SS}^{(k)} & K_{SL}^{(k)} \\ K_{LS}^{(k)} & K_{LL}^{(k)} \end{bmatrix}, \quad M^{(k)} = \begin{bmatrix} M_{SS}^{(k)} & M_{SL}^{(k)} \\ M_{LS}^{(k)} & M_{LL}^{(k)} \end{bmatrix}, \quad k = 1, 2, \ldots n_{ss}. \tag{12}$$

The first task for the substructure-level processors, after the matrices $K^{(k)}$ and $M^{(k)}$ are assembled, is to obtain the matrix

$$\Psi^{(k)} = \begin{bmatrix} I \\ -(K_{LL}^{(k)})^{-1} K_{LS}^{(k)} \end{bmatrix} = \begin{bmatrix} I \\ \Psi_L^{(k)} \end{bmatrix} \tag{13}$$

which requires the decomposition of $K_{LL}^{(k)}$. This decomposition should be saved, as it will be needed in each iteration. Reduced substructure stiffness and mass matrices are then formed based on $\Psi^{(k)}$, and they are given by

$$\hat{K}_0^{(k)} = (\Psi^{(k)})^T K^{(k)} \Psi^{(k)} = [\, K_{SS}^{(k)} + (\Psi_L^{(k)})^T K_{LS} \quad 0 \,] \Psi^{(k)} = K_{SS}^{(k)} + (\Psi_L^{(k)})^T K_{LS} \tag{14}$$

and

$$\hat{M}_0^{(k)} = (\Psi^{(k)})^T M^{(k)} \Psi^{(k)} = (\Psi^{(k)})^T \begin{bmatrix} M_{SS}^{(k)} & M_{SL}^{(k)} \\ M_{LS}^{(k)} & M_{LL}^{(k)} \end{bmatrix} \begin{bmatrix} I \\ \Psi_L^{(k)} \end{bmatrix}. \tag{15}$$

The reduced stiffness and mass matrices $\hat{K}_0^{(k)}$ and $\hat{M}_0^{(k)}$ are sent to the structure-level processor. The lower matrix partition of the matrix product $M^{(k)} \Psi^{(k)}$, which will be denoted by $(M\Psi)_L^{(k)} = M_{LS}^{(k)} + M_{LL}^{(k)} \Psi_L^{(k)}$, must be saved, as it will be required in future calculations.

On the structure level, reduced stiffness and mass matrices for the structure must be assembled. This is straightforward, because the reduced substructure matrices that were just obtained constitute super-element models for the substructures. Reduced matrices for the structure are assembled from these as if they were element matrices and the shared or interface degrees of freedom were inter-element nodal degrees of freedom. The result of this is the $n_S \times n_S$ matrices $\hat{K}_0$ and $\hat{M}_0$, where n_S is the total number of shared degrees of freedom in the structure. The reduced stiffness matrix $\hat{K}_0$ is decomposed for calculations at the present time and later as well, and the m lowest eigenpairs associated with the reduced eigenvalue problem

$$\hat{K}_0 u_{S0} = \hat{M}_0 u_{S0} \lambda_0 \tag{16}$$

are obtained. It should be noted that this eigenvalue problem may be significantly larger than the reduced eigenvalue problems encountered later in this algorithm because its size is determined by the number of shared degrees of freedom. The eigenvectors can be collected in the $n_S \times m$ matrix U_{S0}, which will be assumed to be normalized so that

$$U_{S0}^T \hat{M}_0 U_{S0} = I, \quad U_{S0}^T \hat{K}_0 U_{S0} = \Lambda_0, \tag{17}$$

where I is an identity matrix and Λ_0 is a diagonal matrix of eigenvalues for this reduced eigenvalue problem. Approximate eigenvectors for the entire structure are given by $U_0 = \Psi U_{S0}$, and approximate structure eigenvalues are contained in Λ_0. The approximate eigenvalues and the rows of U_{S0} that pertain to particular substructures are sent to each of the substructure processors, so that iterative refinement can take place.

The refinement of the subspace is represented by $U_0' = K^{-1} M U_0 \Lambda_0$, where the multiplication by Λ_0 serves to keep the vectors in U_0' approximately the same size as in U_0. Recall that the effect of multiplying by $K^{-1} M$ is to divide eigenvectors by their corresponding eigenvalues. Although this refinement is introduced here, its importance will become apparent in a later section. The computation that must be done on the substructure level to carry out the refinement of the subspace includes the following steps. First, a matrix

$$F_0^{(k)} = (M\Psi)_L^{(k)} U_{S0}^{(k)} \Lambda_0 \tag{18}$$

is computed, from which the vectors in $V_0^{(k)'}$ can be obtained:

$$V_0^{(k)'} = (K_{LL}^{(k)})^{-1} F_0^{(k)}. \tag{19}$$

For the next reduced structure stiffness matrix, the following reduced substructure stiffness matrix is calculated:

$$\hat{K}_1^{(k)} = (V_0^{(k)'})^T K_{LL}^{(k)} V_0^{(k)'} = (V_0^{(k)'})^T F_0^{(k)}. \tag{20}$$

For the next reduced structure mass matrix, a matrix $Y_0^{(k)'} = M_{LL}^{(k)} V_0^{(k)'}$ must be obtained first. This matrix is stored, because it will be required later. Then, the matrices

$$\hat{M}_1^{(k)} = (V_0^{(k)'})^T Y_0^{(k)'} \tag{21}$$

and

$$Z_0^{(k)} = ((M\Psi)_L^{(k)})^T V_0^{(k)'} \tag{22}$$

are obtained. The matrices $\hat{K}_1^{(k)}$, $\hat{M}_1^{(k)}$, and $Z_0^{(k)}$ are sent to the structure-level processor.

On the structure level, refinement can be carried out at the same time that the computations described above are being carried out on the substructure level. The refined interface degrees of freedom U_{S0}' are to be obtained, using Eq. (12):

$$U_{S0}' = \hat{K}_0^{-1}(\Psi^T M U_0 \Lambda_0) = \hat{K}_0^{-1} \hat{M}_0 U_{S0} \Lambda_0 = U_{S0}, \tag{23}$$

where the simplification results from the fact that U_{S0} satisfies the eigenvalue problem in Eq. (16), and indicates that the shared degrees of freedom are unchanged in the first iteration. New reduced structure stiffness and mass matrices, based on the refined subspace, are constructed. The reduced stiffness matrix becomes

$$\hat{K}_1 = (U_0')^T K U_0' = (\Psi U_{S0}' + V_0')^T K (\Psi U_{S0}' + V_0') = (U_{S0}')^T \hat{K}_0 U_{S0}' + \sum_{k=1}^{n_{ss}} \hat{K}_1^{(k)}, \tag{24}$$

where it has been noted that $(V_0')^T K\Psi = 0$, which can be seen from Eqs. (6) and (7). In order to form the reduced mass matrix, it is necessary to assemble the matrix

$$Z_0 = \Psi^T M V_0' \tag{25}$$

from the substructure matrices $Z_0^{(k)}$, by superimposing rows of these substructure matrices as the first reduced substructure stiffness and mass matrices $\hat{K}_0^{(k)}$ and $\hat{M}_0^{(k)}$ were superimposed to obtain the matrices $\hat{K}_0$ and $\hat{M}_0$. Then the current reduced structure mass matrix is given by

$$\begin{aligned} \hat{M}_1 &= (U_0')^T M U_0' = (\Psi U_{S0}' + V_0')^T M (\Psi U_{S0}' + V_0') \\ &= (U_{S0}')^T \hat{M}_0 U_{S0}' + U_{S0}^T Z_0 + (U_{S0}^T Z_0)^T + \sum_{k=1}^{n_{ss}} \hat{M}_1^{(k)}. \end{aligned} \tag{26}$$

It should be noted that in this iteration, the first terms of $\hat{K}_1$ and $\hat{M}_1$ need not be computed, because $U_{S0}^T \hat{K}_0 U_{S0} = \Lambda_0$ and $U_{S0}^T \hat{M}_0 U_{S0} = I$. Now the $m \times m$ reduced eigenvalue problem

$$\hat{K}_1 Q_1 = \hat{M}_1 Q_1 \Lambda_1 \tag{27}$$

can be solved. The approximate structure eigenvectors resulting from this are given by

$$U_1 = (\Psi U_{S0}' + V_0') Q_1, \tag{28}$$

so that $U_{S1} = U_{S0}' Q_1$ and $V_1 = V_0' Q_1$, and the approximate eigenvalues are contained in the diagonal matrix Λ_1. The matrix product $Q_1 \Lambda_1$ is calculated and sent to substructure processors.

The next refinement is represented by $U_1' = K^{-1} M U_1 \Lambda_1$, and is nearly identical to the previous one except that V_1 is nonzero, whereas V_0 was a null matrix. Because of this, the substructure-level refinement requires modification of Eq. (18), so that it becomes

$$F_1^{(k)} = \left[(M\Psi)_L^{(k)} U_{S0}^{(k)\prime} + Y_0^{(k)\prime} \right] Q_1 \Lambda_1, \tag{29}$$

and the structure-level refinement of Eq. (23) becomes

$$U_{S1}' = \hat{K}_0^{-1} \left[\hat{M}_0 U_{S0}' + Z_0 \right] Q_1 \Lambda_1. \tag{30}$$

The remaining steps toward obtaining the next reduced eigenvalue problem $\hat{K}_2 Q_2 = \hat{M}_2 Q_2 \Lambda_2$ are unchanged. After this eigenvalue problem is solved, refinement is repeated until convergence takes place.

It should be noted that it is not necessary for substructure processors to wait for the matrix product $Q_1 \Lambda_1$ to be obtained from a reduced eigenvalue problem before calculation of $F_1^{(k)}$ and $V_1^{(k)}$ can begin. Instead, most of the computation for these can be accomplished before $Q_1 \Lambda_1$ becomes available, and when it is available, the necessary postmultiplication can take place. It should also be noted that, although the procedure for substructure-based subspace iteration is more complicated than for the conventional subspace iteration algorithm, the amount of computation that is required is nearly the same for the two approaches. Basically, the substructure-based approach requires some more additions because of the distinction between u_S and v components, but it requires less effort in $K^{-1}F$ computations because K is effectively made block-diagonal.

SUBSPACE AUGMENTATION WITH SUBSTRUCTURE EIGENVECTORS

In the algorithm of the preceding section, it is likely that there will be times when substructure processors are idle. One such time would occur after substructure contributions to the first reduced structure eigenvalue problem have been computed, and before this eigenvalue problem has been solved. In this case, it would probably not be advisable to begin substructure-level refinement before U_{S0} becomes available, because postmultiplication by the rectangular matrix U_{S0} before refinement would reduce the amount of computation required. Also, it may not be possible to divide the computational effort evenly among processors, so that some processors might finish their tasks before others. This raises the question of whether substructure processors can be assigned a task that will make more effective use of this idle time. It can be noted that the eigenvectors of the fixed interface substructure eigenvalue problems $K_{LL}^{(k)} \phi^{(k)} = M_{LL}^{(k)} \phi^{(k)} \gamma^{(k)}$, where $\phi^{(k)}$ and $\gamma^{(k)}$ represent an eigenvector and an eigenvalue, form a complete set for the structure when combined with the vectors in Ψ. A truncated set of these eigenvectors, which are referred to as "normal modes" in the component mode synthesis literature, is appropriate for low-frequency applications such as the present one in which the lowest modes of the structure are sought. A combination of "constraint modes" and "normal modes" has frequently been used in component mode synthesis to obtain approximate lower eigenvalues and eigenvectors for complex structures. If some of these substructure eigenvectors were used to augment the subspace before the reduced eigenvalue problem were solved in a given iteration, the accuracy of the resulting approximate structure eigenpairs could be improved significantly. This suggests that the substructure-level processors should be set to work in any idle time obtaining these eigenvectors using a suitable algorithm. Here the Lanczos algorithm would be appropriate and readily implemented, because the matrices $K_{LL}^{(k)}$ have already been decomposed. For an idea of how quickly substructure eigenpairs would be obtained, the effort required for a vector iteration in all of the substructures is less than for a vector iteration for the entire structure, because the shared degrees of freedom are not included in these computations. Additionally, convergence can be expected to be more rapid, requiring fewer iterations, because the substructure eigenvalues will be more widely spaced. Hence, it is possible that the substructure processors will be able to keep pace with the structure eigenvalue problem as computation progresses through the frequency spectrum, if sufficient idle time is available. Also, it may be advisable to compute substructure eigenpairs even if this requires additional computation time, if their inclusion in reduced subspaces enhances convergence significantly.

In order to include substructure eigenvectors in the subspace for a given reduced eigenvalue problem, additional entries in the reduced matrices $\hat{K}_p$ and $\hat{M}_p$ must be obtained. The following development explains how this can be done at very little computational cost. First, it is necessary to define a matrix containing these eigenvectors:

$$
\Phi = \begin{bmatrix} \Phi^{(1)} & 0 & \cdots & 0 \\ 0 & 0 & \cdots & 0 \\ 0 & \Phi^{(2)} & \cdots & 0 \\ \vdots & \vdots & \ddots & \vdots \\ 0 & 0 & \cdots & \Phi^{(n_{ss})} \end{bmatrix} \tag{31}
$$

where $\Phi^{(k)} = [\phi_1^{(k)} \phi_2^{(k)} \ldots]$ contains a set of eigenvectors for the kth substructure. Additionally, a diagonal matrix of substructure eigenvalues is defined as

$$
\Gamma = \text{block-diag}(\Gamma^{(1)}, \Gamma^{(2)}, \ldots, \Gamma^{(n_{ss})}) \tag{32}
$$

in which $\Gamma^{(k)} = \text{diag}(\gamma_1^{(k)}, \gamma_2^{(k)}, \ldots)$ contains eigenvalues for the kth substructure. It is assumed that all substructure eigenvectors are normalized so that $\Phi^T K \Phi = \Gamma$ and $\Phi^T M \Phi = I$. When the subspace is augmented by adding the vectors in Φ, the reduced structure stiffness and mass matrices become

$$
\hat{K}_p = [U'_p \quad \Phi]^T K [U'_p \quad \Phi] = \begin{bmatrix} (U'_p)^T K U'_p & (\text{sym.}) \\ \Phi^T K U'_p & \Gamma \end{bmatrix} \tag{33}
$$

and

$$
\hat{M}_p = [U'_p \quad \Phi]^T M [U'_p \quad \Phi] = \begin{bmatrix} (U'_p)^T M U'_p & (\text{sym.}) \\ \Phi^T M U'_p & I \end{bmatrix}. \tag{34}
$$

Because $U'_p = K^{-1} M U_p \Lambda_p = K^{-1} M [U'_{p-1} \Phi] Q_p \Lambda_p$, the lower left partition of $\hat{K}_p$ becomes

$$
\Phi^T K U'_p = (\Phi^T M [U'_{p-1} \Phi])(Q_p \Lambda_p), \tag{35}
$$

so that if the lower row partition of the previous reduced mass matrix, $\Phi^T M [U'_{p-1} \Phi]$, is known, $\Phi^T K U'_p$ can be obtained at very little computational cost. Additionally,

$$
\Phi^T K U'_p = \Phi^T K [\Psi U'_{Sp} + V'_p] = \Phi^T K V'_p, \tag{36}
$$

41

because the matrix product $K\Psi$ has zeros in all entries that multiply nonzero entries in Φ, as shown in Eq. (7). Because any vector in V_p' can be represented in terms of a complete set of substructure eigenvectors, it can be easily shown that

$$\Phi^T M V_p' = \Gamma^{-1}\Phi^T K V_p'. \tag{37}$$

Hence,

$$\Phi^T M U_p' = \Phi^T M[\Psi U_{Sp}' + V_p'] = (\Phi^T M \Psi)U_{Sp}' + \Gamma^{-1}\Phi^T K U_p', \tag{38}$$

which indicates that the matrix product $\Phi^T M \Psi$ must be obtained. This matrix product is of the form

$$\Phi^T M \Psi = \begin{bmatrix} (\Phi^{(1)})^T M \Psi \\ (\Phi^{(2)})^T M \Psi \\ \vdots \end{bmatrix} = \begin{bmatrix} (\Phi^{(1)})^T (M\Psi)_L^{(1)} \\ (\Phi^{(2)})^T (M\Psi)_L^{(2)} \\ \vdots \end{bmatrix}, \tag{39}$$

which indicates that rows are contributed independently by different substructures. This indicates that as each substructure eigenvector $\phi_i^{(k)}$ is obtained, the product $(\phi_i^{(k)})^T (M\Psi)_L^{(k)}$ should be computed.

From the above development, it is clear that $\Phi^T K U_p'$ can be obtained very easily if $\Phi^T M U_{p-1}'$ and $Q_p \Lambda_p$ are known, using Eq. (35). Additionally, $\Phi^T M U_p'$ can be obtained very easily given $\Phi^T K U_p'$ and U_{Sp}', using Eq. (38). All that remains to be established is a starting point for this recursive process. This is available in the submatrix $\Phi^T K U_0'$, which is given by

$$\Phi^T K U_0' = \Phi^T M U_0 \Lambda_0 = (\Phi^T M \Psi)U_{S0}\Lambda_0, \tag{40}$$

because $\Phi^T K U_0 = (\Phi^T K \Psi)U_{S0} = 0$. After this starting point, all of the calculations that are required to include Φ in the subspace involve matrices of dimension n_S or less, and can easily be done on the structure level.

SUBSPACE REPETITION

The feature of Lanczos algorithms which is responsible for faster convergence than is observed in the subspace iteration algorithm is the retention of all previous iterates in the reduced subspace. In subspace iteration, only the current iterates are included in the reduced subspace. This section describes how Lanczos-like convergence can be achieved in the context of subspace iteration at an extremely small additional computational cost per iteration, by retaining the subspace from the previous iteration in the current subspace. The procedure for doing this is described in the following.

If the current subspace is to span the set of current and immediately previous iterates, as well as some substructure eigenvectors, this calls for reduced stiffness and mass matrices of the form

$$\hat{K}_p = [U_p \quad U_p' \quad \Phi]^T K [U_p \quad U_p' \quad \Phi] = \begin{bmatrix} U_p^T K U_p & & \text{(sym.)} \\ (U_p')^T K U_p & (U_p')^T K U_p' & \\ \Phi^T K U_p & \Phi^T K U_p' & \Phi^T K \Phi \end{bmatrix} \tag{41}$$

and

$$\hat{M}_p = [U_p \quad U_p' \quad \Phi]^T M [U_p \quad U_p' \quad \Phi] = \begin{bmatrix} U_p^T M U_p & & \text{(sym.)} \\ (U_p')^T M U_p & (U_p')^T M U_p' & \\ \Phi^T M U_p & \Phi^T M U_p' & \Phi^T M \Phi \end{bmatrix}. \tag{42}$$

The matrix partitions $U_p^T K U_p$ and $U_p^T M U_p$ are simply equal to Λ_p and I, respectively, and the matrix partitions outside the first row and column of $\hat{K}$ and $\hat{M}$ are obtained as described in preceding sections of this paper. The remaining four matrix partitions can be obtained at very little computational expense.

The matrix partition $(U'_p)^T K U_p$, upon closer examination, is given by

$$(U'_p)^T K U_p = (K^{-1} M U_p \Lambda_p)^T K U_p = \Lambda_p (U_p^T M U_p) = \Lambda_p, \tag{43}$$

so that it is already available at no additional computational expense. The matrix partition $(U'_p)^T M U_p$ can be written as

$$(U'_p)^T M U_p = (U'_p)^T (K \, K^{-1}) M U_p (\Lambda_p \Lambda_p^{-1}) = \left((U'_p)^T K U'_p \right) \Lambda_p^{-1} \tag{44}$$

so that the columns of $(U'_p)^T K U'_p$ must simply be divided by the most recent eigenvalue estimates to yield $(U'_p)^T M U_p$. The computation associated with this is negligible. Finally, given that $U_p = [U_{p-1} U'_{p-1} \Phi] Q_p$, the lower left partitions of $\hat{K}_p$ and $\hat{M}_p$ can be written in terms of the lower rows of the previous reduced matrices $\hat{K}_{p-1}$ and $\hat{M}_{p-1}$ as

$$\Phi^T K U_p = \left(\Phi^T K \, [\, U_{p-1} \quad U'_{p-1} \quad \Phi \,] \right) Q_p \tag{45}$$

and

$$\Phi^T M U_p = \left(\Phi^T M \, [\, U_{p-1} \quad U'_{p-1} \quad \Phi \,] \right) Q_p, \tag{46}$$

so that they can also be obtained with very little effort. It should be noted that subspace repetition requires that some previous results must be saved for the refinement procedure, and that some straightforward changes must be made to the refinement procedure.

Upon convergence to a structure eigenvector, a further iteration on that eigenvector will simply produce a duplicate copy of that eigenvector. This means that the vectors in the reduced subspace will no longer be linearly independent, and that the reduced stiffness and mass matrices will both become singular. In order to avoid this singularity, and the wasted effort associated with constructing duplicate copies of eigenvectors, it is necessary to incorporate a convergence check in the algorithm. A natural approach to checking to see if an eigenvector has converged is to check for singularity in the reduced mass matrix, because the mass matrix is guaranteed to be positive definite as long as the vectors in the reduced subspace are linearly independent. More specifically, convergence of the ith eigenvector can be checked by determining whether the 2×2 matrix $\hat{m}_{iip} = [u_{ip} u'_{ip}]^T M [u_{ip} u'_{ip}]$ is singular, or nearly singular. The determinant of this 2×2 matrix becomes

$$\det \hat{m}_{iip} = \det \begin{bmatrix} u_{ip}^T M u_{ip} & (\text{sym.}) \\ u_{ip}^T M u'_{ip} & u'^T_{ip} M u'_{ip} \end{bmatrix} = \det \begin{bmatrix} 1 & (\text{sym.}) \\ (k'_{iip}/\lambda_{ip}) & m'_{iip} \end{bmatrix} = m'_{iip} - (k'_{iip}/\lambda_{ip})^2, \tag{47}$$

so that only the ith diagonal elements of the submatrices $(U'_p)^T K U'_p$ and $(U'_p)^T M U'_p$ and the current eigenvalue estimates are involved, and the convergence check is simply whether $\det \hat{m}_{iip} < \epsilon$, where ϵ is a suitable small number.

If this convergence check indicates that a trial vector has converged to an eigenvector, then no further iteration is required for that eigenpair. It would be better, however, to determine whether convergence to an eigenpair will occur when the current reduced eigenvalue problem is solved, rather than whether convergence has already occurred, so that the last refinement of the trial vector is used productively, rather that being discarded. Experience shows that this can be done very effectively by simply using a larger value of ϵ than would be used to test for final convergence. As converged eigenvectors are obtained, new trial vectors can be added to take their place in the set of trial vectors that are undergoing iteration. This suggests a strategy in which the number of approximate eigenpairs obtained from the first reduced eigenvalue problem equals the number of trial vectors that can undergo iteration simultaneously for most effective implementation. This number can be much smaller than the number of eigenpairs that are to be obtained. As trial vectors converge to eigenvectors, they can be removed from the subspace undergoing iteration, and can be replaced by new trial vectors so that the number of trial vectors undergoing iteration remains constant. In this manner, computation can proceed upward through the frequency spectrum. When subspace repetition is employed, the reduced stiffness and mass matrices are frequently nearly singular. Because of this, it is important to ensure that they are as well-conditioned as possible. This is the motivation for multiplying the refined trial vectors by eigenvalue estimates, as mentioned in the second section.

When this approach is taken, the upper left partition in the reduced stiffness matrix remains diagonal, and the eigenvalues on the diagonal can be divided into two groups. The eigenvalues in the first group are associated with eigenvectors that have converged. The eigenvalues in the second group in this upper left partition, equal in number to the number of trial vectors undergoing iteration, are only approximate and have not converged yet. The middle row partition of $\hat{K}_p$ from Eq. (41) has as many rows as the number of trial vectors undergoing iteration. It can be shown rather easily that the partitions $(U_p')^T K U_p$ and $(U_p')^T M U_p$, which are rectangular, consist of null submatrices on the left, and square submatrices on the right that are given by Eqs. (43) and (44), at least in perfect arithmetic. Due to roundoff error, it can be anticipated that the set of trial vectors that have not converged will not remain orthogonal to exact lower eigenvectors, so that the null submatrices in these partitions should be nonzero to reflect this. If these submatrices are nonzero due to roundoff error, solving the reduced eigenvalue problems would enforce orthogonality between the converged eigenvectors and the trial vectors that have not yet converged. Simply setting entries in $\hat{K}_p$ and $\hat{M}_p$ to zero as would be appropriate in perfect arithmetic can be expected to lead to problems of loss of orthogonality. For this reason, occasional explicit orthogonalization will probably be warranted, as in the Lanczos algorithm, although this issue has not been investigated fully at this time.

In comparison with the Lanczos algorithm, the subspace repetition approach presented here has certain advantages. As a rule of thumb, one can expect about one third of the eigenpairs of the subspace generated by the Lanczos procedure to be accurate approximations of the lower eigenpairs of a given structure model. By a theorem by Kaniel[11], which establishes the convergence rate in the Lanczos algorithm, eigenpairs of the Lanczos subspace approximate extremal eigenpairs of the structure model at both ends of the spectrum. This suggests that perhaps another third of the Lanczos eigenpairs approximate the highest eigenpairs of the structure model well, and the remaining eigenpairs are distributed throughout the spectrum between these two extremes. Because the highest eigenpairs of a structure model generated by the finite element method ordinarily have little or no practical significance, it would appear that perhaps half of the Lanczos eigenpairs can be considered extraneous. The subspace repetition approach presented here has the Lanczos feature of retaining previous iterates, but it also provides a way to remove vectors from the subspace that do not significantly enhance convergence to lower structure eigenpairs. In [12] it is pointed out that the subspace generated by the Lanczos procedure for solving some large problems in solid and structural mechanics can become so large that storage limits are exceeded. It would appear that a similar result can be expected in the solution of very large eigenvalue problems if "extraneous" components in the reduced subspace are not removed.

NUMERICAL EXAMPLE

As an example, the algebraic eigenvalue problem associated with a model of a helicopter tail-boom is solved using the method presented in this paper. This model has been used in References [6] and [8]. The model is shown in Figure 1, and consists of an open truss structure

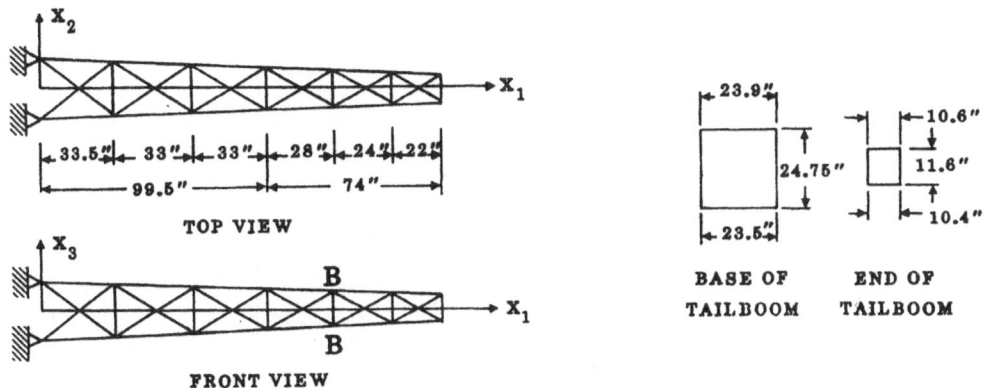

Figure 1. Helicopter tail-boom model.

with 108 truss members and 28 joints. Each joint has three displacement degrees of freedom, for a total of 72 active degrees of freedom in the model. All of the members have an axial stiffness $EA = 10.5 \times 10^6$ and mass per unit length $m = 2.588 \times 10^{-4}$. The model can be partitioned into two substructures by dividing it along the line B-B shown in Fig. 1. Considering that the left substructure has fixed end conditions on both ends and the right substructure is free on the right end when shared degrees of freedom are fixed, this division into substructures results in approximately equal flexibility and numbers of degrees of freedom for the two substructures. The left and right substructures have thirty-six and twenty-four local degrees of freedom, respectively, and there are twelve degrees of freedom that are shared between the two substructures.

The algorithm presented in this paper was implemented on this problem, and six vectors were allowed to undergo iteration simultaneously. To investigate the benefit of calculating substructure eigenvectors to be included in the reduced subspace, different numbers of these per substructure (zero, two, and four) were included in the subspace after the first reduced eigenvalue problem was solved. They were introduced at this time in the iteration process on the assumption that the most significant period of idle time that the substructure processors would encounter would occur while the first reduced eigenvalue problem was being solved by the structure-level processor. The value of ϵ that was used for checking convergence of eigenvectors as in Eq. (47) and following was 10^{-9}. If the determinant of the 2×2 mass matrix was less than ϵ, only a single vector was included in the reduced subspace for representing the eigenvector in question. If this determinant was less than two orders of magnitude smaller than ϵ, the included vector was a linear combination of the present and previous iterates, found by solving the associated 2×2 eigenvalue problem. If the determinant was more than two orders of magnitude smaller than ϵ, it was assumed that the present iterate was sufficiently accurate by itself. This choice of ϵ resulted in eigenvalue accuracy of at least ten decimal places, and usually twelve or more places. Table 1 shows the results of doing this for several iterations, with different numbers of substructure eigenvectors included. The symbols appearing in the table indicate whether trial vectors appeared singly, or with previous iterates, or were found to have converged to eigenvectors in each iteration. As trial vectors converged to eigenvectors, they were replaced in the subspace undergoing iteration with unrefined vectors corresponding to higher eigenpairs from the previous iteration's reduced eigenvalue problem. To keep the number of unconverged trial vectors constant at twelve from iteration to iteration, two new trial vectors were introduced for each converged eigenvector, as long as enough new trial vectors were available. The maximum number of new trial vectors available from the previous iteration was six, because this was the number of repeated vectors in the previous iteration. It was found that substructure eigenvectors were sometimes linearly dependent on the other

Table 1. Convergence histories for the structure of Figure 1.

i	Eigenvalue	0 eigenvectors per substructure					2 eigenvectors per substructure					4 eigenvectors per substructure				
	Iteration	0	1	2	3	4	0	1	2	3	4	0	1	2	3	4
1	0.188023067214D+05	o	•	★	★	★	o	•	★	★	★	o	•	★	★	★
2	0.211263079473D+05	o	•	★	★	★	o	•	★	★	★	o	•	★	★	★
3	0.407509341510D+06	o	•	•	•	★	o	•	•	★	★	o	•	•	★	★
4	0.437043229756D+06	o	•	•	•	★	o	•	•	•	★	o	•	•	★	★
5	0.455316243911D+06	o	•	•	•	★	o	•	•	★	★	o	•	•	★	★
6	0.159300002381D+07	o	•	•	•	★	o	•	•	•	★	o	•	•	•	★
7	0.204466878863D+07	o		o	•	•	o		o	•	•	o		o	•	★
8	0.226197098651D+07	o		o	•	•	o		o	•	•	o		o	•	•
9	0.231171669440D+07	o		o		o	o		o	o	•	o		o	o	•
10	0.562457228264D+07	o		o		o	o		o	o	•	o		o	o	•
11	0.576241794026D+07	o				o	o			o	o	o			o	•
12	0.632615520911D+07	o				o	o			o	o	o			o	o
13	0.999244724618D+07				o						o				o	o
14	0.108173081977D+08				o						o				o	o
15	0.117535707427D+08															o
Reduced evp. dimension		12	12	14	14	16	12	16	18	20	22	12	20	22	18	20
Total no. of ss. evecs.		0	0	0	0	0	0	4	4	4	4	0	8	8	1	1

o = trial vector appearing singly

• = repeated trial vector

★ = converged eigenvector

Table 2. Convergence history using subspace iteration algorithm.

i	Iteration / Eigenvalue	Standard subspace iteration							
		0	1	2	3	4	5	6	7
1	0.188023067214D+05	o	o	★	★	★	★	★	★
2	0.211263079473D+05	o	o	★	★	★	★	★	★
3	0.407509341510D+06	o	o	o	o	o	★	★	★
4	0.437043229756D+06	o	o	o	o	o	★	★	★
5	0.455316243911D+06	o	o	o	o	o	o	★	★
6	0.159300002381D+07	o	o	o	o	o	o	o	★
7	0.204466878863D+07	o			o	o	o	o	o
8	0.226197098651D+07	o			o	o	o	o	o
9	0.231171669440D+07	o						o	o
10	0.562457228264D+07	o						o	o
11	0.576241794026D+07	o							o
12	0.632615520911D+07	o							

vectors in the subspace. When this was true, substructure eigenvectors were removed, starting with the one corresponding to the lowest substructure natural frequency, until the reduced matrices were no longer singular. One would expect that the substructure eigenvectors might stop contributing significantly as accurate structure eigenvalues are obtained that are higher in the frequency spectrum.

From these results, it is clear that convergence is quite rapid, as an average of two or more eigenpairs converge in each iteration, with only six vectors undergoing iteration simultaneously. Including substructure eigenvectors improves convergence somewhat, but not tremendously. For a comparison with subspace iteration without subspace repetition, the same problem was solved by using the basic subspace iteration algorithm with a subspace containing six trial vectors, after the initial reduced eigenvalue problem was solved. The results of doing this are shown in Table 2, where a star indicates that the eigenvalue has converged to the same accuracy as was obtained using subspace repetition. Upon convergence to an eigenpair, a trial vector from the solution of the first eigenvalue problem was refined and then included in the reduced subspace for the next eigenvalue problem. It is clear from these results that subspace repetition enhances convergence significantly.

SUMMARY AND CONCLUSIONS

In this paper, an algorithm is presented for the parallel solution of large structure eigenvalue problems. The algorithm employs one processor for the computation associated with each substructure, and one processor to handle structure-level computation. Computation is carried out simultaneously on both the substructure and structure levels. In idle time, substructure processors can calculate substructure eigenvectors, which can be used to augment the reduced subspace and enhance convergence. Once a substructure eigenvector is computed, along with a vector-matrix product on the substructure level, no further computation is required on the substructure level for inclusion of the eigenvector in the reduced subspace.

Subspace repetition is also employed to accelerate convergence, in which iterates from the preceding reduced eigenvalue problem are included in the reduced subspace. This is similar to what is done in the Lanczos algorithm, in which all prior iterates are retained in the reduced subspace, but the subspace repetition approach provides a way to remove vectors from the Lanczos subspace that do not significantly improve convergence, so that storage requirements are reduced. Employing subspace repetition also does not require any additional effort on the part of substructure processors, because the very small amount of extra computation that is required is done on the structure level. The numerical example presented in this paper demonstrates the method and shows that convergence is significantly faster with this algorithm than with the basic subspace iteration algorithm.

ACKNOWLEDGEMENT

This work was supported by a National Science Foundation Engineering Initiation Award, grant number EET-8709155, monitored by Dr. George K. Lea, whose support is greatly appreciated. Additional support was provided through the ONR-sponsored Basic Research Program under Contract N00014-87-K-0346.

REFERENCES

1. Glowinski, R., Golub, G. H., Meurant, G. A., and Periaux, J., eds., *Proceedings of the First International Symposium on Domain Decomposition Methods for Partial Differential Equations*, SIAM, 1988.

2. Storaasli, O., Bostic, S., Patrick, M., Mahajan, U. and Ma, S., "Three Parallel Computation Methods for Structural Vibration Analysis," *Proc. of 29th Structures, Struc. Dynamics and Materials Conference*, Williamsburg, Va., April 1988, pp. 1401-1410.

3. Hurty, W. C., "Vibrations of Structural Systems by Component-Mode Synthesis," *J. of the Engineering Mechanics Division, ASCE*, Vol. 86, Aug. 1960, pp. 51-69.

4. Hurty, W. C., "Dynamic Analysis of Structural Systems Using Component Modes," *AIAA Journal*, Vol. 3, No. 4, 1965, pp. 678-685.

5. Craig, R. R. Jr. and Bampton, M. C. C., "Coupling of Substructures for Dynamic Analysis," *AIAA Journal*, Vol. 6, No. 7, 1968, pp. 1313-1319.

6. Arora, J. S. and Nguyen, D. T., "Eigensolution for Large Structural Systems with Substructures," *International Journal for Numerical Methods in Engineering*, Vol. 15, No. 3, 1980, pp. 333-341.

7. Hale, A. L. and Meirovitch, L., "A General Procedure for Improving Substructure Representation in Dynamic Synthesis," *J. of Sound and Vibration*, Vol. 84, Sept. 22, 1982, pp. 269-287.

8. Bennighof, J. K., "Component Mode Iteration for Frequency Calculations," *AIAA Journal*, Vol. 25, No. 7, 1987, pp. 996-1002.

9. Bathe, K.-J., *Finite Element Procedures in Engineering Analysis*, Prentice-Hall, Englewood Cliffs, N. J., 1982.

10. Parlett, B. N., *The Symmetric Eigenvalue Problem*, Prentice-Hall, Englewood Cliffs, N. J., 1980.

11. Kaniel, S., "Estimates for Some Computational Techniques in Linear Algebra," *Math. Comp.*, Vol. 20, 1966, pp. 369-378.

12. Ferencz, R. M., "A Comparison of Element-by-Element Preconditioned Iterative Methods for Solid and Structural Mechanics," *Proc. of Two-Day Symposium on Solution of Super Large Problems in Computational Mechanics*, Mystic, Conn., October 1988.

PARALLEL SPECTRAL ELEMENT METHODS FOR THE INCOMPRESSIBLE NAVIER-STOKES EQUATIONS

Paul F. Fischer and Anthony T. Patera

Massachusetts Institute of Technology
Room 3-264, 77 Massachusetts Avenue
Cambridge MA 02139

Abstract

We present a parallel spectral element method for solution of the unsteady incompressible Navier-Stokes equations in general three-dimensional geometries. The approach combines high-order spatial discretizations with iterative solution techniques in a way which exploits with high efficiency currently available medium-grained distributed-memory parallel computers. Emphasis is placed on the development of algorithm constructs which allow for solution of physically relevant problems; we specifically address the problem of parallel solution in domains of general topology. The success of the procedure is demonstrated by several examples of moderate Reynolds number Navier-Stokes calculations on the Intel vector hypercube.

1 Introduction

The solution of incompressible fluid dynamics problems by numerical simulation has advanced rapidly in recent years due to simultaneous improvements in algorithms and computers. However, despite these advances, the large number of degrees-of-freedom required to resolve even relatively simple three-dimensional laminar flows, let alone transitional or turbulent flows, has prevented computational fluid dynamics from addressing many problems of fundamental and practical importance. In essence, large-scale fluid mechanics calculations are still too costly in terms of human and computational resources to assume the role of "primary means of analysis."

A promising approach to reducing the costly nature of fluid dynamics calculations is to solve problems not on a single (expensive) computer, but rather to distribute the work amongst many less powerful (and less expensive) processors. The potential increase in efficiency due to the economies of parallel processing derive not only from decreases in direct costs, but also from improvements in productivity and creativity brought about by a more local and interactive computing environment. Unfortunately, the availability of parallel processors does not necessarily imply their efficient usage, and care must be taken in developing numerical algorithms that are appropriate for parallel implementation.

The purpose of the present paper is to illustrate a spectral element algorithm for the Navier-Stokes equations which exploits with high parallel efficiency the highly economical parallel computers currently available. Our work builds extensively on past work on parallel partial differential equation solution in the choice of an iterative solver, as well

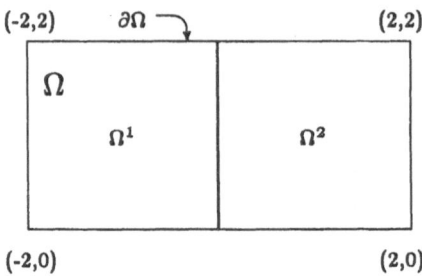

Figure 1. Computational domain Ω consisting of $K = 2$ subdomains.

as in the underlying strategy of load balancing, communication, and topological embeddings. In particular, our schemes are founded on the following well-developed precepts: use of iterative solvers that exploit sparsity and minimize non-concurrent operations, e.g. [1,2]; geometry-based distribution of work amongst processors, e.g. [3,4,5,6]; exploitation of nearest-neighbor sparsity and substructuring to minimize communication, e.g. [7,8]; efficient embedding of vector reduction operations to allow for more general and implicit solution algorithms, e.g. [9,10].

The methods used represent an extension of these well-established ideas in the following ways. First, the spectral element discretizations [11] employed are high-order, leading not only to improved accuracy but also to a more efficient, work-intensive "native" medium-grained parallelism. Second, the discretizations, solvers, and parallel constructs are built upon the general foundation of locally-structured/globally-unstructured representations, thus allowing for efficient implementation in arbitrary geometries. Third, the equations solved are the full equations describing viscous fluid flow [6], as opposed to (second-order elliptic) subsets of the Stokes problem; all potentially non-concurrent hazards are therefore addressed. Lastly, the methods are implemented on a fast vector parallel processor, thus providing a useful and economical fluid mechanics analysis tool. Numerous, physically relevant, examples are presented to illustrate this point.

2 Spectral Element Discretizations

Our numerical methods for the Navier-Stokes equations are premised upon a 'layered' approach, in which the discretizations and solvers are constructed on the basis of a hierarchy of nested operators proceeding from the highest to the lowest derivatives. For incompressible viscous flow equations the linear self-adjoint elliptic Laplace operator represents the 'kernel' of our Navier-Stokes algorithm insofar that it involves the highest spatial derivatives. This operator governs the continuity requirements, conditioning and stability of the system. The fully discretized Navier-Stokes equations are typically solved at each time step by performing a series of elliptic solves and preconditioning steps.

The spectral element method [11,12] is a generalized variational scheme which exploits the rapid convergence rates of spectral methods while retaining the geometric flexibility of the finite element techniques. It is based upon a macro- (spectral) element discretization in which the degrees-of-freedom internal to elements are coefficients of global (elemental) expansion functions with C^0 continuity imposed across element boundaries. With an appropriate choice of interpolants and quadrature formulae, it is can be shown that the error for problems having smooth solutions will decrease exponentially fast as the order of the expansion, N, is increased.

As the parallel implementation of the spectral element method is intrinsically tied to the discretization, we briefly outline the basis of the discretization by considering the solution to a two-dimensional Poisson equation on the domain shown in Figure (1).

$$-\nabla^2 u = f \quad in \ \Omega,$$

$$u = 0 \quad on \ \partial\Omega. \tag{1}$$

Equation (1) can be equivalently expressed as: Find $u \in \mathcal{H}_0^1(\Omega)$ such that

$$\int_\Omega \nabla\phi\nabla u \, d\Omega = \int_\Omega \phi f \, d\Omega \quad \forall\phi \in \mathcal{H}_0^1(\Omega), \tag{2}$$

where the space $\mathcal{H}_0^1$ is the space of all functions which are zero on the boundary and have a square integrable first derivative. The variational form has the significant advantage that it reduces the required level of continuity on the solution from C^1 to C^0, which in turn has implications as regards parallel communication.

Discretization of the variational statement proceeds by restricting the admissible solutions and trial functions in (2) to a finite-dimensional subspace, X_h, of the infinite-dimensional space, $\mathcal{H}_0^1$. For the spectral element method we choose the space X_h to be:

$$X_h = \{\phi|_{\Omega^k} \in \mathbf{P}_N(\Omega^k)\} \cap \mathcal{H}_0^1(\Omega) \tag{3}$$

where $\mathbf{P}_N(\Omega^k)$ is the space of all polynomials of degree $\leq N$ in each spatial direction on element k. The spectral element method is thus characterized by the discretization pair $h = (K, N)$, where K is the number of subdomains (elements) and N is the order of the polynomial approximations. For reasons of efficiency (tensor products, [13]) the subdomains are taken to be quadrilaterals in $\mathbf{R}^2$ and hexahedra in $\mathbf{R}^3$. The spectral element discretization of (1) thus corresponds to numerical quadrature of the variational form (2) restricted to the subspace X_h: Find $u_h \in X_h$ such that:

$$\int_\Omega \nabla\phi_h\nabla u_h \, d\Omega = \int_\Omega \phi_h f_h \, d\Omega \quad \forall\phi_h \in X_h(\Omega), \tag{4}$$

where f_h is the interpolant of f in the space X_h.

While (4) is a statement of the type of solution which we seek, it does not indicate the form in which our solution will be represented, i.e., the choice of basis functions to be used for the polynomials in X_h. The Legendre spectral element method employs a tensor product form of Lagrangian interpolants based on a local, elemental, mapping of $\mathbf{x} \in \Omega^k \rightarrow \mathbf{r} = (r, s) \in [-1, 1]^2$. We consider for illustration the case where the elemental decomposition consists of the union of squares with sides of length 2. Within each element, u_h has the form:

$$u_h(x, y)|_{\Omega^k} = \sum_{p=0}^{N}\sum_{q=0}^{N} u_{pq}^k h_p(r)h_q(s) \ , \tag{5}$$

where $u_{pq}^k = u^k(r_p, s_q)$ are the unknown grid values of the approximate solution in element k. The interpolants, $h_i(\xi)$, satisfy:

$$h_i(\xi) \in \mathbf{P}_N[-1, 1] \tag{6}$$
$$h_i(\xi_j) = \delta_{ij} \ ,$$

where the grid points, ξ_j, are chosen to be the Gauss-Lobatto-Legendre quadrature points [14,15].

With the explicit representation of functions given by (5), it is straight forward to evaluate the discrete variational form (4). Integration is performed using Gauss-Lobatto quadrature:

$$\int_\Omega f(x, y)d\Omega \rightarrow \sum_{k=1}^{K}\left\{\sum_{i=0}^{N}\sum_{j=0}^{N} f_{ij}^k \rho_{ij}\right\} \ , \tag{7}$$

where ρ_{ij} is the quadrature weight associated with the point $\mathbf{r}_{ij}$. The derivatives at the quadrature points are computed as:

$$\left.\frac{\partial u}{\partial x}\right|_{\mathbf{x}_{ij}^k} \quad \rightarrow \quad \left.\frac{\partial u^k}{\partial r}\right|_{\mathbf{r}_{ij}} = (D_{ip}\,u_{pj}^k), \tag{8}$$

$$D_{ij} \equiv \left.\frac{dh_j(r)}{dr}\right|_{r=r_i} \quad ,$$

where, for notational convenience in this and the following two equations, we use (.) to imply summation over the repeated index within the parentheses. The variational statement (4) therefore takes the form:

$$\sum_{\Omega^k}\left\{\sum_{i=0}^{N}\sum_{j=0}^{N}\rho_{ij}\left[(D_{ip}\,\phi_{pj}^k)(D_{iq}\,u_{qj}^k) + (D_{jp}\,\phi_{ip}^k)(D_{jq}\,u_{iq}^k) - \phi_{ij}^k\,f_{ij}^k\right]\right\} = 0 \ , \tag{9}$$

where it remains to specify what are the admissible values of ϕ_{ij}^k.

Since Eq. (9) holds for arbitrary $\phi_h \in X_h$, the requisite discrete system of equations can be generated by setting $\phi_{ij}^k = \delta_{ii'}\delta_{jj'}\delta_{kk'}$, for all (i', j', k') corresponding to unique points in the domain interior. Note that the outermost summation in equation (9) implies that in the case where ϕ_{ij}^k has a physical counterpart in another element, k', (i.e. ϕ_{ij}^k lies on the interface between elements k and k'), the contributions to the integral (sum) from the adjacent elements must be added together. We refer to this operation as direct stiffness summation and denote it by $\sum_{\Omega^k}'$. The final system to be solved for u is therefore:

$$\sum_{\Omega^k}'\left\{\sum_{p=0}^{N}D_{pi}(D_{pq}u_{qj}^k)\rho_{pj} + \sum_{p=0}^{N}D_{pj}(D_{pq}u_{iq}^k)\rho_{ip}\right\} = \sum_{\Omega^k}' f_{ij}^k\rho_{ij} \tag{10}$$

Although the discrete Laplacian in (10) is never evaluated in standard matrix form, it is convenient to express equation (10) as:

$$\mathbf{A}\mathbf{u} = \mathbf{B}\mathbf{f} \ , \tag{11}$$

where $\mathbf{A}$ is the global stiffness matrix and $\mathbf{B}$ is the mass matrix.

3 Iterative Solution Procedures

The natural choice of solution algorithm in a parallel environment is an iterative procedure, given that such techniques can be both highly local and concurrent. In this section we examine the computational complexity of iterative spectral element solvers.

At the heart of any iterative solver is the evaluation of matrix-vector products of the form $\mathbf{A}\mathbf{u}$. It is clear from equation (10) that for discretizations in $\mathbf{R}^d$ these products can be efficiently calculated in $O(KN^{d+1})$ operations using sum-factorization methods [13]. Additional computational efficiency can be gained by exploiting the regular structure of the spectral (intra-element) operators and recognizing that the inner most kernel of the matrix-vector product, $(D_{ip}\,u_{pj})$, is exactly equivalent to a *matrix-matrix* product which can be rapidly evaluated on many vector architectures. In addition to the $O(KN^{d+1})$ computational effort, the direct stiffness summation will require $O(KN^{d-1})$ operations, corresponding to the number of degrees-of-freedom lying on the element interfaces. It thus follows that the number of clock cycles required to evaluate the left side of (10) on a single processor is

$$Z_1^e = c_1 K N^{d+1} + c_2' K N^d + c_3 K N^{d-1}, \tag{12}$$

where the constants c_1, c_2', and c_3 depend only (weakly) on spatial dimension. The $O(KN^d)$ contribution to Z_1^e is only present in the case of complex geometry or non-separable coefficients. It should also be noted that only $O(KN^d)$ storage is required to evaluate $\mathbf{A}u$.

The proper choice of spectral element basis is directly reflected in the "good" computational complexity estimate Z_1^e. First, the sum-factorization (10) and the operation count (12) applies to general-geometry isoparametric spectral element discretizations of non-separable equations [12], due to the tensor product spaces (3), tensor product quadratures (7), and tensor product bases (5) described in Section 2. Second, the direct stiffness summation contribution to Z_1^e is only $O(KN^{d-1})$, rather than $O(KN^{d+1})$, due to our choice of basis in which the number of test functions which are nonzero on the elemental boundary is minimal. Although the fact that the direct stiffness summation work is small does not appear particularly important in the single-processor estimate (12), in the parallel case the direct stiffness contribution will be the leading-order *communication* term.

We next consider the work estimate for Jacobi (diagonal) preconditioned conjugate gradient solution [16] of the multi-dimensional elliptic equation (11). In its simplest form, the conjugate gradient algorithm requires evaluation of (global) matrix-vector products, scalar-vector multiplies, vector-vector additions, and vector inner products. In serial processor applications, the majority of the computational effort is typically associated with the matrix-vector products, as the computation for the remaining operations scales only as the total number of degrees-of-freedom, $O(KN^d)$. Denoting N_ϵ^A as the total number of $\mathbf{A}$ iterations required to reach a specified tolerance, ϵ, the single processor operation count for conjugate gradient solution of (10) will be:

$$Z_2^e = N_\epsilon^A \left\{ c_1 K N^{d+1} + c_2 K N^d + c_3 K N^{d-1} \right\} , \tag{13}$$

where c_2 accounts for the additional conjugate gradient operations. Note that the condition number of the preconditioned $\mathbf{A}$ system is $O(K_1^2 N^2)$ [17], which implies that $N_\epsilon^A \sim O(K_1 N)$ [16], where K_1 is the number of spectral elements in a *single* spatial direction. This convergence rate, though respectable for a high-order method, is clearly not order-independent, and will deteriorate significantly for large problems. The convergence rate can be improved by the use of recently developed spectral element multigrid algorithms [17,18].

4 Parallel Implementation

Due to the heterogeneous structure of the discretization, the natural parallel implementation of the spectral element method is to to distribute the data and work amongst independent processors on an element by element basis. This decomposition is ideally suited to medium-grained architectures which consist of a large number of processors, each capable of supporting a large number of degrees-of-freedom. In this section we review some basics of distributed memory parallel processing, outline a spectral element implementation which allows for efficient parallel solutions for domains of *general topology*, and examine the computational complexity of parallel Navier-Stokes calculations.

4.1 Distributed Memory Parallel Processing

Distributed memory parallel computers offer tremendous potential as a scalable parallel architecture because both the memory and the primary memory-processor bandwidth grow in direct proportion to the number of processors in the system. Distributed memory systems are comprised of independent computers, or nodes, each containing a processor which is directly connected to *local* memory containing both data and source code. Processors typically access non-local data via a slow, indirect, network and data transfer protocol, where the network is characterized by its topology, e.g., ring, mesh, or hypercube. This direct/indirect memory hierarchy favors algorithms in which the majority of the computational effort is localized within a processor.

During execution, each node runs asynchronously, following locally resident source code and processing locally resident data. In practice, distributed memory machines are typically programmed in what has been termed "single program, multiple data" fashion where the nodes have identical copies of the source code but different sets of data [19]. Data is transferred between processors via a message passing protocol whereby data packets are explicitly sent from one processor to another. The data is identified by a label so that the receiving processor can discriminate incoming messages. Issuing a receive-wait command will cause a processor to halt execution until the desired data is received. Processor synchronization is inherent in this type of protocol since receipt of data implies that the data is in fact ready for processing by the receiving node.

We have implemented our algorithms on the Intel iPSC/1-VX and iPSC/2-VX hypercubes which are typical of the class of architectures for which the parallel spectral element method is well suited. The iPSC is a distributed memory, message passing, parallel processor consisting of $M = 2^D$ independent processor/memories, or nodes, arranged on a D dimensional hypercube communication network. The iPSC/2-VX is an upgraded version of Intel's original iPSC hypercube which incorporates an improved message router allowing data to be transferred between non-nearest-neighbor processors with minimal degradation in data transfer rate. Each node has an Intel 80386/80387 processor/coprocessor which achieves a floating point execution rate of roughly 0.1 MFLOPS. The message transfer rates are $\Delta(1) \approx 300\mu\text{sec}$ and $\Delta(\infty) \approx 1.4\mu\text{sec}$, where $\Delta(n)$ corresponds to the time-per-word required to transfer n words from one processor to another. On the VX system used for the present calculations, each node has an attached vector processor which achieves 3-4 MFLOPS on standard vector operations and 10-12 MFLOPS on matrix-matrix products. The nodes have 2-8 Mbytes of memory with an additional 1 Mbyte of fast memory on the vector board. The programmer is responsible for vector calls and explicit transfer of data between the standard and vector memory.

4.2 Parallel Work Decomposition for Problems of General Topology

The underlying strategy in the development of the spectral element algorithms is to treat each element as a "virtual parallel processor" such that each element is an independent and indivisible unit. The only types of operations permitted to transcend element boundaries are interface operations, such as direct stiffness summation, and vector reduction operations, such as inner product evaluation. These operations are sufficient to effect almost any operation associated with second-order equations as a consequence of the variational approach and the solution and trial functions being in $\mathcal{H}_0^1$. It is clear from equation (10) that the character of a given spatial derivative operator is determined by the matrix-vector products within the braces; direct stiffness summation then assembles the elemental contributions. Strict decoupling of intra- and inter-element operations in this manner allows for independent development of general spectral element algorithms and parallel communication algorithms which can be readily combined into a productive tool.

Implementation of the spectral element method on a distributed memory parallel processor follows by mapping elements or groups of elements onto separate processors and constructing the requisite inter-processor vector reduction and direct stiffness summation operators. Iterative solution of the linear system (11) differs from the serial case in that elemental data and matrix coefficients are distributed so that elemental residuals, $A^k u^k$, are computed simultaneously for distinct elements on distinct processors. The global residual calculation is completed via the inter-processor direct stiffness operation. The remaining vector operations for the conjugate gradient routine are computed in the same distributed manner, with requisite inter-processor communication to complete the inner products. [20]

4.3 Mapping and Communication

The element to processor map is constructed to minimize processor load imbalance and communication overhead [21]. Since the spectral element method is inherently a macro-element decomposition, load balance is achieved by ensuring that the number of elements on any given pair of processors differs by no more than one. The communication time associated with vector reduction operations is fixed for any problem topology (see below) and hence does not influence the mapping strategy. In contrast, the overhead associated with element interface communication is directly influenced by the element to processor mapping; it is governed by the number of "exposed" faces on each processor, i.e., those interfaces for which adjacent elements are assigned to different processors, and by the distance on the network which separates two, physically adjacent, elements. Our current decomposition algorithm employs a nested-dissection scheme [22,23] to minimize the number of exposed element interfaces and thereby reduce the data communication traffic. In practice, network-dependent element to processor mapping optimizations which seek to ensure that only "nearest neighbor" processors are communicating have been found to affect the solution times by at most twenty percent [20], and are therefore not of primary concern.

The inner product is a standard vector reduction, which is evaluated by first performing intra-element/intra-processor sums, and then evoking an inter-processor vector reduction which sums and redistributes the contributions from each processor. For hypercube multiprocessors consisting of M processors, it is well known that vector reductions can be effected in $O(\log M)$ communication cylcles using a binary tree embedding [9]. On the Intel machine, the inter-processor vector reduction software is provided so that optimal performance may be obtained without explicit knowledge of the network topology. In this rapidly evolving technology, manufacturer provision of such basic parallel communication kernels is important in order that applications programmers are spared redundant, architecture specific, development efforts.

The direct stiffness summation procedure imposes function continuity across element interfaces and is in essence the operation which binds the subdomains together. Calculation of the global matrix vector product, $\mathbf{Au}$ (or $\mathbf{Bf}$), is completed by summing residuals of coincident nodes and redistributing them back to the original element based data structrures, as illustrated in Figure 2. The operation is inherently complex due to the unstructured nature of the global element assembly; an arbitrary number of elements may be joined at a single element vertex. The direct stiffness implementation must therefore be sufficiently robust to treat all admissible element assemblies, yet also be efficient so that, as a costly, communicative operation, it is not the rate limiting step in parallel computation. As simultaneous attainment of these goals is non-trivial, we analyse the requirements and restrictions of the operation in a parallel context.

We first consider the simple spectral element mesh shown in Figure 2a in which data is given on each element. In Figures 2c through 2f we show diagrammatically how direct stiffness summation can be performed by local directional splitting of the operation into a sequence ψ of d=2 element exchanges. It can be seen that the nodal values at the vertices are, indeed, correct; that is the sequence ψ ensures that the contributions from each element at the shared corner are summed together. The advantages of this splitting method over, say, a bi-directional parallel edge pass followed by vertex-specific operations are: the splitting mehtod is algorithmically clean; the splitting method avoids costly short messages; the splitting method avoids non-nearest-neighbor communication and contention. These advantages are even clearer in three space dimensions.

For many spectral (or substructured finite) element decompositions it is difficult, if not impossible, to find an edge pass sequence ψ which results in correct nodal values at all vertices. However, it is often relatively simple to find a sequence ψ for which the number of vertices with incorrect values, denoted "special" nodes, is small. This suggests the following strategy: first, a vector reduction (sum) operation is performed to accumulate the contributions of all elments to the residuals associated wtih the special nodes; second, a standard d-pass sequence is performed, ψ; third the results of the vector reduction are redistributed to the special nodes. The summaton strategy is illustrated diagrammatically in Figure 3.

This summation algorithm is both general and relatively efficient, and is guaranteed to effect the direct stiffness operatoin by virtue of the special node treatment. Furthermore, the regularity and special nodes associated with a candidate sequence ψ can readily be

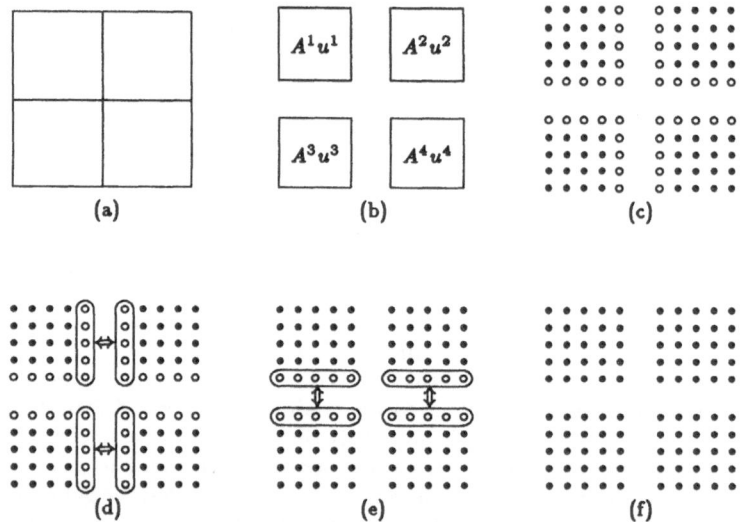

Figure 2. Computation of residual vector for regular geometry in $\mathbf{R}^2$: (a) four element mesh; (b) simultaneous (parallel) computation of incomplete residual, $\tilde{r}^k = A^k \tilde{u}^k$; (c) nodal content of $\tilde{r}^k$ is denoted by circles - solid circles indicate correct residuals, open circles indicate values requiring contributions from neighboring elements; (d) and (e) bi-directional exchange and sum sequence, ψ, which effects the completed residual at all nodes including the corner nodes; (f) completed residual, $\tilde{r}^* = \sum' A^k \tilde{u}^k$.

determined by computing the result of a multiplicative ψ for a field comprising (locally) distinct primes, and comparing with the correct result of such a direct stiffness operation. (Direct stiffness multiplication is defined as in Figure 2 with arrows implying multiplication, not addition.) The analysis of our parallel direct stiffness method is however, not complete, in that no studies have been perfromed to determine minimally irregular sequences.

4.4 Parallel solution costs

The serial processor conjugate gradient work estimate (13) is modified for the parallel case to account for the additional communication costs. It is pertinent to characterize the relevant hardware parameters: δ - the time required to execute a single floating point operation, $\Delta(n)$ - the time-per-word required to transfer n words of data from one processor to another, and $\sigma(n) \equiv \Delta(n)/\delta$. Generally, $\sigma(1) > \sigma(n)$ for $n > 1$, because of the startup costs associated with any inter-processor data transfer. (Note that this first order estimate does not account for *where* data is being transferred). To leading order, the total solution time on M processors can be expressed as:

$$\tau_{sol} = \delta N_e^A \left\{ \frac{c_1 K N^{d+1}}{M} + c_3 \sigma(N^{d-1}) \tilde{K} N^{d-1} + c_4 \sigma(1) log(M) \right\} \quad , \qquad (14)$$

where the c_3 term accounts for direct stiffness summation and the c_4 term accounts for vector reduction. We introduce $\tilde{K} \leq 2dK/M$ as the maximum number of element edges per processor for which inter-processor communication is required in order to complete the direct stiffness summation.

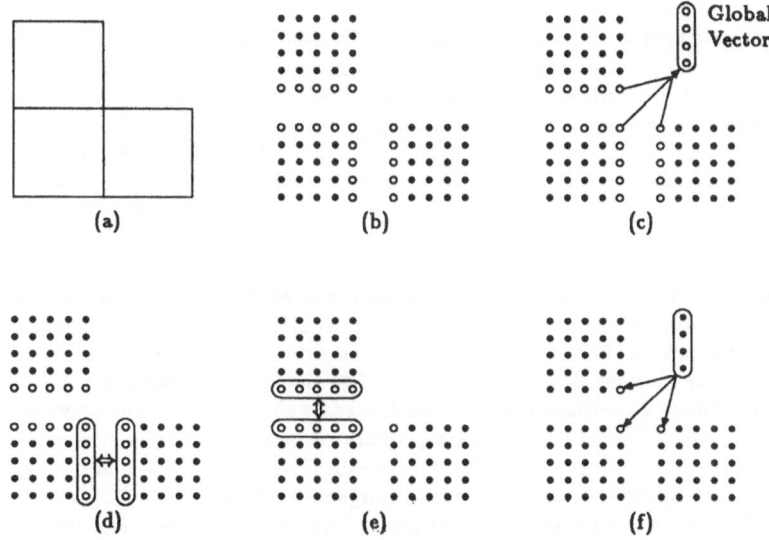

Figure 3. Direct stiffness summation for irregular geometry in $\mathbf{R}^2$: (a) three element mesh; (b) nodal content of incomplete residual, $\tilde{r}^k$; (c) $O(\log M)$ map-and-sum vector reduction to gather special nodes onto a global data structure; (d) and (e) standard bi-directional exchange and sum sequence, ψ; (f) final map ($O(\log M)$ fan out) which overwrites local data at special nodes with correct, complete residual.

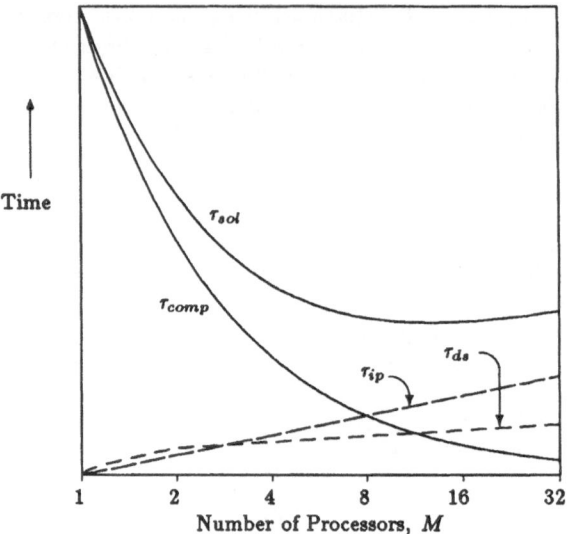

Figure 4. Illustration of computing costs associated with the parallel conjugate gradient algorithm, measured in wall clock time. The growth of the inner-product time illustrates the limiting nature of *global* operations, resulting in a minimum solution time for a number of processors $M_0 < \infty$.

The estimate (14) is based upon the assumptions that the computational load (c_1 term) is balanced and that direct stiffness summation is M independent. The first assumption only holds in the case where K is a multiple of M. The second assumption is justified to the extent that disjoint element-element/processor-processor pairs can exchange data simultaneously, at a rate which is independent of the location of the processors in the network. Because direct stiffness summation is inherently local (though not intra-element), it will not be strongly dependent upon the system size. Note finally that the c_4 term associated with vector reductions grows as $O(\log M)$ and is strongly M dependent due to the global nature of the operation.

For illustration, we plot in Figure 4 a caricature of equation (14) in the form of solution time versus number of processors for some fixed problem. The τ_{comp} curve, corresponding to the total computing effort (operations) required to solve the problem, decreases in direct proportion to the number of processors assigned to the task, assuming the task is M partitionable. The parallel overhead costs are illustrated by the τ_{ds} and τ_{ip} curves which represent the direct stiffness summation and inner-product communication times, respectively. The total solution time, τ_{sol}, is the sum of the three cost terms: τ_{comp}, τ_{ds}, and τ_{ip}. Note that, in general, it is not possible to span the entire range of processors, $M = 1, 2, \ldots, M_{max}$, for a single problem (K, N, d) due to the coarse grained nature of the spectral element discretization. The lower bound is controlled by the ratio $K N^d / M$ which is limited by the memory on each processor. The upper bound is $M \leq K$, since elements are indivisible entities.

It is clear from Figure 4 that the inner-product time can become dominant as the number of processors is increased, and that there results a minimum obtainable solution time at a finite number of processors. Such minima have been observed in spectral element test problems on the iPSC/1-VX [20]. Two approaches to avoid operating in the "speed-down" regime of Figure 4 are to increase K, i.e., increase the amount of work on each processor, or to decrease σ. Both of these options are addressed by Intel's second generation hypercube, the iPSC/2-VX, which has increased memory, allowing for a larger KN^d/M, and improved communications, resulting in a decreased Δ. The vector speed on both generations of hypercubes is the same, so for vector dominated calculations, a decrease in Δ corresponds to a decrease in σ.

We next analyse the parallel efficiency, $\eta \equiv \tau_{comp}/\tau_{sol}$. From equation (14) we have, to leading order:

$$\eta \approx 1 - \frac{c_3}{c_1}\frac{M\tilde{K}}{KN^2}\sigma(N^{d-1}) - \frac{c_4}{c_1}\frac{M}{KN^{d+1}}\sigma(1)\log(M) \tag{15}$$

We note that the constants governing the degradation in efficiency in (15) are directly proportional to σ, thus implying that a straightforward approach to improving system performance is to decrease Δ/δ. While this certainly yields better processor utilization, it does so at a cost, since a reduced Δ implies a faster, more expensive, network. The difficulty is that efficiency only measures processor utilizaton and does not account for the cost of the associated network. In addition, parallel efficiency is strongly dependent upon the choice of algorithm and the scalar-vector mix of the particular application. It cannot be taken as a measure of "goodness" of any particular algorithm-architecture coupling, as it relatively easy to obtain high efficiency on a system with slow processors, implying that the reduction in σ is brought about by an increase in δ rather than a decrease in Δ. Moreover, used as a performance measure it can in fact be misleading, since any algorithm or architecture improvements which directly decrease τ_{comp} will reduce τ_{sol}, but will also yield reduced efficiency.

It is nonetheless of interest to analyse the efficiency versus number of processors for a fixed algorithm-architecture coupling and for a fixed ratio, K/M [24]. Such an analysis addresses the question, "What performance will be obtained if the problem size is doubled and the number of processors is doubled?" This is readily computed from (15), yielding:

$$\eta\bigg|_{K/M} \approx \eta_0 - \frac{c_4}{c_1}\frac{M}{KN^{d+1}}\sigma(1)\log(M) \tag{16}$$

where η_0 is the efficiency obtained on a smaller system for some $M > 1$. From (16), it is clear that there is only a small logarithmic degradation in the efficiency as the problem size and machine size grow simultaneously. Such performance has been observed empirically in our experience with the iPSC/2-VX. A typical three-dimensional flow calculation with $N = 10$ and $K/M = 2$ runs at roughly 75 percent efficiency and 10 to 11 MFLOPS on four processors. Similar performance is obtained for larger systems in which $K/M = 2$: 16 processors yields 44 MFLOPS, 32 - 80 MFLOPS, and 64 - 160 MFLOPS.

5 Results

We conclude with the presentation of several example calculations which illustrate the generality of the parallel spectral element implementation, its capabilty in handling laminar and transitional flows, and its utility as a fluid mechanics analysis tool. Except where indicated, the calculations were performed on a four node Intel iPSC/2-VX hypercube.

The first calculation is a steady three-dimensional Stokes flow consisting 312,000 degrees-of-freedom ($K = 128$, $N = 10$). The physical problem is that of a spiral groove thrust

(a)   (b)

Figure 5. Mesh geometry for spiral groove thrust bearing. Line intersections indicate location of Gauss-Lobatto quadrature points for $N = 5$: (a) 32 element lower level comprising inlet grooves, (b) thin gap region consisting of 96 elements.

bearing with ambient pressure at the inner and outer radii. The elemental geometry in Figure 5 is comprised of two levels of elements stacked upon one another. The lower layer corrsponds to the bearing grooves which have a depth of .04, normalized with respect to the outer radius. The upper layer corresponds to the gap between the thrust plates having a depth of 0.01. The flow is driven by imposing counter-clockwise plane rotation on the upper bearing surface. Lift is generated by viscous pumping in the channels, and the computed lift is in good agreement with Reynolds equation solutions for similar configurations [25]. The main point of this calculation is to illustrate the scalability of the parallel algorithms described previously. The Stokes problem was run on a 64 node Intel iPSC/2-VX at a sustained execution rate of 160 MFLOPS for the duration of the 16 minute solution time. Previous calculations on a 16 node machine had achieved 44 MFLOPS [20], indicating roughly a ten percent degradation in performance per processor as the number of processors was increased from 16 to 64. As access to the 64 node machine was limited, extensive parameter studies were not feasible; consequently, it is not clear whether the observed performance degradation is due solely to the $\log M$ term in equation (15) or to variance resulting from non-optimal mapping. In either case, the performance on 64 nodes is quite good and indicates that this particular algorithm/architecture coupling can be leveraged further with the addition of more processors.

The next example we consider is the classical problem of external startup flow past a cylinder of diameter D at a Reynolds number of $R = U_\infty D/\nu = 100$. The cylinder is initially in quiescent fluid at $t = 0$, with an external uniform flow U_∞ imposed abruptly at $t = 0_+$. The problem is treated with the spectral element Navier-Stokes discretizations based on consistent approximation spaces for the velocity and the pressure, and the Uzawa conjugate gradient-based iterative solvers [26]. Figure 6a shows the good comparison between the numerical prediction of the recirculation zone length at early times with the experimental observations of [27]. At later times the familiar unsteady von Karman vortex street forms as shown in Figure 6b. The startup calculation was performed on an eight node iPSC/1-VX hypercube.

The third problem is again the cylinder startup problem, but now at a Reynolds number $R = 1000$. This calculation is based on a fractional step method with conjugate gradient

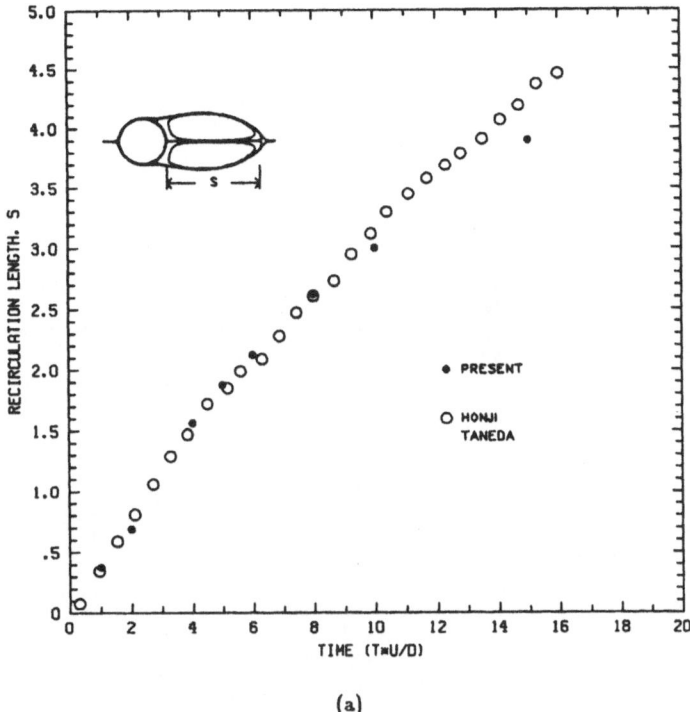

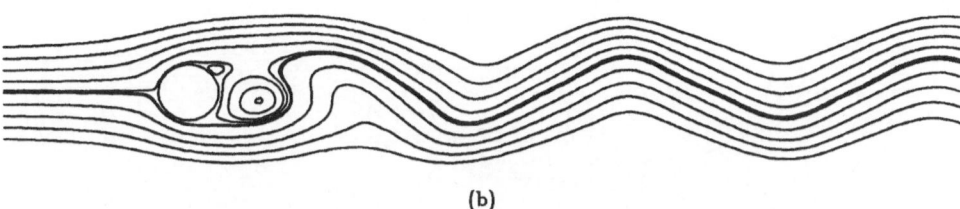

(b)

Figure 6. Startup flow past a cylinder at a Reynolds number $R = U_\infty D/\nu = 100$. (a) Comparison of early time nondimensional recirculation zone length with experimental measurements of [27]. (b) Instantaneous streamlines showing the von Karman vortex street at a nondimensional time of 110.

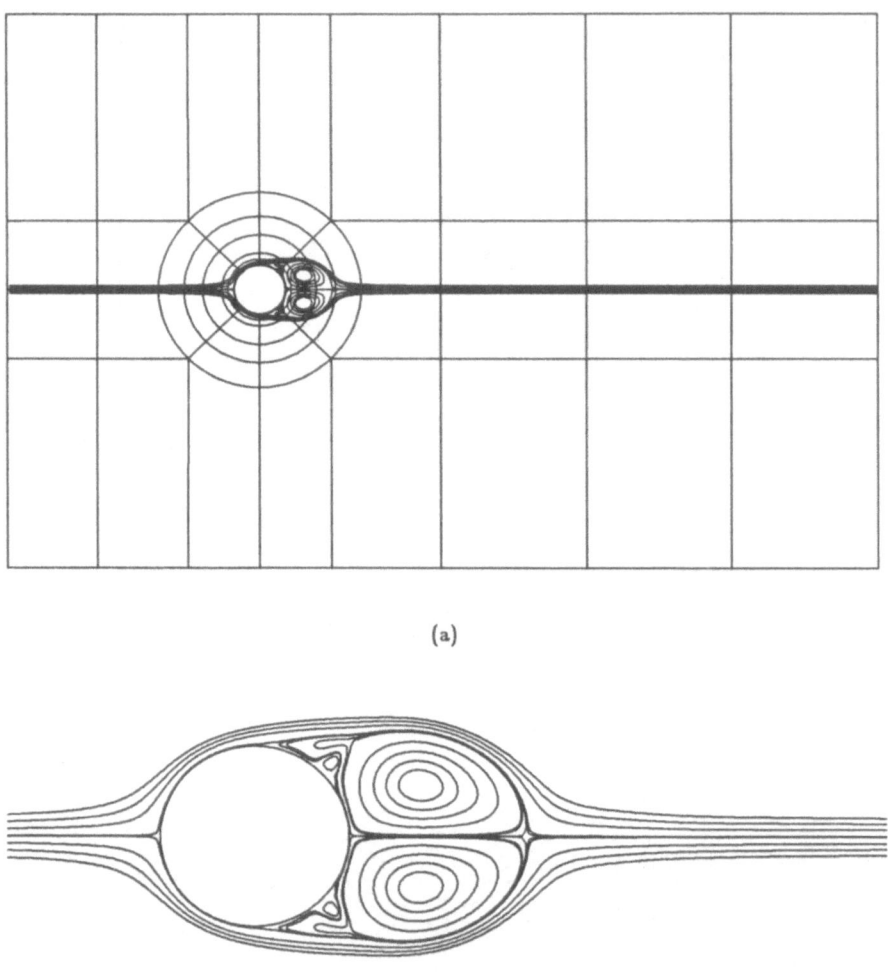

(a)

(b)

Figure 7. Startup flow past a cylinder at a Reynolds number $R = U_\infty D/\nu = 1000$. (a) Spectral element mesh $(K = 68,\ N = 10)$ and instantaneous streamlines at a nondimensional time of $U_\infty t/D = 3$. (b) Enlarged view reveals small secondary vortices which are also observed experimentally [29].

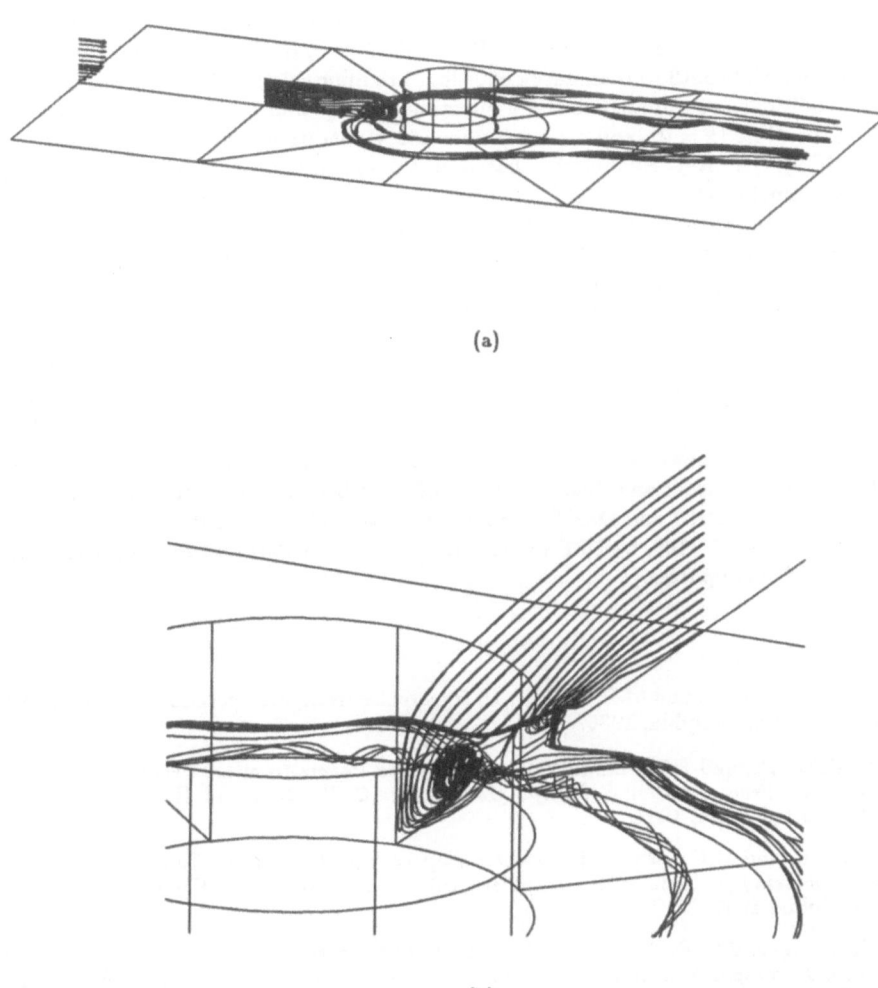

(a)

(b)

Figure 8. Three-dimensional flow past a cylinder mounted on a flat plate at a Reynolds number $R = 1000$ based on maximum inflow velocity, U_*, and cylinder diameter, D. (a) Mesh, inflow velocity profile, and three-dimensional horseshoe vortex structures shown at an early time $U_* t/D = 4$. (b) Detailed flow structure close to the cylinder reveals the presence of two horseshoe vortices at time $U_* t/D = 9$, in good qualitative agreement with the experiments of [30].

solution of the elliptic subproblems [28]. Figure 7a shows the spectral element discretization of the computational domain consisting of $K = 52$ elements, together with instantaneous streamlines at an early non-dimensional time of $U_\infty t/D = 3$. Figure 7b shows a more detailed close-up of the vortex structure which forms behind the cylinder; this result is in excellent agreement with the experimental results provided by [29], in particular as regards the placement of the small secondary vortex.

The last example is a three-dimensional flow past a cylinder mounted on a flat plate. In the flow direction the velocity profile $u = U[1 - 16(1/2 - z/H)^4], z \leq H/2$, is imposed at the inflow plane, while outflow boundary conditions are imposed at the exit plane. Periodic boundary conditions are imposed in the y-direction, no-slip wall boundary conditions are imposed on the plate ($z = 0$), and symmetry boundary conditions are assumed on the top plane ($z = H/2$). The Reynolds number based on the symmetry plane inflow velocity (U) and cylinder diameter, D, is $R = 1000$; the geometric ratio D/H is unity for this calculation. Figure 8a shows the three-dimensional horseshoe vortex structure that forms around the cylinder at a non-dimensional time $Ut/D = 9$, while Figure 8b shows the more detailed flow structure close to the cylinder. The computational results are in good qualitative agreement with the experimental results of [30].

Acknowledgements

We would like to acknowledge the significant contributions to this work by Einar Rønquist of M.I.T., and David Scott and Justin Rattner of Intel Scientific Computers. This work was supported by the ONR and DARPA under contracts N00014-85-K-0208, N00014-87-K-0439, and N00014-88-K-0188, by NSF under Grants DMC-8704357 and ASC-8806925, and by Intel Scientific Computers.

References

[1] T.F. Chan, Y. Saad, and M.H. Schultz, in *Hypercube Multiprocessors 1986* (M.T. Heath, ed), SIAM, Philadelphia, 1986.

[2] O.A. McBryan, and E.F. van de Velde, in *Selected Papers from the Second Conference on Parallel Processing for Scientific Computing* (C.W. Gear and R.G. Voigt, eds.), SIAM, Philadelphia, 1987.

[3] T.F. Chan, and D.C.Resasco, in *Selected Papers from the Second Conference on Parallel Processing for Scientific Computing* (C.W. Gear and R.G. Voigt, eds.), SIAM, Philadelphia, 1987.

[4] R. Glowinski and M.F. Wheeler in *Proceedings of the First International Conference on Domain Decomposition Methods for Partial Differential Equations, Paris* (R. Glowinski, G. Golub, G. Meurant, and J. Periaux, eds.), pp. 144-172, SIAM, Philadelphia, 1987.

[5] O.B. Widlund, in *Proceedings of the First International Conference on Domain Decomposition Methods for Partial Differential Equations, Paris* (R. Glowinski, G. Golub, G. Meurant, and J. Periaux, eds.),pp. 113-127, SIAM, Philadelphia, 1987.

[6] D.E. Keyes and W.D. Gropp, in *Proceedings of the Second International Conference on Domain Decomposition Methods for Partial Differential Equations, Los Angeles* SIAM, Philadelphia, 1988

[7] L.Adams and R.G. Voigt, in *Large Scale Scientific Computation* (S. Parter, ed.), pp 301-321, Academic Press, Orlando, Florida, 1984.

[8] W.D. Gropp and D.E. Keyes, SIAM J. of Sci. and Stat. Comput. 9 (1988) 312-326.

[9] Y. Saad and M.H. Schultz, *Topological Properties of Hypercubes*, Research Report YALEU/DCS/RR-389, Yale University, New Haven, 1985.

[10] H.X.Lin, and H.J. Sips, in *Proc. 1986 Int. Conf. on Parallel Processing*, 503-510, 1986.

[11] A.T. Patera, A spectral element method for fluid dynamics; Laminar flow in a channel expansion, *J. Comput. Phys.*, **54**, 1984, p.468.

[12] Y. Maday and A.T. Patera, Spectral element methods for the Navier-Stokes equations, in *State of the Art Surveys in Computational Mechanics* (Edited by A.K. Noor), ASME, New York, to appear.

[13] S.A. Orszag, Spectral Methods for Problems in Complex Geometries, *J. Comput. Phys.*, **37**, 1980, p. 70.

[14] A.H. Stroud, and D. Secrest, *Gaussian Quadrature Formulas*, Prentice Hall, 1966.

[15] C. Canuto, M. Hussaini, A. Quarteroni, and T. Zang, *Spectral Methods in Fluid Dynamics*, Springer-Verlag, 1987.

[16] G.H. Golub and C.F. van Loan, *Matrix Computations*, Johns Hopkins University Press, 1983.

[17] E.M. Rønquist and A.T. Patera, Spectral element multigrid. I. Formulation and numerical results, *J. Sci. Comput*, **2**, 4, 1987, p. 389.

[18] Y. Maday and Muñoz, Spectral Element Multigrid. II. Theoretical Justification, *J. Sci. Comput.*, to appear.

[19] M.T. Heath, The Hypercube: A Tutorial Overview, in *Hypercube Multiprocessors 1986*, (M.T. Heath, ed.), SIAM, Philadelphia, 1986.

[20] P.F. Fischer and A.T. Patera, Parallel Spectral Element Solution of the Stokes Problem, *J. Comput. Phys.*, submitted.

[21] B. Nour-Omid, A. Raefsky, and G. Lyzenga, Solving Finite Element Equations on Concurrent Computers, in *Parallel Computations and Their Impact on Mechanics*,(A.K. Noor, Ed.), ASME, N.Y., 1987.

[22] R.F. Lucas, *Solving Planar Systems of Equations on Distributed-Memory Multiprocessors* Ph.D. Thesis. , Dept. of Electrical Engineering, Stanford University, 1987.

[23] G. Fox, M. Johnson, G. Lyzenga, S. Otto, J. Salmon, D. Walker (1988). *Solving Problems on Concurrent Processors. Volume 1 : General Techniques and Regular Problems.* Prentice-Hall, Englewood Cliffs, New Jersey.

[24] J.L. Gustafson, G.R. Montry, and R.E. Benner, Development of Parallel Methods for a 1024-Processor Hypercube, *SIAM J. on Sci. and Stat. Comput.*, **9**, 4, July 1988, pp. 609-638.

[25] W.A. Gross, "Fluid Film Lubrication", John Wiley & Sons, Inc., New York (1980).

[26] E.M. Rønquist, *Optimal Spectral Element Methods for the Unsteady Three-dimensional Incompressible Navier-Stokes Equations*, Ph.D. Thesis, Massachusetts Institute of Technology, 1988.

[27] H. Honji and A. Taneda, Unsteady Flow Past a Circular Cylinder, *J. Phys. Soc. Japan*, **27**, 6, 1969, p. 1668.

[28] G.E. Karniadakis, E.T. Bullister, and A.T. Patera, A Spectral Element Method for Solution of the Two- and Three-Dimensional Time-Dependent Incompressible Navier-Stokes Equations, in: *Finite Element Mehtods for Nonlinear Problems, Europe-US Symposium, Trondheim, Norway 1985*, Bergan, Bathe, Wunderlich, eds., Springer, Berlin Heidelberg, 1986.

[29] M. Van Dyke, *An Album of Fluid Motion*, The Parabolic Press, Stanford, California, 1982, p. 36.

[30] C.J. Baker, The laminar horseshoe vortex, *J. Fluid Mech.*, **95**, 1979, p. 347.

A PROJECT FOR DEVELOPING A LINEAR ALGEBRA LIBRARY FOR HIGH-PERFORMANCE COMPUTERS

J. Demmel, J. Dongarra, J. DuCroz, A. Greenbaum,
S. Hammarling, and D. Sorensen

Mathematics and Computer Science Division
Argonne National Laboratory
Argonne, Illinois 60439-4844

1. Introduction and Objectives

Argonne National Laboratory, the Courant Institute for Mathematical Sciences, and the Numerical Algorithms Group, Ltd., are developing a transportable linear algebra library in Fortran 77. The library is intended to provide a uniform set of subroutines to solve the most common linear algebra problems and to run efficiently on a wide range of high-performance computers.

To be effective, the new library must satisfy several criteria. First, it must be highly efficient, or at least "tunable" to high efficiency, on each machine. Second, the user interface must be uniform across machines. Otherwise much of the convenience of portability will be lost. Third, the programs must be widely available. NETLIB [4] has demonstrated how useful and important it is for these codes to be available easily, and preferably on line. We intend to distribute the new library in a similar way, for no cost or a nominal cost only. In addition, the programs must be well documented.

The new linear algebra library will be based on the successful EISPACK [9, 7] and LINPACK[1] libraries, with the following further developments:

- integration of the two sets of algorithms into a unified, systematic library;
- incorporation of recent algorithmic improvements; and
- restructuring of the algorithms to make as much use as possible of the Level 1, 2, and 3 Basic Linear Algebra Subprograms (BLAS)

We expect that the library will become a central part of the high-performance scientific

†Work supported in part by the Applied Mathematical Sciences subprogram of the Office of Energy Research, U. S. Department of Energy, under Contract W-31-109-Eng-38.

programming environment, much as conventional libraries for serial machines are essential to conventional scientific computing.

2. Outline of the Project

2.1 Contents

The new library will provide approximately the same functionality as LINPACK and EISPACK together, namely, solution of systems of simultaneous linear equations, least-squares solution of overdetermined systems of equations, and solution of matrix eigenvalue problems. The associated matrix factorizations (LU, Cholesky, QR, SVD, Schur, generalized Schur) will also be provided, as will related computations such as reordering of the factorizations and condition numbers. Dense and band matrices will be provided, but not general sparse matrices. In all areas, similar functionality will be provided for real and complex matrices.

Many of the algorithms of LINPACK and EISPACK will be carried over with only minor modifications to their numerical behavior, although often with extensive restructuring to improve performance. Some algorithms may be deleted, for example, to eliminate duplication; others may be extended to provide additional functionality. Where the state of the art is sufficiently clear, new algorithms will be added.

Test software will be developed, building on that already developed for EISPACK, LINPACK, and the BLAS. Timing programs will also be provided.

2.2 Use of the BLAS

We intend to code the routines in the new library so that as much of the computation as possible is performed by calls to Level 2 or Level 3 BLAS.

In some algorithms (e.g., computing eigenvalues of a symmetric tridiagonal matrix), use of the BLAS is not feasible. In the majority of the algorithms, however, there is certainly scope for using Level 2 and 3 BLAS, and a considerable amount of experience has already been accumulated[8, 2, 3, 5, 6].

To exploit Level 3 BLAS, one usually must restructure the algorithm into a "block" form, in which the original matrices are partitioned into submatrices or blocks, and the algorithm is expressed in terms of basic matrix-matrix operations on the blocks. The performance of the algorithms depends on the dimensions chosen for the blocks. It will therefore be necessary to investigate the appropriate blocking strategy for each of our target machines, and then develop a mechanism whereby the routines can determine good block dimensions automatically (possibly via a machine-specific enquiry function).

It is likely that this project will reveal the need for a few additional basic routines whose performance may need to be optimized for different architectures and may be regarded as extensions to the current sets of BLAS (e.g., applying sequences of plane rotations to a matrix).

2.3 Target Machines

The library will be designed to perform efficiently on machines with a modest number of processors (say, 1-20), each having a powerful vector-processing capability. These machines include all of the most powerful computers currently available and in use for general-purpose scientific computing: CRAY-2, CRAY X-MP, CYBER 205, ETA-10, Fujitsu/Amdahl VP, IBM 3090/VF, NEC SX, Alliant FX/8, Convex C-1, Scientific Computer Systems SCS-40, Ardent, Stellar, Sequent Symmetry, Encore Multimax, and BBN Butterfly.

We do not claim that the strategy of using Level 2 or Level 3 BLAS will necessarily attain optimal performance on all these machines; indeed, some algorithms can be structured in several different ways, all calling Level 3 BLAS, but with different performance characteristics. In such cases we shall choose the structure that provides the best "average" performance over the range of target machines.

We hope that the library will also perform well on a wider class of parallel machines, including the IBM RP3, Intel iPSC, NCUBE, and FPS T-Series.

On conventional serial machines, the performance of the library is expected to be at least as good as that of the current LINPACK and EISPACK codes. Thus the library will be suitable across the whole range of machines from personal computers to supercomputers to experimental architectures.

2.4 Programming Language and Style

The software will be developed in standard Fortran 77, with extensions to the standard only where necessary.

Single- and double-precision versions will be prepared; conversion between different precisions will be performed automatically by software tools.

Routines for complex matrices will use the COMPLEX data type (like LINPACK, but unlike EISPACK); hence the availability of a double-precision complex (COMPLEX*16 or DOUBLE COMPLEX) data type will be assumed as an extension to Fortran 77. Routines for real and complex matrices will be written to maintain a close correspondence between the two and to permit automatic transformation, as far as possible; however, in some algorithms (e.g., unsymmetric eigenvalue problems) the correspondence will necessarily be weaker.

We realize that Fortran 8X is likely to have a number of features that would improve the design and coding of the library—optional arguments, dynamic allocation of workspace, and array features, to name a few. However, if we are to begin development and testing of the library now on our range of target machines, there seems no reasonable choice other than Fortran 77.

2.5 User Interface and Documentation

The user interface to the routines will be similar to that of LINPACK. Routine names and arguments will follow a systematic scheme.

We intend to provide a set of top-level driver routines that can solve a complete problem (e.g., compute all eigenvalues and eigenvectors of a symmetric matrix). Each routine will call a number of lower level routines which perform the parts of the solution (e.g., reduction to tridiagonal form, accumulation of the orthogonal transformation matrix, compute eigenvectors and eigenvectors of a tridiagonal matrix). Both levels of routines will be accessible to users and will be documented in a guide, similar in many respects to the LIN-PACK Users' Guide.

3. Organization

Argonne National Laboratory and the Courant Institute has received funding from the National Science Foundation; the Numerical Algorithms Group in the United Kingdom has agreed to cooperate in this research effort. The three institutions plan to work in a coordinated but independent fashion. Argonne will serve as a center for the project and be responsible for collecting, editing, testing material, and distributing the software. We intend to meet at least twice a year to coordinate activities and discuss developments.

We shall use test sites to help verify that the software is working correctly. We also plan to distribute working notes and to report at conferences on the status, progress, and future directions throughout the project life.

Information may be obtained from any of the following: Jack Dongarra and Danny Sorensen, Argonne National Laboratory, Argonne, IL 60439-4844; James Demmel and Anne Greenbaum, Courant Institute, 251 Mercer Street, New York. NY 10012; Jeremy Du Croz and Sven Hammarling, Numerical Algorithms Group, Ltd., NAG Central Office, Mayfield House, 256 Banbury Road, Oxford OX2 7DE, England.

References

1. J.J. Dongarra, J. Bunch, C. Moler, and G. Stewart, *LINPACK Users' Guide,* SIAM Pub., Philadelphia (1979).

2. J.J. Dongarra, J. DuCroz, S. Hammarling, and R. Hanson, "An Extended Set of Fortran Basic Linear Algebra Subprograms," Argonne National Laboratory Report, ANL-MCS-TM-41 (Revision 3) (November 1986).

3. J.J. Dongarra and S. C. Eisenstat, "Squeezing the Most out of an Algorithm in Cray Fortran," *ACM Trans. Math. Software* **10, 3**, pp. 221-230 (1984).

4. J.J. Dongarra and E. Grosse, "Distribution of Mathematical Software via Electronic Mail," Argonne National Laboratory Report, ANL-MCS-TM-48, (to appear in CACM) (March 1985).

5. J.J. Dongarra, L. Kaufman, and S. Hammarling, "Squeezing the Most out of Eigenvalue Solvers on High-Performance Computers," *Linear Algebra and Its Applications* **77**, pp. 113-136 (1986).

6. J.J. Dongarra and D.C. Sorensen, "Linear Algebra on High-Performance Computers," pp. 3-32 in *Proceedings Parallel Computing 85*, ed. U. Schendel, North Holland (1986).

7. B.S. Garbow, J.M. Boyle, J.J. Dongarra, and C.B. Moler, *Matrix Eigensystem Routines - EISPACK Guide Extension*, 1977.

8. C. Lawson, R. Hanson, D. Kincaid, and F. Krogh, ''Basic Linear Algebra Subprograms for Fortran Usage,'' *ACM Transactions on Mathematical Software* **5**, pp. 308-323 (1979).

9. B.T. Smith, J.M. Boyle, J.J. Dongarra, B.S. Garbow, Y. Ikebe, V. Klema, and C. Moler, *Matrix Eigensystem Routines - EISPACK Guide, Second Edition,,*

A COMPARISON OF ELEMENT-BY-ELEMENT PRECONDITIONED ITERATIVE METHODS FOR SOLID AND STRUCTURAL MECHANICS

Robert M. Ferencz *

Methods Development Group
Lawrence Livermore National Laboratory
Livermore, CA

ABSTRACT

Past work with element-by-element (EBE) preconditioned conjugate gradient iterative solution strategies has shown these techniques can be effective for large-scale, three-dimensional calculations in solid and structural mechanics. Significant gains over the profile storage direct solution method traditionally used in implicit finite element codes have been observed for a variety of real engineering analyses, especially in solid mechanics. Structural mechanics applications have proved less successful due to the ill-conditioned linear systems engendered by standard structural discretizations. This lack of robustness has recently motivated reconsideration of Lanczos-based algorithms as alternative iterative drivers. In this paper we compare the relative strengths of the conjugate gradient and Lanczos drivers when coupled with EBE preconditioning. The performance of the two methods is characterized, and compared with direct solution, using a model problem and a number of real engineering meshes. These examples suggest that the conjugate gradient method remains superior for large-scale computation.

INTRODUCTION

Until recent times a general lack of computational capacity has limited implicit finite element calculations in solid and structural mechanics to moderately sized two-dimensional simulations. Direct methods of solving linear systems of equations have proved quite adequate for this task, and their simplicity and robustness has led to almost universal adoption of these techniques. However, the increasing demand for three-dimensional simulation, as well as ever larger two-dimensional applications, requires us to confront the unfavorable rates of cost growth for both storage and operations with direct methods. These difficulties have long inspired the study and development of iterative linear equation solvers by numerical analysts. The computational mechanics community is returning to this topic in search of cost-effective methods for large-scale analysis. Furthermore, iterative methods may prove to be the most natural avenue for parallel solution of systems of linear equations, thus the availability of low-cost multiprocessors adds further impetus to this research.

---

* Work performed under the auspices of the U.S. Department of Energy by the Lawrence Livermore National Laboratory under Contract W-7405-Eng-48.

Element-by-element (EBE) methods were initially viewed as a generalization of the global operator splitting techniques of finite difference methods to the element (local) data structure which is intrinsic to finite element implementations. Specifically, Hughes et al. (1983a) introduced EBE methods as a means of implicit time-integration for the heat conduction problem. Based upon this idea, Ortiz et al. (1983) developed an EBE time-stepping scheme for structural dynamics. Both these algorithms are unconditionally stable and have formal second-order accuracy, but display unacceptable spatial truncation errors in some test problems. Due to that experience, subsequent strategies have relied upon applying EBE procedures to solve the systems of linear equations arising from standard time-integration schemes (Hughes et al. 1983b). In particular, EBE methods have been identified as effective preconditioners when coupled with the conjugate gradient algorithm as the iterative driver (Hughes et al. 1983c; Hughes et al. 1984; Nour-Omid and Parlett 1985; Winget and Hughes 1985).

The efficiency of the EBE preconditioned conjugate gradient (EBE/PCG) algorithm for large-scale, vectorized calculations in linear and nonlinear solid mechanics was first documented in Hughes et al. (1987). Subsequent extensions to shell and contact/impact problems are presented in Hughes and Ferencz (1988a and 1988b). Ferencz (1989) has illustrated considerable success with EBE/PCG for nontraditional applications of shell elements, such as sheetmetal forming simulation and other contact problems. However, the inability of EBE/PCG to cope with the ill-conditioning associated with traditional structural element discretizations has long been apparent. Muller (1985) attempted to enhance EBE/PCG performance for such structural applications by both reformulating the fundamental element discretization as well as defining more general element factors for the EBE preconditioner. Recently, Nour-Omid et al. (1988) proposed the use of a Lanczos algorithm with partial reorthogonalization as another means of solving ill-conditioned problems.

The present paper is motivated by the observations of Nour-Omid et al. (1988), which we seek to extend into the range of large-scale engineering applications. We present an overview of the conjugate gradient and Lanczos algorithms as well as a representative EBE preconditioner. Before discussing numerical examples, the concept of a fractal mesh dimension is reviewed, as it proves useful in interpreting some of our results. A model problem and several engineering meshes of considerable size then illustrate our experience comparing the relative performance of EBE/PCG, EBE/Lanczos and direct solution.

ORIGIN OF THE LINEAR SYSTEM

For simplicity in comparing the candidate iterative drivers, here we restrict our attention to linear problems in solid and structural mechanics. However, these implementations have been developed in NIKE3D, a general purpose, large deformation finite element code with a variety of nonlinear material models (Hallquist 1984). NIKE3D's nonlinear strategy is summarized in (Hughes et al. 1987). For our present purpose, we simply state that NIKE3D generates symmetric, positive definite linear systems with N_{eq} unknowns, which we denote by the generic form

$$\mathbf{A}\mathbf{x} = \mathbf{b}. \tag{1}$$

The matrix $\mathbf{A}$ and vector $\mathbf{b}$ arise from a straightforward Galerkin finite element spatial discretization, and in the case of transient problems also reflect the common Newmark (1959) time-integration algorithm. To date, the production version of NIKE3D has relied upon a direct method utilizing a Crout factorization to solve linear systems. The so-called FISSLE package is a vectorized variable band, active column, out-of-core solver developed by S. J. Sackett and R. L. Taylor (see Taylor et al. 1981).

ITERATIVE LINEAR EQUATION SOLVING

The conjugate gradient (Hestenes and Stiefel 1952) and Lanczos (1952) algorithms are closely related, and their mathematical structure in perfect arithmetic is identical; for background see Golub (1983). Both are characterized as *Krylov space* methods, as at iteration j each computes an approximate solution $\mathbf{x}_j$ which resides in the Krylov space

$$\mathcal{K}_j(\mathbf{A}, \mathbf{q}_1) \equiv \text{span}\{\mathbf{q}_1, \mathbf{A}\mathbf{q}_1, \mathbf{A}^2\mathbf{q}_1, \dots, \mathbf{A}^{j-1}\mathbf{q}_1\} \tag{2}$$

where the generating vector $\mathbf{q}_1$ is usually taken parallel to $\mathbf{b}$. Lanczos algorithms are based on the notion of computing an orthonormal basis for the Krylov space using the recurrence

$$\beta_{j+1}\mathbf{q}_{j+1} = \bar{\mathbf{q}}_{j+1} = \mathbf{A}\mathbf{q}_j - \alpha_j\mathbf{q}_j - \beta_j\mathbf{q}_{j-1}, \tag{3}$$

where $\alpha_j = \mathbf{q}_j^T\mathbf{A}\mathbf{q}_j$, $\beta_{j+1} = \|\bar{\mathbf{q}}_{j+1}\|_2$ and $\mathbf{q}_0 \equiv \mathbf{0}$. This relation may be written in the matrix form

$$[\mathbf{0}_{j-1}\,|\,\beta_{j+1}\mathbf{q}_{j+1}] = \mathbf{A}\mathbf{Q}_j - \mathbf{Q}_j\mathbf{T}_j, \tag{4}$$

where $\mathbf{T}_j$ is the tridiagonal matrix

$$\mathbf{T}_j = \begin{bmatrix} \alpha_1 & \beta_2 & & & & & \\ \beta_2 & \alpha_2 & \beta_3 & & & & \\ & \beta_3 & \alpha_3 & \ddots & & & \\ & & \ddots & \ddots & \beta_{j-1} & & \\ & & & \beta_{j-1} & \alpha_{j-1} & \beta_j & \\ & & & & \beta_j & \alpha_j \end{bmatrix}_{j \times j}, \tag{5}$$

and

$$\mathbf{Q}_j = [\mathbf{q}_1, \mathbf{q}_2, \dots, \mathbf{q}_j]_{N_{eq} \times j}. \tag{6}$$

These basis vectors are also frequently referred to as *Lanczos vectors*. Due to the orthonormality of the basis vectors $\mathbf{Q}_j^T\mathbf{Q}_j = \mathbf{I}$, and so from (4) it follows that

$$\mathbf{0} = \mathbf{Q}_j^T\mathbf{A}\mathbf{Q}_j - \mathbf{T}_j \tag{7}$$

i.e., $\mathbf{T}_j$ is the projection of $\mathbf{A}$ onto the basis $\mathbf{Q}_j$. Using the basis $\mathbf{Q}_j$ generated by the Lanczos process, we wish to construct an approximate solution

$$\mathbf{x}_j = \mathbf{Q}_j\mathbf{y}_j, \tag{8}$$

with the associated residual

$$\mathbf{r}_j = \mathbf{b} - \mathbf{A}\mathbf{Q}_j\mathbf{y}_j. \tag{9}$$

A Galerkin approximation to $\mathbf{x}$ will result from making the residual orthogonal to the Krylov space, i.e.,

$$\mathbf{Q}_j^T\mathbf{r}_j = \mathbf{Q}_j^T(\mathbf{b} - \mathbf{A}\mathbf{Q}_j\mathbf{y}_j) \equiv \mathbf{0}. \tag{10}$$

Hence $\mathbf{x}_j$ is defined by the reduced system

$$\mathbf{T}_j\mathbf{y}_j = \mathbf{Q}_j^T\mathbf{b}, \tag{11}$$

which is typically simplified by taking $\beta_1\mathbf{q}_1 = \mathbf{b}$ so that

$$\mathbf{T}_j\mathbf{y}_j = \beta_1\mathbf{e}_1 = \beta_1 \begin{bmatrix} 1 \\ 0 \\ 0 \\ \vdots \\ 0 \end{bmatrix}_{j \times 1}. \tag{12}$$

In its simplest form, the Lanczos algorithm consists of constructing a basis for the Krylov space using (3) until the residual is sufficiently small, then solving the reduced system (12) for $\mathbf{y}_j$ and constructing $\mathbf{x}_j$. Available computational simplifications include monitoring the residual norm without computation of the residual vector, and accumulating the current approximation after each iteration; see Golub (1983) for further discussion. The CG algorithm can be equated to the Lanczos process by recognizing that its successive search directions ($\mathbf{p}_j$ in the notation of Box 1) are linear combinations of the Lanczos vectors. A more compelling development of CG stems from the minimization of the quadratic form $\frac{1}{2}\mathbf{x}^T\mathbf{A}\mathbf{x} - \mathbf{x}^T\mathbf{b}$ (e.g., Golub 1983).

As the CG and Lanczos methods are mathematically equivalent, they are equally dependent upon preconditioning to speed convergence. Preconditioning with a matrix $\mathbf{B}$ that is a computationally efficient approximation to $\mathbf{A}$ is typically introduced into the CG algorithm by defining an auxiliary vector $\mathbf{z}_j = \mathbf{B}^{-1}\mathbf{p}_j$. This leads to the standard PCG implementation which has been used in our previous work. If this same strategy is applied to Lanczos, the orthonormality condition on the basis vectors is modified by using $\mathbf{B}^{-1}$ as a metric:

$$(\mathbf{q}_i, \mathbf{q}_j)_{B^{-1}} \equiv \mathbf{q}_i \cdot \mathbf{B}^{-1}\mathbf{q}_j = \mathbf{q}_i \cdot \mathbf{z}_j = \begin{cases} 1, & \text{if } i = j; \\ 0, & \text{if } i \neq j. \end{cases} \tag{13}$$

This change leads to the preconditioned Lanczos algorithm presented in Nour-Omid et al. (1988). That algorithm suffers from a serious shortcoming when coupled with reorthogonalization strategies for the basis vectors: reorthogonalizing $\mathbf{q}_{j+1}$ would require j passes through the preconditioner during the *modified* Gram-Schmidt process. Using the unmodified Gram-Schmidt algorithm would permit holding $\mathbf{z}_{j+1} = \mathbf{B}^{-1}\mathbf{q}_{j+1}$ fixed, but is less stable numerically. Another possible remedy is to write the auxiliary vectors $\{\mathbf{z}_i\}_{i=1}^j$ to secondary storage and recall them to compute $\mathbf{q}_i \cdot \mathbf{B}^{-1}\mathbf{q}_{j+1} = \mathbf{z}_i \cdot \mathbf{q}_{j+1}$. This remedy would double the number of N_{eq}-length vectors written to secondary storage, which is unacceptable for large problems.

Shakib (1989) recognized these difficulties with reorthogonalization in connection with a preconditioned GMRES algorithm applied to nonsymmetric systems arising from fluid dynamics simulations. To avoid operational or storage penalties analogous to those noted above, Shakib decomposes his nonsymmetric preconditioner $\mathbf{B}$ into $\mathbf{LU}$ and uses GMRES to solve the linear system

$$\mathbf{L}^{-1}\mathbf{A}\mathbf{U}^{-1}\,\mathbf{U}\mathbf{x} = \mathbf{L}^{-1}\mathbf{b}, \tag{14}$$

or

$$\mathbf{L}^{-1}\mathbf{A}\mathbf{U}^{-1}\,\hat{\mathbf{x}} = \hat{\mathbf{b}}. \tag{15}$$

Shakib refers to this as "wrap-around" preconditioning. Application of this strategy to the CG and Lanczos methods with the symmetric preconditioner $\mathbf{B} = \mathbf{U}^T\mathbf{U}$ results in the algorithms of Boxes 1 and 2. Each method now computes an approximation from the Krylov space $\mathcal{K}_j(\mathbf{U}^{-T}\mathbf{A}\mathbf{U}^{-1}, \mathbf{U}^{-T}\mathbf{b})$. Raefsky (1988) has confirmed that wrap-around preconditioning was used for the actual implementation associated with Nour-Omid et al. (1988).

REMARK 1. The title *modified* PCG for Box 1 refers to the operations in (18)–(19), which arise from NIKE's treatment of nonzero displacement boundary conditions. These degrees of freedom are maintained as "active" diagonal equations in the system, which are then solved trivially by a direct method. Although a suitably defined preconditioner $\mathbf{B} = \mathbf{U}^T\mathbf{U}$ will replicate this behavior during the solution of $\mathbf{U}^T\hat{\mathbf{r}}_0 = \mathbf{r}_0$, the subsequent line search may scale the boundary displacements from the correct values. The present initialization serves to contract the search space to the remaining

Box 1. Algebraic definition of the modified conjugate gradient algorithm with wrap-around preconditioning.

Given the linear system $\mathbf{A}\mathbf{x} = \mathbf{b}$, where $\mathbf{A}$ is symmetric positive definite, and the preconditioner $\mathbf{B} = \mathbf{U}^T\mathbf{U}$

Step 1. Solve uncoupled equations and initialize:

$$\mathbf{r}_0 = \mathbf{b} \tag{16}$$
$$\hat{\mathbf{x}}_0 = \mathbf{0} \tag{17}$$

for $l = 1, 2, \ldots, N_{eq}$
    if $A_{kl} = 0$ for all $k < l$ then

$$\hat{x}_l = r_l/A_{ll} \tag{18}$$
$$r_l = 0 \tag{19}$$

    endif
continue

$$\hat{\mathbf{r}}_0 = \mathbf{U}^{-T}\mathbf{r}_0 \tag{20}$$
$$\hat{\mathbf{p}}_1 = \hat{\mathbf{r}}_0 \tag{21}$$

Step 2. Iterate for $j = 1, 2, \ldots, j_{\max}$:
    Perform line search to update solution and residual:

$$\alpha_j = \frac{\hat{\mathbf{r}}_{j-1} \cdot \hat{\mathbf{r}}_{j-1}}{\hat{\mathbf{p}}_j \cdot \mathbf{U}^{-T}\mathbf{A}\mathbf{U}^{-1}\hat{\mathbf{p}}_j} \tag{22}$$

$$\hat{\mathbf{x}}_j = \hat{\mathbf{x}}_{j-1} + \alpha_j\hat{\mathbf{p}}_j \tag{23}$$

$$\hat{\mathbf{r}}_j = \hat{\mathbf{r}}_{j-1} - \alpha_j\mathbf{U}^{-T}\mathbf{A}\mathbf{U}^{-1}\hat{\mathbf{p}}_j \tag{24}$$

    Check convergence (δ_L is a user-defined tolerance):
    if $\|\hat{\mathbf{r}}_j\|_2 \leq \delta_L \|\hat{\mathbf{r}}_0\|_2$ then

$$\mathbf{x}_j = \mathbf{U}^{-1}\hat{\mathbf{x}}_j \tag{25}$$

    return
endif
Compute new conjugate search direction:

$$\beta_{j+1} = \frac{\hat{\mathbf{r}}_j \cdot \hat{\mathbf{r}}_j}{\hat{\mathbf{r}}_{j-1} \cdot \hat{\mathbf{r}}_{j-1}} \tag{26}$$

$$\hat{\mathbf{p}}_{j+1} = \hat{\mathbf{r}}_j + \beta_{j+1}\hat{\mathbf{p}}_j \tag{27}$$

---

unspecified degrees of freedom. Care must be taken however, so that the preconditioner does not reintroduce non-zero components for the specified degrees of freedom into subsequent search directions $\mathbf{p}_j$ for $j > 1$. The identical initialization of uncoupled equations is included in the Lanczos algorithm of Box 2. With this modification the algorithms are also applicable to systems with blocks of diagonal equations, as would arise from a mixed implicit-explicit time-integration scheme as discussed in Hughes et al. (1979). □

As noted, the CG and Lanczos algorithms yield identical approximations in perfect arithmetic. In our experience the algorithms as discussed so far behave very similarly in actual calculations with finite precision arithmetic, i.e., the iterations to converge differ by less than one percent. Thus as the linear system becomes more ill-

Box 2. Algebraic definition of the Lanczos algorithm with wrap-around preconditioning and accumulated solution. (...) denotes an optional operation.

Given the linear system $\mathbf{Ax} = \mathbf{b}$, where $\mathbf{A}$ is symmetric positive definite, and the preconditioner $\mathbf{B} = \mathbf{U}^T\mathbf{U}$

Step 1. Solve uncoupled equations:

$$\hat{\mathbf{x}}_0 = \mathbf{0} \tag{28}$$

for $l = 1, 2, \ldots, N_{eq}$

if $A_{kl} = 0$ for all $k < l$ then

$$\hat{x}_l = b_l/A_{ll} \tag{29}$$
$$b_l = 0 \tag{30}$$

endif

Step 2. Initialize:

$$\hat{\mathbf{q}}_0 = \mathbf{0} \tag{31}$$
$$\hat{\mathbf{b}} = \mathbf{U}^{-T}\mathbf{b} \tag{32}$$
$$\beta_1 = \|\hat{\mathbf{r}}_0\|_2 = \|\hat{\mathbf{b}}\|_2 \tag{33}$$
$$\hat{\mathbf{q}}_1 = \beta_1^{-1}\hat{\mathbf{b}} \tag{34}$$
$$d_1 = \alpha_1 = \hat{\mathbf{q}}_1 \cdot \mathbf{U}^{-T}\mathbf{A}\mathbf{U}^{-1}\hat{\mathbf{q}}_1 \tag{35}$$
$$p_1 = \beta_1/\alpha_1 \tag{36}$$
$$\hat{\mathbf{c}}_1 = \hat{\mathbf{q}}_1 \tag{37}$$
$$\hat{\mathbf{x}}_1 = \hat{\mathbf{x}}_0 + p_1\hat{\mathbf{c}}_1 \tag{38}$$

Step 3. Iterate for $j = 1, \ldots, j_{\max}$:

Compute next basis vector:

$$\bar{\mathbf{q}}_{j+1} = \mathbf{U}^{-T}\mathbf{A}\mathbf{U}^{-1}\hat{\mathbf{q}}_j - \alpha_j\hat{\mathbf{q}}_j - \beta_j\hat{\mathbf{q}}_{j-1} \tag{39}$$

(Monitor orthogonality of $\bar{\mathbf{q}}_{j+1}$)

$$\beta_{j+1} = \|\bar{\mathbf{q}}_{j+1}\|_2 \tag{40}$$
$$\|\hat{\mathbf{r}}_j\|_2 = \beta_{j+1}|p_j| \tag{41}$$

Check convergence:

if $\|\hat{\mathbf{r}}_j\|_2 \leq \delta_L \|\hat{\mathbf{r}}_0\|_2$ then

$$\mathbf{x}_j = \mathbf{U}^{-1}\hat{\mathbf{x}}_j \tag{42}$$

return

endif

Normalize basis vector and update solution:

$$\hat{\mathbf{q}}_{j+1} = \beta_{j+1}^{-1}\bar{\mathbf{q}}_{j+1} \tag{43}$$

(Write $\hat{\mathbf{q}}_j$ to secondary storage)

$$\alpha_{j+1} = \hat{\mathbf{q}}_{j+1} \cdot \mathbf{U}^{-T}\mathbf{A}\mathbf{U}^{-1}\hat{\mathbf{q}}_{j+1} \tag{44}$$
$$u_{j+1} = \beta_{j+1}/d_j \tag{45}$$
$$d_{j+1} = \alpha_{j+1} - \beta_{j+1}^2/d_j \tag{46}$$
$$p_{j+1} = -p_j\beta_{j+1}/d_{j+1} \tag{47}$$
$$\hat{\mathbf{c}}_{j+1} = \hat{\mathbf{q}}_{j+1} - u_{j+1}\hat{\mathbf{c}}_j \tag{48}$$
$$\hat{\mathbf{x}}_{j+1} = \hat{\mathbf{x}}_j + p_{j+1}\hat{\mathbf{c}}_{j+1} \tag{49}$$

conditioned, both algorithms degrade due to the effects of machine arithmetic which lead to a loss of conjugacy (orthogonality) between the search (basis) vectors. As this occurs, the projection of $\mathbf{A}$ upon the sequence of Lanczos vectors will no longer be tridiagonal. However, the Lanczos algorithm permits the addition of auxiliary algorithms to actively maintain orthogonality of the basis vectors; this is indicated by the parenthetical operations in Box 2. The implementations of Nour-Omid et al. (1988) and this paper are both based upon the *partial reorthogonalization* strategy of Simon (1984). Simon developed a simple recurrence to estimate the inner products of the basis vectors and showed that a reduced notion of orthogonality was sufficient for $\mathbf{T}$ to remain tridiagonal. Specifically, define a computer's *unit roundoff* ϵ as the smallest number such that in machine arithmetic $1 + \epsilon > 1$. Then the *semiorthogonality* condition $|\mathbf{q}_i \cdot \mathbf{q}_j| < \sqrt{N_{eq}}\epsilon$ for $i \neq j$ is sufficient to maintain the desired structure of $\mathbf{T}$; this is a considerable relaxation of the orthogonality condition $|\mathbf{q}_i \cdot \mathbf{q}_j| < N_{eq}\epsilon$ for $i \neq j$. Simon's recurrence circumvents the need to recall the previous basis vectors from secondary storage and monitor orthogonality with dot product calculations at each iteration. Instead, the historical pool of basis vectors is only recalled from storage when the recurrence predicts that the latest vector $\bar{\mathbf{q}}_{j+1}$ violates semiorthogonality with respect to a subset of that pool. At that point $\bar{\mathbf{q}}_{j+1}$ and $\mathbf{q}_j$ are reorthogonalized with respect to the recalled basis vectors. Henceforth, reference to the Lanczos algorithm will mean Lanczos with partial reorthogonalization. Further discussion of our implementation is contained in Ferencz (1989).

REMARK 2. Although the operational structure of the CG and Lanczos algorithms of Boxes 1 and 2 are different, their cost in the absence of any reorthogonalization is nearly the same. The major costs per iteration are identical: the two-phase preconditioning $\mathbf{v}' = \mathbf{U}^{-1}\mathbf{v}$ and $\mathbf{v}''' = \mathbf{U}^{-T}\mathbf{v}''$, and the intervening matrix-vector product $\mathbf{v}'' = \mathbf{A}\mathbf{v}'$, where $\mathbf{v}$ denotes a dummy vector argument to these operations. In addition, PCG uses three dot products and three SAXPYs, while Lanczos requires two dot products and four and one half SAXPYs.$\square$

THE CROUT EBE PRECONDITIONER

The selection of the preconditioner $\mathbf{B} = \mathbf{U}^T\mathbf{U}$ is crucial to the performance of iterative algorithms. We desire $\mathbf{B}^{-1}$ to approximate $\mathbf{A}^{-1}$ in the sense that the condition number $\kappa(\mathbf{U}^{-T}\mathbf{A}\mathbf{U}^{-1}) \ll \kappa(\mathbf{A})$ while retaining a computationally efficient structure. A simple preconditioner with some useful properties is nodal block-diagonal scaling, i.e., $\mathbf{B}$ is a block-diagonal matrix where each block contains the coefficients of $\mathbf{A}$ describing the coupling between degrees of freedom at an individual node. This preconditioner may be implemented by using the Cholesky decomposition $\mathbf{W}^T\mathbf{W} = $ block-diag($\mathbf{A}$) to transform the system of equations:

$$\mathbf{W}^{-T}\mathbf{A}\mathbf{W}^{-1} \, \mathbf{W}\mathbf{x} = \mathbf{W}^{-T}\mathbf{b}, \tag{50}$$

or

$$\tilde{\mathbf{A}}\tilde{\mathbf{x}} = \tilde{\mathbf{b}}. \tag{51}$$

This transformed system is solved iteratively, then the solution to the original system is recovered: $\mathbf{x} = \mathbf{W}^{-1}\tilde{\mathbf{x}}$. In addition to automatically embedding block-diagonal scaling into the iterative driver, this transformation nondimensionalizes the equation system. This ensures that the residual norm used to monitor convergence is well-defined for structural mechanics, where the untransformed residual has mixed dimensions. Therefore, both of our iterative algorithms are posed in this transformed system. In subsequent discussion of the numerical examples, the term block-diagonal preconditioning will mean using the algorithms of Boxes 1 and 2 to solve the transformed system (51) with no further preconditioning, i.e., taking $\mathbf{B} = \mathbf{I}$.

Element-by-element preconditioners are motivated by a desire to maintain the element-based data structure of current finite element codes. For the *Crout EBE preconditioner*, we utilize the product decomposition

$$\mathbf{B}_{\text{Crout}} = \prod_{e=1}^{N_{el}} \mathcal{L}_p^e \times \prod_{e=1}^{N_{el}} \mathcal{D}_p^e \times \prod_{e=N_{el}}^{1} \mathcal{U}_p^e, \tag{52}$$

where

$$\mathcal{L}_p^e = \mathcal{L}_p^e(\bar{\mathbf{A}}^e, \mathbf{P}^e) = (\mathbf{P}^e)^T L_p[\mathbf{P}^e \bar{\mathbf{A}}^e (\mathbf{P}^e)^T] \mathbf{P}^e,$$

$$\mathcal{D}_p^e = \mathcal{D}_p^e(\bar{\mathbf{A}}^e, \mathbf{P}^e) = (\mathbf{P}^e)^T D_p[\mathbf{P}^e \bar{\mathbf{A}}^e (\mathbf{P}^e)^T] \mathbf{P}^e,$$

$$\mathcal{U}_p^e = \mathcal{U}_p^e(\bar{\mathbf{A}}^e, \mathbf{P}^e) = (\mathbf{P}^e)^T U_p[\mathbf{P}^e \bar{\mathbf{A}}^e (\mathbf{P}^e)^T] \mathbf{P}^e,$$

and N_{el} is the number of elements in the mesh. The reversed element order in the third product of (52) results in **B** being symmetric. Denoting the eth element's contribution to the transformed global matrix $\tilde{\mathbf{A}}$ by $\tilde{\mathbf{A}}^e$, then $\bar{\mathbf{A}}^e$ is the *Winget regularized element matrix*:

$$\bar{\mathbf{A}}^e = \mathbf{I} + (\tilde{\mathbf{A}}^e - \tilde{\mathbf{W}}^e), \tag{53}$$

$$\tilde{\mathbf{W}}^e = \text{diag}(\tilde{\mathbf{A}}^e). \tag{54}$$

The regularization ensures $\bar{\mathbf{A}}^e$ is positive definite, hence the Crout factorization $L_p[\cdot] D_p[\cdot] U_p[\cdot]$ is well-defined. The permutation matrix $\mathbf{P}^e$ is a notational convenience intended to indicate that the element matrix is factored in the local ordering of the element equations. A fully vectorized implementation of the Crout EBE preconditioner is operational within NIKE3D and described in Hughes et al. (1987). Further details of the implementation may be found in Ferencz (1989). Other EBE preconditioners have been defined (see Hughes and Ferencz (1988b) for a summary and attribution), but restriction to this single example is sufficient for the present purpose of comparing CG and Lanczos.

FRACTAL DIMENSION OF A FINITE ELEMENT MESH

Prior to discussing our computational experience with these EBE preconditioned iterative algorithms, we wish to review the concept of the fractal dimension of a mesh introduced by Hughes et al. (1987). Traditionally, a mesh is said to be n-dimensional if it is composed of n-dimensional elements. Hence a mesh consisting of a long strip of three-dimensional continuum elements is labelled a three-dimensional mesh. Such nomenclature obscures a key measure of the relative efficiency of iterative and direct solution algorithms.

Definition. Given a mesh containing three-dimensional, eight-node brick continuum elements, consider three meshes with equal numbers of active degrees of freedom: a strip of elements connected on end, a square with one element through the thickness, and a cube. We refer to these as the one-, two- and three-dimensional reference meshes. For each reference mesh we compute the total profile storage, or "fill", needed to hold the associated global matrix **A**. Then the fill for the given mesh determines a *fractal dimension* d_f:

If $fill \leq fill_{2D}$

$$d_f = \frac{fill - fill_{1D}}{fill_{2D} - fill_{1D}} + 1, \tag{55}$$

If $fill > fill_{2D}$

$$d_f = \frac{fill - fill_{2D}}{fill_{3D} - fill_{2D}} + 2. \tag{56}$$

For simplicity, the fill for each reference mesh is computed using its *peak* bandwidth. NIKE's four-node shell element has the same number of degrees of freedom as the eight-node continuum element, and in the general case identical nodal connectivity within an element, so it is consistent to use the same reference meshes to define the fractal dimension of a model containing these shell elements.

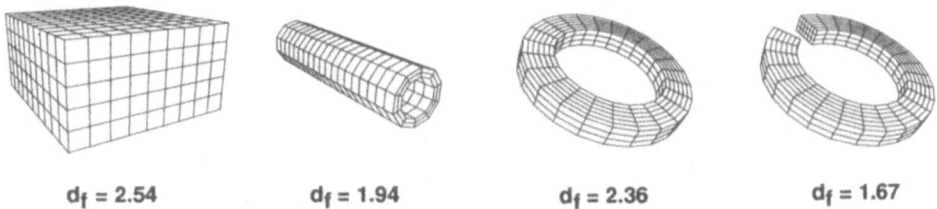

$d_f = 2.54$ $d_f = 1.94$ $d_f = 2.36$ $d_f = 1.67$

Figure 1. Fractal dimension of four simple meshes.

Given that the number of equations N_{eq} is constant, the fractal dimension reflects the change in mean bandwidth of **A** as a function of mesh configuration. Figure 1 presents the fractal dimension for a number of simple meshes. Note the resulting drop in d_f if the annulus is "cut" and opened into a helix. As d_f approaches 1.0, two effects combine to make direct solution more efficient than EBE preconditioned iteration. First, the bandwidth of **A** decreases, reducing the operations for direct factorization which are proportional to the bandwidth squared. Second, for the EBE preconditioner discussed here, the number of iterations scales with the largest number of nodes along an edge, increasing the cost of iteration. These trends are illustrated in Figure 2, which plots the ratio of Crout EBE/PCG to direct equation solving CPU for three series of linear problems. Each mesh is a rectangular parallelepiped constructed of eight-node continuum elements with unit aspect ratios. The number of global degrees of freedom is kept as constant as possible within each series, but the number of elements along each of the three principal directions of the mesh is varied to alter the fractal dimension. For these well-conditioned problems, Crout EBE/PCG is more efficient than FISSLE for $d_f > \approx 1.35$. The relative behavior of Crout EBE/Lanczos is equivalent for these simple examples.

NUMERICAL EXAMPLES

The following calculations summarize our experience in comparing EBE/PCG to EBE/Lanczos as well as direct solution using NIKE3D. Computations were performed on either a CRAY X-MP/48 with a 9.5 nanosecond clock period or an X-MP/416 with an 8.5 nanosecond clock period. Each of these machines has a 134×10^6 word Solid-state Storage Device (SSD) providing high-speed secondary storage. We are currently using the CFT 1.14 FORTRAN compiler, which utilizes both the bi-directional memory and hardware gather/scatter capabilities of the X-MP. Unless noted otherwise, all computations were performed with a linear iteration tolerance of $\delta_L = 10^{-4}$. The CPU and I/O costs presented are for equation solving only and include assembly of the global equations for the direct method, but exclude element formation for all

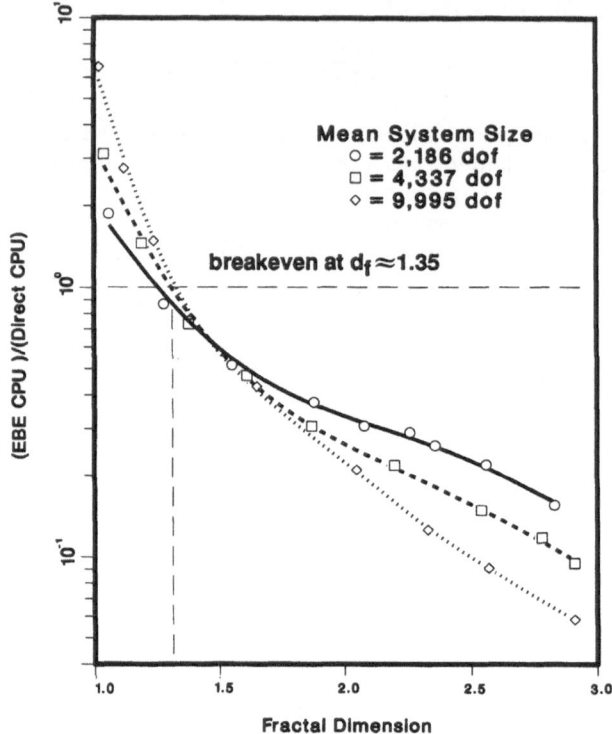

Figure 2. Relative efficiency of Crout EBE/PCG versus direct solution with FISSLE for three series of well-conditioned problems.

methods. The I/O costs include I/O-related system charges, which can vary with the level of machine usage. Storage requirements are stated in units of 64-bit words.

Timing data for some of these examples have previously appeared in Hughes et al. (1987) and Hughes and Ferencz (1988a,b). Those data were gathered on the X-MP/48 when it was still running under the Livermore Time Share System (LTSS). The X-MP/48 has more recently become the testbed for the Network Livermore Time Share System (NLTSS), which will support distributed processing in the future. We have found that NLTSS increases I/O cost 20 to 30 percent due to added system overhead, so for large, out-of-core solutions we have migrated to the X-MP/416 which continues to run under LTSS. Therefore, the large-scale calculations reported here reflect new calculations on the X-MP/416 and should not be confused with timings from the slower X-MP/48 reported in earlier papers.

Plane-Strain Cantilever Beam

We first consider a simple plane-strain cantilever beam subjected to a transverse end load. A nearly identical problem was examined in Nour-Omid et al. (1988), and here we compare them to verify our implementation of the Lanczos strategy in NIKE3D. The mesh consists of a single layer of continuum brick elements, and because the deformation is antisymmetric, it models only half the physical domain. The end load and reaction forces are defined in accordance with standard beam bending theory. The kinematic boundary conditions are illustrated in Figure 3 for a mesh with 4×16 elements, however all numerical results reported are for a mesh containing 16×64

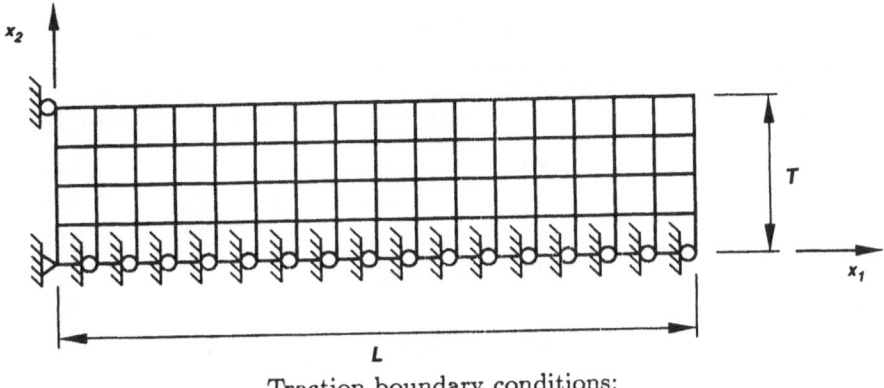

Traction boundary conditions:

$$\sigma_{22}(x_1, \pm T) = \sigma_{12}(x_1, \pm T) = 0, \qquad x_1 \in]0, L[$$

$$\sigma_{11}(L, x_2) = 0; \ \sigma_{12}(L, x_2) = \frac{P}{2I}(T^2 - x_2^2), \qquad x_2 \in]0, T[$$

$$\sigma_{11}(0, x_2) = \frac{PLx_2}{I}; \ \sigma_{12}(0, x_2) = -\frac{P}{2I}(T^2 - x_2^2), \qquad x_2 \in]0, T[$$

Figure 3. Kinematic and traction boundary conditions for the plane-strain cantilever beam problem. P is the magnitude of the end load and I is the area moment of inertia $I = 2T^3/3$.

elements. The length of the beam is held constant at $L = 16$, but the half-thickness T is varied to alter the condition number of the problem. Rather than refer to the thickness, we will index the results by the planar element aspect ratio which equals $4/T$. An elastic modulus of 1.0 and a Poisson's ratio of 0.3 are specified. The plane-strain assumption is enforced by setting all through-width displacements to zero. The number of global degrees of freedom is reduced to 2143 by defining rigid nodal constraints between each pair of nodes having identical x_1 and x_2 coordinates. These constraints simply assign the same global equation number to each physically equivalent pair of nodal x_1 or x_2 displacements and *do not* involve a penalized constraint.

Table 1 provides a brief comparison of the NIKE3D results with some of the data reported by Nour-Omid et al. (1988). The condition numbers listed are those for the Crout EBE preconditioned systems (data for the unpreconditioned systems are plotted in Figure 4). The iteration counts are for convergence to a linear tolerance of $\delta_L = 10^{-6}$. With NIKE3D, the PCG and Lanczos iterations differ by less than one percent up to a condition number of $O(10^6)$. A much more rapid divergence between the methods is reported in Nour-Omid et al. (1988), which led us to conjecture that those calculations were performed in 32-bit arithmetic. This hypothesis was subsequently confirmed by Raefsky (1988). In general, the added numerical precision on the CRAY leads to more rapid convergence for either method. The only exception occurs for the Lanczos solution of the most ill-conditioned problem. This apparent contradiction again appears to arise from the difference in numerical precision. With 32-bit arithmetic, the unit roundoff ϵ is $O(10^{-7})$, hence it would in general be impossible to compute a meaningful matrix-vector product when the condition number of $\tilde{\mathbf{A}}$ exceeds $\epsilon^{-1} \approx 10^7$ (Wilkinson 1963). Therefore, it is doubtful that the 32-bit calculation had converged to the real solution of the linear system. Nevertheless, the

Table 1. Condition number and iteration counts for linear solution of the cantilever beam problem with Crout EBE preconditioning.

Aspect Ratio	Condition Number	Conjugate Gradient		Lanczos	
		Nour-Omid*	NIKE3D	Nour-Omid*	NIKE3D
1	5.1×10^4	182	133	146	133
2	4.3×10^5	344	188	199	185
4	4.8×10^6	850	319	329	311
8	6.3×10^7	2,714	702	586	559
16	8.7×10^8	—	1,835	—	1,018
40	2.4×10^{10}	>6,000	5,954	886	1,368

* Data from Nour-Omid et al. (1988).

data in Table 1 support the conclusion that the NIKE3D implementation is in reasonable agreement with the methodology proposed in Nour-Omid et al. (1988).

REMARK 3. We wish to emphasize that the results presented in Nour-Omid et al. (1988) are for a cantilever beam in pure bending with 2142 degrees of freedom. Selected executions of that problem with NIKE3D indicate that the same iteration counts are attained as reported for the present example. However, we feel that the definition given here has more consistent kinematic boundary conditions and therefore serves as a better model problem. □

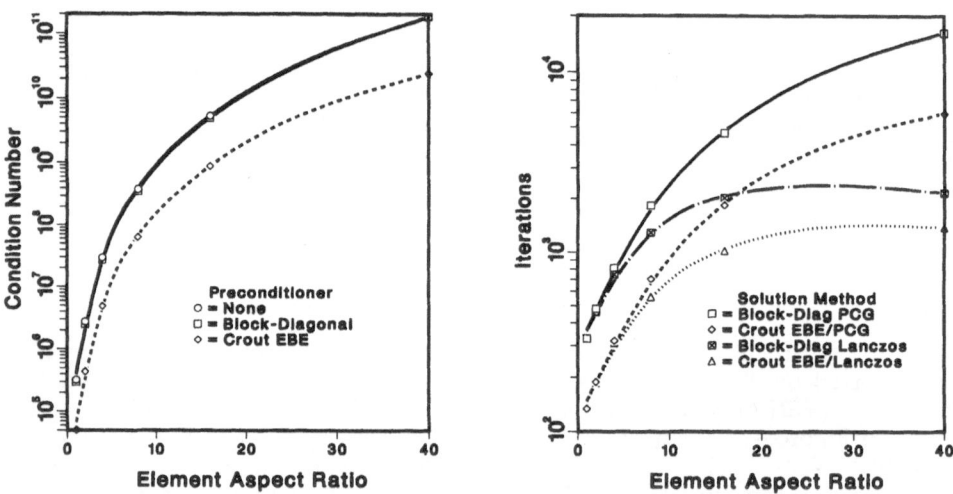

Figure 4. Condition numbers and iteration counts for the plane-strain cantilever beam problems with the PCG and Lanczos algorithms.

The range of condition number κ spanned by the examples is illustrated in Figure 4. We remind the reader that the change in element aspect ratio is a reflection of decreasing beam half-thickness T, so not only does the maximum eigenvalue change proportional to $\approx T^{-1}$, but the minimum eigenvalue grows proportional to $\approx T^3$. As a result, κ is growing proportional to $\approx T^{-4}$. Block-diagonal preconditioning has essentially no effect upon condition number, as we would expect for a uniform grid. Crout EBE preconditioning reduces the condition number by a factor of 5.5 for the first five analyses, but for the highest aspect ratio attains a reduction factor of 7.5. Figure 4 also shows the iteration performance of the four algorithms examined. The block-diagonal PCG iterations grow catastrophically with κ to a maximum of 16,252. Crout EBE/PCG behaves similarly, peaking at a maximum of 5,954 iterations. For the lower range of condition number, the Lanczos methods perform identically with their PCG counterparts. At higher values of κ, the explicit maintenance of semiorthogonality between the basis vectors retains the theoretical finite termination property of the Lanczos algorithm. This "cap" of N_{eq} iterations is quite clear for the block-diagonal Lanczos algorithm: the problems contain 2143 degrees of freedom, and the final two solutions require 2,017 and 2,138 iterations. The Crout EBE/Lanczos algorithm displays a similar flattening, but the number of iterations would continue to grow for even larger κ until the N_{eq} limit is reached.

The CPU data plotted in Figure 5 is a direct reflection of the trends in iteration count: PCG cost grows relentlessly while Lanczos flattens as the N_{eq} iteration limit is approached. Clearly Lanczos offers a decisive cost advantage relative to PCG for these highly ill-conditioned problems. However, it is essential to note that this mesh has a fractal dimension of only 1.12, and that these problems may be solved with our direct method *in less than one second* of X-MP CPU. Thus for this problem, Lanczos is superior to PCG in a regime where both are vastly inferior to direct solution.

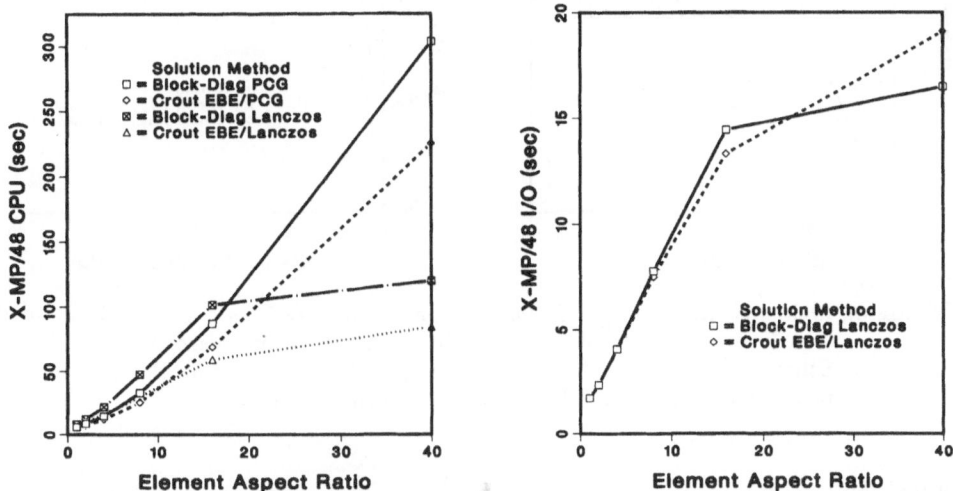

Figure 5. CPU and I/O costs for the plane-strain cantilever beam problems with the PCG and Lanczos algorithms.

The I/O data in Figure 5 is presented to show that I/O remains a reasonable percentage of CPU cost for Lanczos when the SSD is utilized. The I/O is actually excessive for this problem due to our current implementation. The Lanczos algorithm was coded in anticipation of large problems where the element matrices would be stored out of core. The reorthogonalization module relies on using the element buffer area in central memory as a work space for the recalled Lanczos vectors. As a result, all Lanczos solutions must be performed with the out-of-core element storage option selected. For relatively small problems such as this, it would be more efficient to keep the element data in core during successive iterations with a copy stored on the SSD. These data then need only be retrieved after a reorthogonalization phase has overwritten the element storage in central memory with Lanczos vectors.

Further insight into the level of effort required by the Lanczos algorithm is provided by Figure 6. The term *reorthogonalization cycle* is intended to signify a reorthogonalization of q_j and q_{j+1} with respect to all previous Lanczos vectors. A *reorthogonalization operation* refers to the effort to reorthogonalize q_j or q_{j+1} with respect to a single stored Lanczos vector, i.e., one dot product and one SAXPY. The data illustrate a conclusion noted by Nour-Omid et al. (1988): EBE preconditioning can be a useful addition to a Lanczos-based solution strategy. As shown, EBE preconditioning, when it successfully improves conditioning, reduces the frequency of reorthogonalization cycles. This in turn reduces the total number of reorthogonalization operations. In Figure 6 we see nearly an order of magnitude reduction in this cost due to EBE preconditioning. Finally, because EBE speeds convergence, fewer basis vectors need to be retained in secondary storage.

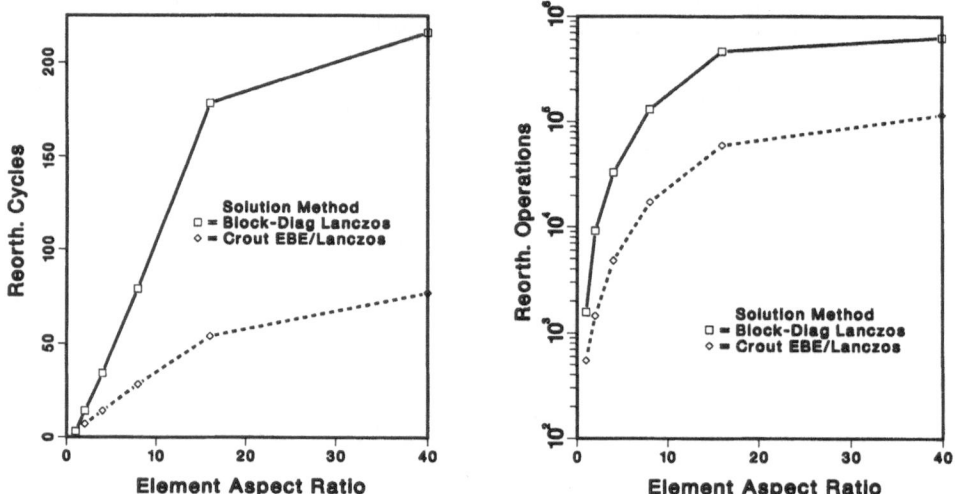

Figure 6. Incidence and cost of Lanczos reorthogonalizations for the plane strain cantilever beam problems. Each *cycle* denotes an entire modified Gram-Schmidt reorthogonalization process, and one *operation* denotes one dot product and one SAXPY with N_{eq}-length vectors.

Tunnel Intersection

The stress field about intersecting tunnels in geological material is under study at Los Alamos National Laboratory. The mesh in Figure 7 was developed by R. Rosinsky of L.L.N.L. as a prototype to evaluate our iterative algorithms. Triple symmetry is used to model two 130 inch radius tunnels in a $600 \times 600 \times 600$ inch block of rock subjected to a uniform hydrostatic pressure of 100 psi. The mesh contains 11,713 nodes, 10,240 elements, 34,272 degrees of freedom with $d_f = 3.02$. To study the effect of mesh distortion, a series of analyses were performed on meshes with evenly graded mesh transitions from the tunnel intersection outward. The measure of distortion was taken as the largest to smallest element length along an edge, which is nearly equivalent to the maximum element aspect ratio in the mesh. The mesh in Figure 7 corresponds to an aspect ratio of ten. As shown in Figure 8, this series of meshes spans a range of condition number from nearly 10^4 to 10^6 for the unpreconditioned system.

PCG and Lanczos performance is compared for the block-diagonal and Crout EBE preconditioners in Figures 9 and 10. Due to the high fractal dimension of the

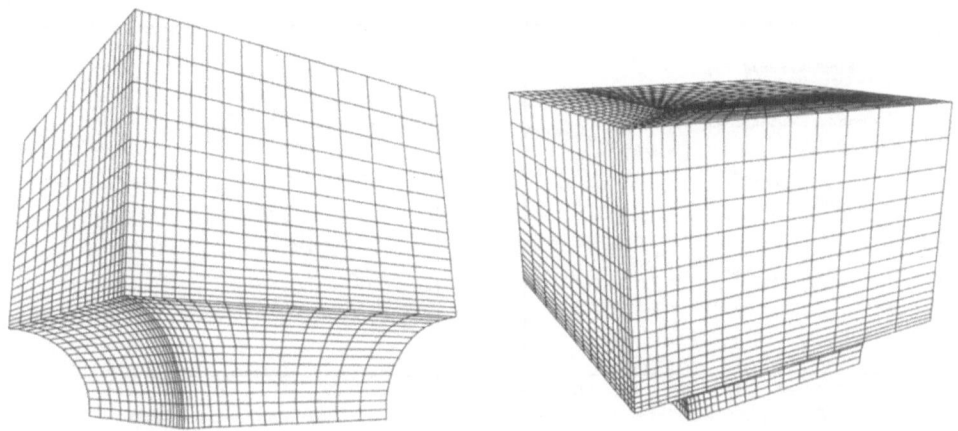

Figure 7. Two views of mesh for the tunnel intersection problem with a maximum element aspect ratio of 10.

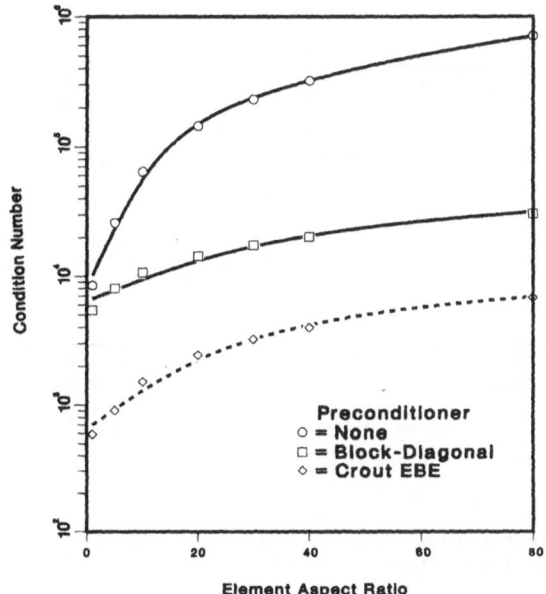

Figure 8. Condition number versus element aspect ratio for the tunnel intersection problem.

problem, both iterative methods use no more than seven percent of the direct solution CPU for all meshes. The Lanczos algorithm provides a minor reduction in iterations for either preconditioner as the maximum element aspect ratio increases. The cost of reorthogonalization can more than offset this advantage. Although Crout EBE CPU decreases slightly with Lanczos for higher aspect ratios, block-diagonal CPU shows a steady increase. These notions are made more precise in Figure 10, which plots relative performance (Lanczos to PCG) with the two preconditioners. For block-diagonal preconditioning, Lanczos attains a near uniform iteration ratio of 0.98, but the CPU ratio rises from 1.05 to nearly 1.10. With Crout EBE, Lanczos provides a

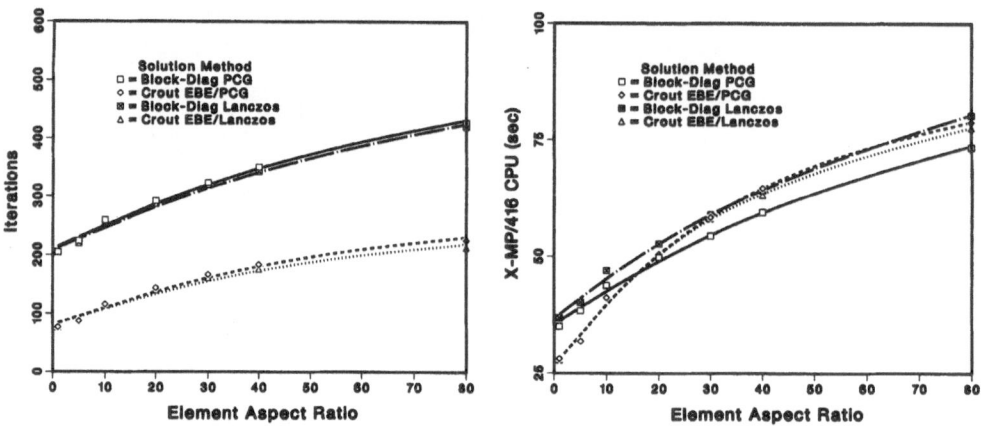

Figure 9. Iteration and X-MP/416 CPU performance versus element aspect ratio for the tunnel intersection problem with the PCG and Lanczos algorithms.

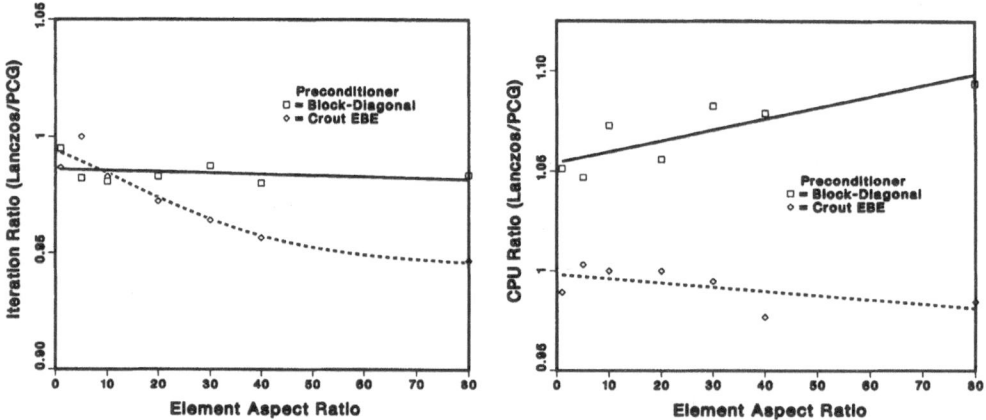

Figure 10. Relative performance of the PCG and Lanczos algorithms versus element aspect ratio for the tunnel intersection problem.

steady decrease in iteration ratio from about 1.0 to 0.95. This is mirrored by the CPU ratio decreasing from 1.0 to 0.98.

The difference in relative performance of the Lanczos algorithm using these two preconditioners is at first unexpected. Yet it appears this behavior may be explained with reference to the plot of condition number as a function of aspect ratio and preconditioner contained in Figure 8. The somewhat ironic conclusion is that the relative success of Lanczos with Crout EBE stems from the more rapid degradation of this preconditioner as the mesh is distorted. This performance is reflected by the change in condition number κ as the maximum element aspect ratio increases from 1 to 80. With block-diagonal preconditioning, κ increases from 5444 to 30,491: a factor of 5.6. Crout EBE is quite successful for the uniform mesh, as κ has an initial value of 588. However, as the mesh distorts κ increases to 6761: a factor of 11.5. The more rapid growth in ill-conditioning with Crout EBE provides a greater opportunity for Lanczos to attain a net reduction in cost. This behavior is reflected directly by the reorthogonalization strategy. Consider a cost measure $(n_c|n_o)$, where n_c is the number of reorthogonalization cycles and n_o the number of reorthogonalization operations. For the uniform mesh, block-diagonal Lanczos has a cost of (6|1452), which increases to (13|6407) for the highest aspect ratio. In contrast, Crout EBE Lanczos goes from a cost of (2|166) to (11|2741), and this faster rise in reorthogonalization cost appears coincident with the greater relative benefit.

Stiffened Cylindrical Shell

The analysis of thin shell structures is a frequent engineering task and typically among the most difficult problems to solve iteratively. A scale model of a ring-stiffened cylindrical shell has been proposed by Dr. Rembert Jones of the U.S. Navy's David Taylor Research Center for experimental and computational study of shell buckling due to external pressure loads (Jones 1986). This problem geometry is representative of a common type of engineering shell structure and as such is a useful example for comparing the performance of our iterative methods for linear shell analysis. A series of meshes was generated to determine the performance trends with problem size; the second coarsest mesh is pictured in Figure 11. The axial symmetry of the problem permits modelling only half the length of the shell. Due to the anticipated buckling mode, all meshes for the original buckling study were limited to a $360°/(2 \cdot 7) \approx 25.71°$ arc about the generating axis. These same meshes were used for the linear analyses presented here. Symmetry boundary conditions must be imposed on the bottom ($\theta = 0°$) and top ($\theta = 25.71°$) edges of the mesh. As with the axial symmetry conditions at the midlength, the symmetry conditions at the bottom edge may be enforced with nodal boundary conditions. Symmetry conditions at the top edge are applied through penalized constraints.

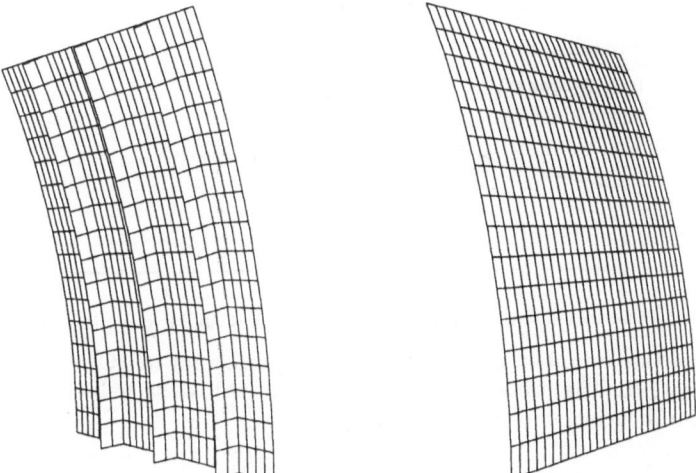

Figure 11. Inner and outer views of mesh 2 for the stiffened cylindrical shell problem. The stiffeners are also modelled with shell elements.

Table 2 summarizes the size characteristics of the four meshes considered; Meshes 2 through 4 each double the lineal resolution of the preceeding mesh. Figure 12 contains a plot of the associated condition numbers. The large, nearly constant condition numbers for the original systems stem from the penalized symmetry constraint. Figure 12 also plots the CPU ratio for PCG iteration relative to direct solution. Due to the relatively low fractal dimension and high condition number, PCG uses more CPU than

Table 2. Mesh size and characteristics for the stiffened cylindrical shell.

	Mesh 1	Mesh 2	Mesh 3	Mesh 4
No. Nodes	162	595	2,277	8,905
No. Elements	136	544	2,176	8,704
No. Equations	875	3,382	13,292	52,696
Storage[a]	55,212	415,339	3,365,969	27,047,389
Storage[b]	92,641	368,861	1,472,101	5,881,781
Storage[c]	222,141	1,329,349	8,570,029	64,163,557
Fractal Dimension	1.70	1.78	1.86	1.89
$\max(h^e/t)$	13.2	6.60	3.30	1.65

[a] Direct solution, [b] Crout EBE/PCG, [c] Crout EBE/Lanczos.

direct solution. Although the costs are nearly equal by Mesh 4, that mesh represents a much finer spatial discretization than normally used in shell analysis. The improving performance of Crout EBE/PCG with mesh size results from increased accuracy of the EBE preconditioner as the element length-to-thickness ratio h^e/t decreases.

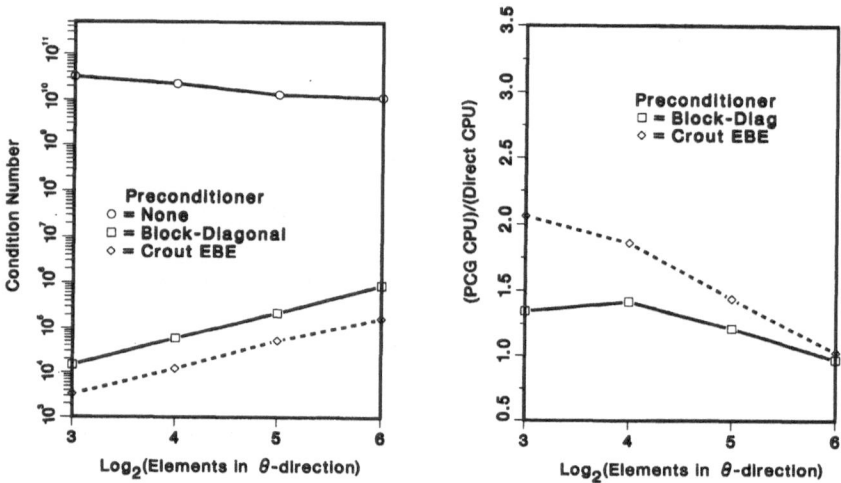

Figure 12. Condition number and relative CPU performance (Crout EBE/PCG to Direct) for the stiffened cylindrical shell. Abscissa values 3–6 correspond to Meshes 1–4 described in Table 2.

The poor performance of the EBE/PCG methods for linear analysis of thin shells such as these first motivated our consideration of adding a Lanczos algorithm to NIKE3D. The performance plots in Figure 13 illustrate that Lanczos has no effect upon cost trends as the mesh is refined, and the relative cost data in Figure 14 sound a familiar refrain: iteration counts are reduced by less than five percent while CPU cost grows ten or more percent. The largest problem also underscores another pitfall of the Lanczos algorithm, as we could not access sufficient SSD storage to hold the basis vectors generated with block-diagonal preconditioning. Based upon the relative performance for Mesh 3, we project that the Lanczos solution for Mesh 4 would require 2064 iterations. This translates to a data base of approximately 109 million words, or essentially four times the data for direct solution. Indeed, block-diagonal Lanczos

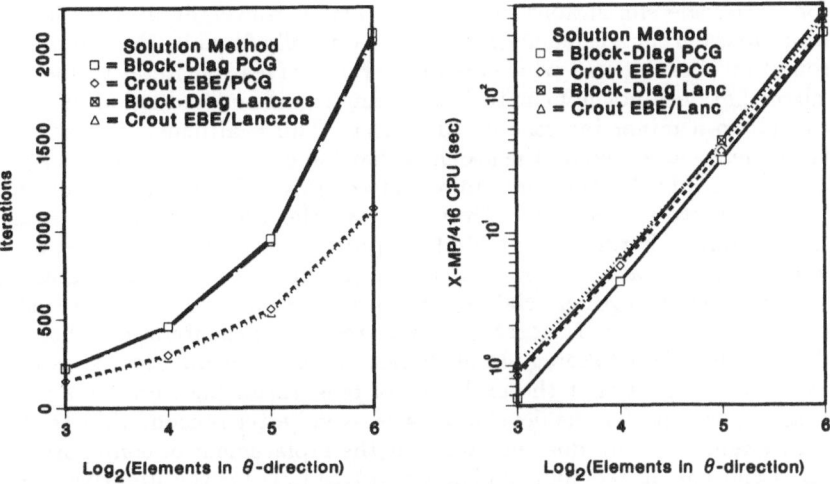

Figure 13. Iteration and X-MP/416 CPU performance for the stiffened cylindrical shell using the block-diagonal transformation with the PCG and Lanczos algorithms.

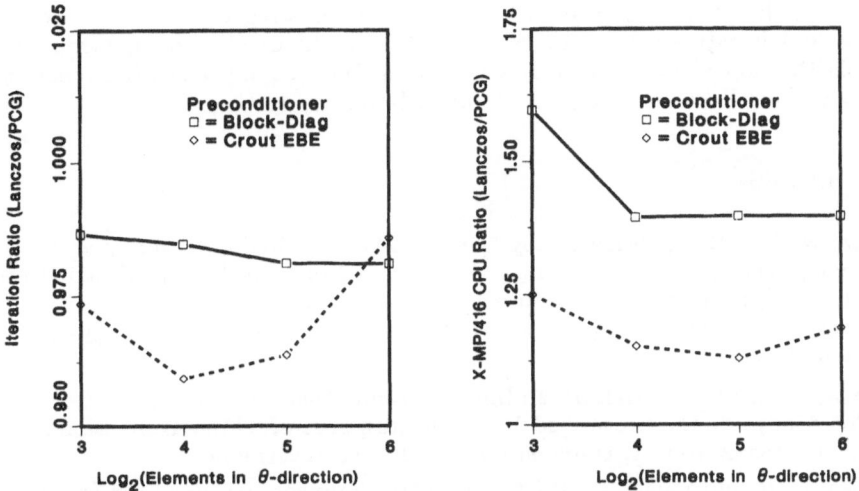

Figure 14. Relative performance of the PCG and Lanczos algorithms for the stiffened cylindrical shell using the block-diagonal transformation.

solution for all the meshes uses at least three times the storage of the direct method, while Crout EBE/Lanczos requires at least double the storage (see Table 2).

CONCLUSIONS

The numerical examples presented support several conclusions concerning the performance of a Lanczos-based iterative algorithm. First, the use of partial reorthogonalization as developed by Simon (1984) does preserve the theoretical finite termination property of the Lanczos algorithm. However, this property is of little practical interest for large-scale simulations, as we must attain convergence in far fewer than N_{eq} iterations in order for a calculation to be economically feasible. Second, for moderately ill-conditioned problems—which are in the range of feasible iterative solution, the Lanczos algorithm does not appear to lead to any significant reduction in iterations, and may result in a minor increase in CPU cost. Third, although this minor CPU differential might be accepted on the grounds that Lanczos is a more robust algorithm than PCG, this added robustness is attained at the price of accruing a large auxiliary data set of basis vectors. For linear stress analysis, the size of available memory and fast secondary storage is the current limit on problem size (see Hughes et al. 1987). Hence retention of the basis vectors is a substantial burden and totally unacceptable when it exceeds the storage associated with direct solution. Preliminary efforts at retaining only a subset of the Lanczos vectors or restarting the iteration when the vector pool hits a pre-assigned storage limit do not appear fruitful. Finally, as a general matter, we cannot expect that the exchange of one reasonable iterative strategy for another is any substitute for the development of even better preconditioners. In total, the experience reported here does not warrant the replacement of conjugate gradient with Lanczos and partial reorthogonalization in NIKE3D for the iterative solution of large-scale problems.

ACKNOWLEDGEMENT

The author would like to express his appreciation to: Prof. Thomas J. R. Hughes, Stanford University, for his guidance during the course of our EBE research; Dr. Arthur Raefsky, JPL and Stanford, for many interesting conversations concerning Lanczos and iterative solvers in general; Dr. Gerald L. Goudreau, Group Leader, Methods Development Group, L.L.N.L., for his interest and encouragement, and to the Engineering Research Program at L.L.N.L. for support.

REFERENCES

Ferencz, R. M., 1989, "Element-by-element Preconditioning Techniques for Large-scale, Vectorized Finite Element Analysis in Nonlinear Solid and Structural Mechanics," Ph.D. thesis, Stanford University, Stanford.

Golub, G. H. and Van Loan, C. F., 1983, "Matrix Computations," Johns Hopkins, Baltimore.

Hallquist, J. O., 1984, "NIKE3D: an Implicit, Finite Deformation, Finite Element Code for Analyzing the Static and Dynamic Response of Three–Dimensional Solids," UCID-18822, Rev. 1, University of California, Livermore.

Hestenes, M. R. and Stiefel, E., 1952, Method of Conjugate Gradient for Solving Linear Systems, *Journal of Research of the National Bureau of Standards*, 49:409.

Hughes, T. J. R., Pister, K. S., and Taylor, R. L., 1979, Implicit-explicit Finite Elements in Nonlinear Transient Analysis, *Computer Methods in Applied Mechanics and Engineering*, 17/18:159.

Hughes, T. J. R., Levit I., and Winget, J. M., 1983a, Implicit, Unconditionally Stable Algorithms for Heat Conduction Analysis, *Journal of the Engineering Mechanics Division, ASCE*, 109:576.

Hughes, T. J. R., Levit I., and Winget, J. M., 1983b, An Element-by-element Solution Algorithm for Problems of Structural and Solid Mechanics, *Computer Methods in Applied Mechanics and Engineering*, 36:241.

Hughes, T. J. R., Winget, J. M., Levit, I., and Tezduyar, T., 1983c, New Alternating Direction Procedures in Finite Element Analysis Based Upon EBE Approximate

Factorizations, in: "Computer Methods for Nonlinear Solids and Structural Mechanics," AMD-Vol. 54, S. N. Atluri and N. Perrone eds., ASME, New York.

Hughes, T. J. R., Raefsky, A., Muller, A., Winget, J. M., and Levit, I., 1984, A Progress Report on EBE Solution Procedures in Solid Mechanics, in: "Numerical Methods for Nonlinear Problems, Vol. 2," C. Taylor et al. eds., Pineridge Press, Swansea.

Hughes, T. J. R., Ferencz, R. M., and Hallquist, J. O., 1987, Large-scale Vectorized Implicit Calculations in Solid Mechanics on a CRAY X-MP/48 Utilizing EBE Preconditioned Conjugate Gradients, *Computer Methods in Applied Mechanics and Engineering* 61:215.

Hughes, T. J. R. and Ferencz, R. M., 1988a, Fully Vectorized EBE Preconditioners for Nonlinear Solid Mechanics: Applications to Large-scale Three-dimensional Continuum, Shell and Contact/Impact Problems, in: "Proceedings for the 1st International Symposium on Domain Decomposition Methods for Partial Differential Equations," R. Glowinski et al. eds., SIAM, Philadelphia.

Hughes, T. J. R. and Ferencz, R. M., 1988b, Implicit Solution of Large-scale Contact and Impact Problems Employing an EBE Preconditioned Iterative Solver,in: "IMPACT: Effects of Fast Transient Loadings," W. J. Ammann et al. eds., Balkema, Rotterdam.

Jones, R., 1986, private communication to Thomas J.R. Hughes.

Lanczos, C., 1952, Solution of Systems of Linear Equations by Minimized Iteration, *Journal of Research of the National Bureau of Standards*, 45:33.

Muller, A. M., 1985, "Element-by-element Iterative Procedures in Structural Finite Element Analysis," Ph.D. thesis, Stanford University, Stanford.

Newmark, N. M., 1959, A Method of Computation for Structural Dynamics, *Journal of the Engineering Mechanics Division, ASCE* 85:67.

Nour-Omid, B. and Parlett, B. N., 1985, Element Preconditioning using Splitting Techniques, *SIAM Journal on Scientific and Statistical Computing* 6:761.

Nour-Omid, B., Parlett, B. N., and Raefsky, A., 1988, Comparison of Lanczos with Conjugate Gradient using Element Preconditioning, in: "Proceedings for the 1st International Symposium on Domain Decomposition Methods for Partial Differential Equations," R. Glowinski et al. eds., SIAM, Philadelphia.

Ortiz, M., Pinsky, P. M., and Taylor, R. L., 1983, Unconditionally Stable Element-by-Element Algorithms for Dynamics Problems, *Computer Methods in Applied Mechanics and Engineering* 36:223.

Raefsky A., 1988, private communication.

Shakib, F., 1989, "Finite Element Analysis of the Compressible Euler and Navier-Stokes Equations," Ph.D. thesis, Stanford University, Stanford.

Simon, H. D., 1984, The Lanczos Algorithm with Partial Reorthogonalization, *Mathematics of Computation* 42:115.

Taylor, R. L., Wilson, E. L. and Sackett, S. J., 1981, Direct Solution of Equations by Frontal and Variable Band, Active Column Methods, in: "Nonlinear Finite Element Analysis in Structural Mechanics," W. Wunderlich et al. eds., Springer-Verlag, Berlin.

Wilkinson, J. H., 1963, "Rounding Errors in Algebraic Processes," Prentice-Hall, Englewood Cliffs.

Winget, J. M. and Hughes, T. J. R., 1985, Solution Algorithms for Nonlinear Transient Heat Conduction Analysis Employing Element-by-Element Iterative Strategies, *Computer Methods in Applied Mechanics and Engineering* 52:711.

TRANAIR TECHNOLOGY: SOLUTIONS FOR LARGE PDE PROBLEMS

John E. Bussoletti and Forrester T. Johnson, *
David P. Young, Robin G. Melvin,
Richard H. Burkhart and Michael B. Bieterman, †
Satish S. Samant and Gautam SenGupta ‡

The Boeing Company
M/S: 7C-36
PO Box 3707
Seattle WA 98124-2207

ABSTRACT

TRANAIR is a series of computer codes for the solution of second order partial differential equations (PDEs) in three dimensions. The technology used in the TRANAIR codes can be viewed as a novel combination of boundary element methods and finite element field methods for the solution of three dimensional boundary value problems. This combination of techniques allows the solution of larger systems of equations than conventional boundary element or finite element techniques permit. TRANAIR codes have been applied to a wide variety of three dimensional computational problems which are purely elliptic (linear potential flow), problems which are elliptic but have embedded regions of hyperbolic character (the full potential equation, describing transonic flow) and indefinite elliptic problems (acoustic and electromagnetic scattering problems). Many applications have involved extremely complex and realistic geometries. Performance of the approach is very similar regardless of the type of problem. In this paper we discuss the TRANAIR technology in the context of its applications to the solution of large PDE problems.

INTRODUCTION

In aerodynamics, as in most engineering disciplines, many phenomena are described by solutions to boundary value problems for partial differential equations. Representative of these equations are the Prandtl-Glauert equation, describing linear potential flow:

$$\tilde{\nabla}^2\Phi = (1 - M_\infty^2)\Phi_{xx} + \Phi_{yy} + \Phi_{zz} = 0, \tag{1}$$

the full potential equation, describing transonic potential flow:

$$\vec{\nabla} \cdot \rho\vec{\nabla}\Phi = 0, \tag{2}$$

(where ρ is a non-linear function of $\vec{\nabla}\Phi$), and time-harmonic linearizations of these equations, describing various acoustic phenomena:

$$\tilde{\nabla}^2\Phi + k_o^2\Phi = 0, \tag{3}$$

---

*Boeing Advanced Systems
†Boeing Computer Services
‡Boeing Commercial Airplanes

(where $k_o^2 = \omega^2/c_o^2$ is a constant wavenumber), and

$$\vec{\nabla} \cdot \rho \vec{\nabla} \Phi + k^2 \Phi = 0, \tag{4}$$

(where $k^2 = \omega^2/c^2$ and both ρ and c may vary with position in space). These equations describe steady and (linearized) time-harmonic fluid flow in various physical regimes which characterize the cruise flight conditions of most aircraft. The equations arise through simplifications of the fundamental Navier-Stokes equations, which express the physical conditions of conservation of mass, momentum and energy.

Maxwell's equations, which describe electromagnetic phenomena, are closely related to Equation 4 in a mathematical sense. This system can be described by the vector equation:

$$\vec{E} - \alpha \vec{\nabla} \times \beta \vec{\nabla} \times \vec{E} = 0 \tag{5}$$

where

$$\alpha^{-1} = \iota \omega \epsilon + \sigma, \tag{6}$$

$$\beta^{-1} = -\iota \omega \mu,$$

ϵ characterizes the local dielectric permittivity, σ characterizes the local conductivity, and μ characterizes the local magnetic permeability. The material properties ϵ, σ and μ may vary from point to point in space and may also be discontinuous across boundaries.

Boundary conditions for these equations are usually expressed as conditions on the function Φ itself or on its normal derivative, $\hat{n} \cdot \vec{\nabla} \Phi$, but may be more general, involving linear combinations of the two or might involve conditions on the discontinuity of the function or its normal derivative across a boundary.

The solutions of these equations have significant engineering and economic impact. For example, design engineers in aerodynamics evaluate various aircraft configurations by solving equations like these and by examining detailed predictions of pressure distributions over a configuration surface. In the process they develop the necessary insight to create a configuration surface geometry which satisfies engineering and economic constraints while delivering the required aircraft performance. Design engineers and their management have readily acknowledged that the very existence of some extremely successful aircraft would not have been possible without these kinds of computational tools.

Because of the economic success of such a design process, there is an ongoing interest in finding ways to extend these computational capabilities. These extensions are aimed at both reducing the cost of obtaining solutions to a given problem and also at enlarging the size of problems that can be solved on current computer systems. The motivation for reducing solution cost is pretty universal and should be obvious in the context of the economic aspects of engineering design. The interest in ever larger problem sizes, however, comes from a variety of physical motivations that are somewhat application dependent.

BOUNDARY INTEGRAL AND FIELD APPROACHES

Numerical solutions of Equations 1 through 5 have been found by two approaches. In the first, (primarily associated with solutions of Equations 1 and 3), Green's theorem is used to convert the partial differential equation into an integral equation:

$$\Phi(\vec{r}) = \int G_o(|\vec{r} - \vec{r'}|)\sigma(\vec{r'})dS' + \int \mu(\vec{r'})\hat{n} \cdot \vec{\nabla} G_o(|\vec{r} - \vec{r'}|)dS' \tag{7}$$

where $G_o(|\vec{r} - \vec{r'}|)$ is the Green's function for the partial differential equation and σ and μ are unknown singularity distributions confined to boundary surfaces. These singularity distributions are sources for the field and characterize the discontinuities in the field (μ) and/or its normal derivative (σ) across the boundary. These techniques are generally called boundary integral methods, and in various engineering disciplines go by names like panel methods and moment methods.

The unknown parameters are often polynomial-based splines of various orders. The number and size of the set of parameters is determined by the ability of the local polynomial approximation to represent the characteristic length scale of the solution variation. The field Φ and its derivatives are computed in terms of a set of parameters characterizing μ and σ at locations on the boundary surface. Appropriate boundary conditions are invoked on the surface to define a discrete system of equations whose solution provides values for the singularity parameters. The matrix which emerges from this process is a dense matrix and may or may not be diagonally dominant, depending on the nature of the boundary value problem.

The advantage of this approach is that, after the continuous equation is discretized, the entire three dimensional field is characterized by a two dimensional set of parameters. This has the effect of reducing the number of solution parameters to be determined. The disadvantage of the approach is that the matrix in the governing equations is a dense matrix. Thus the solution cost scales asymptotically as $\mathcal{O}(p^3)$ where p represents the number of parameters characterizing μ or σ. In addition, Equation 7 is useful only if the Green's function is known or easily computed. This function typically is known only for constant coefficient partial differential equations. For Equations 2 and 4, an additional term must be added to Equation 7 to include the effects of the local variation in the coefficients ρ or k^2. This requires a volumetric parameterization of the source distribution in addition to the surface parametrization of the solution. In such a case, the advantages of the reduction of dimensionality in the problem is lost.

In the case of linear potential flow solutions, increasing modeling complexity creates larger problem sizes which can saturate computer resources. Computational costs for traditional boundary element approaches to these problems[1] scale asymptotically as $\mathcal{O}(p^3)$, where p represents the number of surface elements describing the configuration. This scaling usually forces the aerodynamics designer to compromise the fidelity of the configuration representation in some region in order to obtain higher accuracy in a region of greater interest. Occasionally this compromises overall solution accuracy.

For acoustic and electromagnetic scattering problems, an additional burden is placed on configuration modeling through the need to resolve the wavelength of the oscillatory phenomena. This means that for a given wavelength, no surface patch can be larger than about $\lambda/2$, where λ represents the wavelength of the driving field (assuming polynomial basis functions). Indeed for most applications, accuracy dictates that patches be no larger than $\lambda/6$ to $\lambda/10$. The overall effect of this added constraint is usually a frustrating limit on the maximum frequency for which calculations can be performed.

The second solution approach involves a discrete approximation to the partial differential equation in which the fundamental unknowns are parameters describing the fields themselves. Finite difference, finite volume and finite element schemes all work in this manner. The entire volume of space is filled with a computational grid. Parameters describing the field are introduced at locations on this grid and linear combinations of the field parameters are used to approximate the partial derivatives. For example,

$$(1 - M_\infty^2)(\frac{(\Phi_{i+1,j,k} - \Phi_{i,j,k})}{\Delta x} - \frac{(\Phi_{i,j,k} - \Phi_{i-1,j,k})}{\Delta x})/\Delta x +$$
$$(\frac{(\Phi_{i,j+1,k} - \Phi_{i,j,k})}{\Delta y} - \frac{(\Phi_{i,j,k} - \Phi_{i,j-1,k})}{\Delta y})/\Delta y + \tag{8}$$
$$(\frac{(\Phi_{i,j,k+1} - \Phi_{i,j,k})}{\Delta z} - \frac{(\Phi_{i,j,k} - \Phi_{i,j,k-1})}{\Delta z})/\Delta z = 0$$

approximates Equation 1. The matrix for the resulting equations which result is quite sparse and may or may not be diagonally dominant. The number of field parameters which are introduced depends on how well local polynomials in the field unknowns can approximate the solution variation. Offsetting the sparse nature of the matrix is the fact that the field variation must be described over the full three dimensional volume of space. For problems with a boundary condition at infinity, an additional approximation is required to permit the computational grid to be truncated at an appropriate point, usually far removed from the configuration.

The advantage of this approach is that the technique can readily be used for problems with varying coefficients. The sparsity of the matrix helps offset the additional number of parameters which are required to represent the field over three dimensional space. But it is still the case that with this approach, the requirements of characterizing the solution over a three dimensional volume greatly expands the size of the problem.

A field method with equivalent resolution to a boundary element surface parameterization would have $N = p^{3/2}$ grid points. However, many simple preconditioned iterative methods provide solutions at costs like $\mathcal{O}(N^{7/6}) = \mathcal{O}(p^{7/4})$[2]. Thus field methods asymptotically offer a very competitive alternative to boundary element methods. However, with field methods, one needs to generate a three dimensional grid of points about a configuration. The quality of the grid can have a big impact on the accuracy of the solution. However, the most reliable way to identify such an impact is to solve the problem again on a different grid.

With our development efforts in TRANAIR[3, 4, 5, 6, 7, 8], we seek to combine the best features of both the boundary integral approach and the field method approach (finite element methods, in particular). The basic idea is to characterize the solution in terms of sources for the field, rather than in terms of the field itself. For many problems this results in a parameterization which is quite close to a two dimensional surface of parameters embedded in a three dimensional grid. The equations themselves, however are formulated in terms of the values of the field. This allows us to maintain the sparse nature of the discrete equations in an iterative solution process. In this manner, we can extend our ability to solve large problems on existing computers.

The key to this combination lies in recognizing that the integrals occurring in Equation 7 are convolution integrals. As such each has the property that when Fourier transformed, it becomes a product of the individual Fourier transforms of the Green's function and the source distribution(s). Thus Fourier transforms permit a rapid and efficient conversion from source parameters to the fields which the sources generate. In addition, such fields have the correct behavior at infinity as long as that behavior is built into the Green's function, even though the computational grid is only defined in the near field of the configuration. These properties carry over directly into the discrete version of the problem when the Fourier transform is replaced by the Discrete Fourier Transform over a uniform grid of points. Computationally there are a variety of techniques which implement a Fast Fourier Transform (FFT) in a manner which permits these convolution integral to be performed in $\mathcal{O}(N \log N)$ operations, where N is the total number of points in the computational grid.

These attractive properties of the FFT have been recognized for quite some time (see Bojarski[9] and Mittra[10]). Unfortunately, all FFT algorithms are limited to specialized distributions of grid spacings, usually uniform Cartesian grids. These are not particularly appropriate for general aircraft configurations, since Cartesian grids have a uniform length scale associated with them. With such a grid, the grid spacing must be selected so that the highest local solution gradient is accurately represented. This means that the solution is represented at the finest length scale over the complete computational field, a tremendous waste of resources.

In addition, it is not enough to have a convenient way to evaluate the field generated by source distributions. One must also be able to solve a large system of equations for the sources which generate the field. The development of TRANAIR has been a consistent exploration of ways to circumvent the limitations of Cartesian grids and to find efficient ways of solving the system of equations which result in this formulation. In effect, the essence of TRANAIR is an embodiment of the philosophical idea of combining boundary element and field methods effectively.

TRANAIR TECHNOLOGY

To bring this combination of boundary integral and field methods together the TRANAIR family of codes makes use of a variety of numerical techniques. These include discrete Green's functions for constant coefficient operators on Cartesian grids, Discrete Fourier Transforms and Convolutions, Finite Element discretizations, configuration geometries defined by two dimensional surface patches (panels), oct-tree based descriptions of hierarchically refined grids, and sparse matrix solvers with nested dissection orderings used as preconditioners for the GMRES algorithm. In this section we present essential features of the ways these techniques are used in TRANAIR.

Source Parameterization

Consider the full potential equation, Equation 2. The density ρ is a nonlinear function of the velocity $\vec{\nabla}\Phi$. In most fluid flow applications, both the density and the velocity are slowly varying functions in regions removed from configuration boundaries and shock locations. In such a region $\mathcal{R}$, the full potential equation is well-approximated by linear potential flow, Equation 1. Assuming a uniform incident flow field $\vec{V}_\infty$, we write Equation 2 as

$$L(\Phi + \vec{V}_\infty \cdot \vec{r}) = 0 \tag{9}$$

and its limiting form (Equation 1) as

$$T\Phi = 0 \tag{10}$$

Given a solution to Equation 9, one may apply to it the operator T and obtain

$$T\Phi = Q \tag{11}$$

This defines the sources Q which generate the solution Φ. If one has a Green's function T^{-1} for T so that

$$\Phi = T^{-1}Q \tag{12}$$

one can add and subtract $T\Phi$ to Equation 9. Then by defining

$$\Delta L(\Phi + \vec{V}_\infty \cdot \vec{r}) = L(\Phi + \vec{V}_\infty \cdot \vec{r}) - T\Phi \tag{13}$$

Equation 9 becomes

$$(Q + \Delta L(T^{-1}Q + \vec{V}_\infty \cdot \vec{r}) = 0 \tag{14}$$

Outside region $\mathcal{R}$, $L = T$. Thus $\Delta L = 0$, and Equation 14 states that outside $\mathcal{R}$, Q vanishes. Thus, the problem on an infinite domain can be characterized by a finite set of sources located within $\mathcal{R}$, given the Green's function T^{-1}.

The usefulness of the Q (source) parameterization of the solution over the Φ (field) parameterization depends on how easily Φ can be computed from Q. As mentioned above, if Q is provided on a regularly spaced Cartesian grid, the Discrete Fourier Transform performs the required operation in a few, $\mathcal{O}(N \log_2 N)$, operations provided we have an appropriate Green's function on the grid. Buneman[11] has indicated how to compute a Green's function for discrete versions of the Laplace equation. We have developed techniques for obtaining discrete Green's functions for both the Laplace and Helmholtz equations[3, 4, 5, 6]. In addition, Hockney[12] and James[13] have described techniques for computing the convolution efficiently.

For fluid dynamics applications using the full potential equation, if aircraft configurations contain lifting surfaces, sources Q are not confined to a finite region. In these cases the lifting surfaces in the configuration generate a discontinuity surface in Φ which extends to infinity. The discrete representation of this surface is provided by two sheets of sources of opposite sign extending to infinity. For these applications we have made extensions to the work by Buneman[11], Hockney[12] and James[13] by formulating a discrete Green's function with downstream sources[3, 5]. That is, along one coordinate direction (by convention the x coordinate direction), whatever values the sources have in the next-to-last plane in the grid are extended to infinity to define the discrete Green's function for the problem. We are in the process of extending this to the Helmholtz Green's function.

Discretization

Use of a Cartesian mesh is essential to obtaining the efficient evaluation of the convolution integral, but this is in conflict with the representation of general aircraft configurations. It can also prove wasteful in that a uniform Cartesian grid alone must be specified with a resolution equal to the shortest length scale expected in the solution. In the most recent implementation of TRANAIR we have resolved these conflicts.

The success of linear panel methods in aerodynamic applications has resulted in a general capability within the engineering community to represent aircraft (or undersea) configurations with panels (discrete surface patches). For compatibility with this technology, configuration boundaries are provided to TRANAIR through a panel description of the object[1]. The panel definition of the configuration is embedded into the Cartesian grid. The continuous differential equation is then discretized with finite element techniques using an appropriate variational principle[5, 6]. The variational integral is supplemented by additional surface integral terms to correctly represent the required boundary conditions on the paneled surfaces wherever the boundary conditions differ from the "natural" impermeable surface conditions.

This use of finite element techniques allows independent discretizations of the configuration and the field and, in addition, removes from the user the burden of generating a three dimensional field grid. The manner in which the finite element discretization has been implemented differs from traditional approaches in one fashion. For any grid cell which contains a portion of the boundary of the configuration, multiple basis functions are introduced by adding multiple unknown parameters (denoted by ψ) at nodes of the cell, so that each disconnected region in the cell (defined by cell face boundaries and those panel boundaries which lie within and/or intersect a cell, see Figure 1), has an independent basis function characterized by appropriate ϕ and ψ unknowns. For the full potential equation and for acoustics applications, we use a trilinear basis function. For electromagnetic applications a special "vector bilinear" basis function[3, 6] is used.

To allow accurate and efficient representation of varying length scale effects, local mesh refinement of a global Cartesian grid has recently been introduced into the TRANAIR code for the full potential equation[20, 7]. A hierarchical data representation based on oct-trees is used to describe the local refinements[14, 15]. This data structure is extremely compact in its storage requirements. A complete definition of the refined grid requires typically $10N/7$ storage, where N is the total number of all finite element grid cells. Refinement of the global Cartesian grid is based on three criteria. First, since the local curvature of the configuration surface is one parameter that is suggestive of the length scale characterizing solution gradients and since the local panel size of the configuration provides some indication of local curvature, the global grid may be automatically refined in the neighborhood of the boundary based on local panel size. Second, since in a typical application designers have a greater interest in a high resolution solution on some sub-portion of the full configuration, the user can specify a "region of interest" for grid refinement by defining coordinates of a volume of space enclosing the region of interest. Third, since another characteristic of solutions to the full potential equation is shocks, the location of which are not known prior to the solution of the system of equations, a solution adaptive refinement of the grid may be invoked to deal with such situations.

A typical grid automatically generated by panel size for an F-16 configuration is illustrated in Figure 2. This grid contains about 220,000 finite element regions. A similar grid for a 747 wing-body configuration is shown in Figure 3. This grid consists of 120,000 regions.

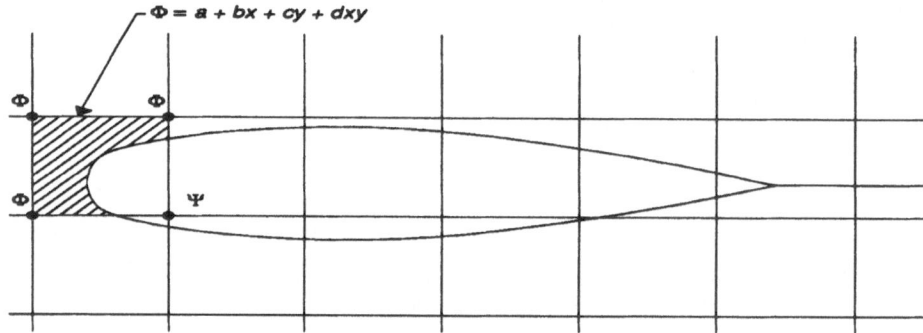

Figure 1. Multiple Basis Functions in Boundary Boxes

In any fluid flow problem where the configuration generates a net lift, there is a discontinuity in the potential field along a surface embedded in the field. This surface is called a wake. In order to correctly describe wake surfaces, the user specifies each wake by defining a surface which extends from the trailing edge of any lifting potion of the configuration, along the streamwise direction, out through the back of the computational grid. Additional unknowns are introduced on the leading edges of these wake surfaces. These unknowns are called doublet parameters, and are denoted by μ. These parameters are used to impose the correct Kutta condition at trailing edges of wings and may also be used to impose the correct jump conditions across volumetric regions with differing total pressure, total temperature and/or total swirl. The wake surfaces convect the appropriate discontinuities in potential to the end of the computational grid, and then convolution of sources on the grid with the exterior Poisson Green's function automatically extends wake sheets to infinity by using these distributions of "downstream sources".

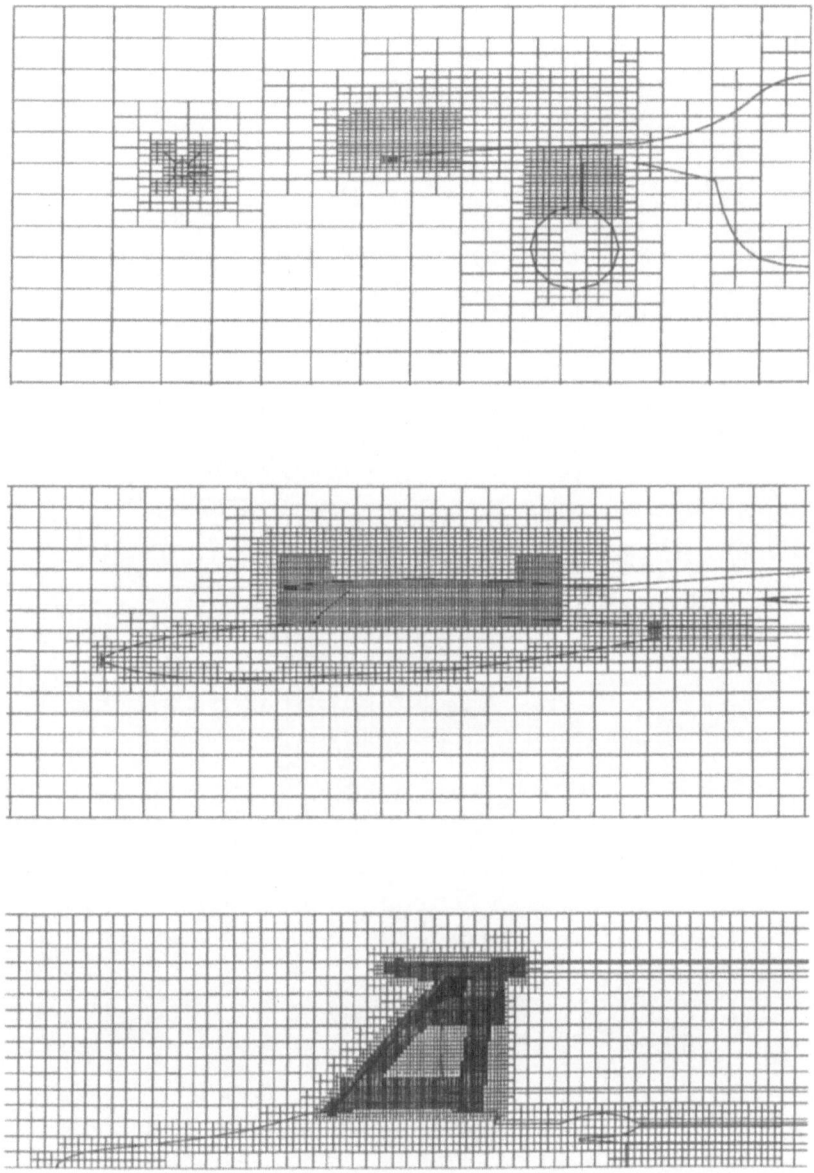

Figure 2. Computational Grid for F-16 Configuration Case

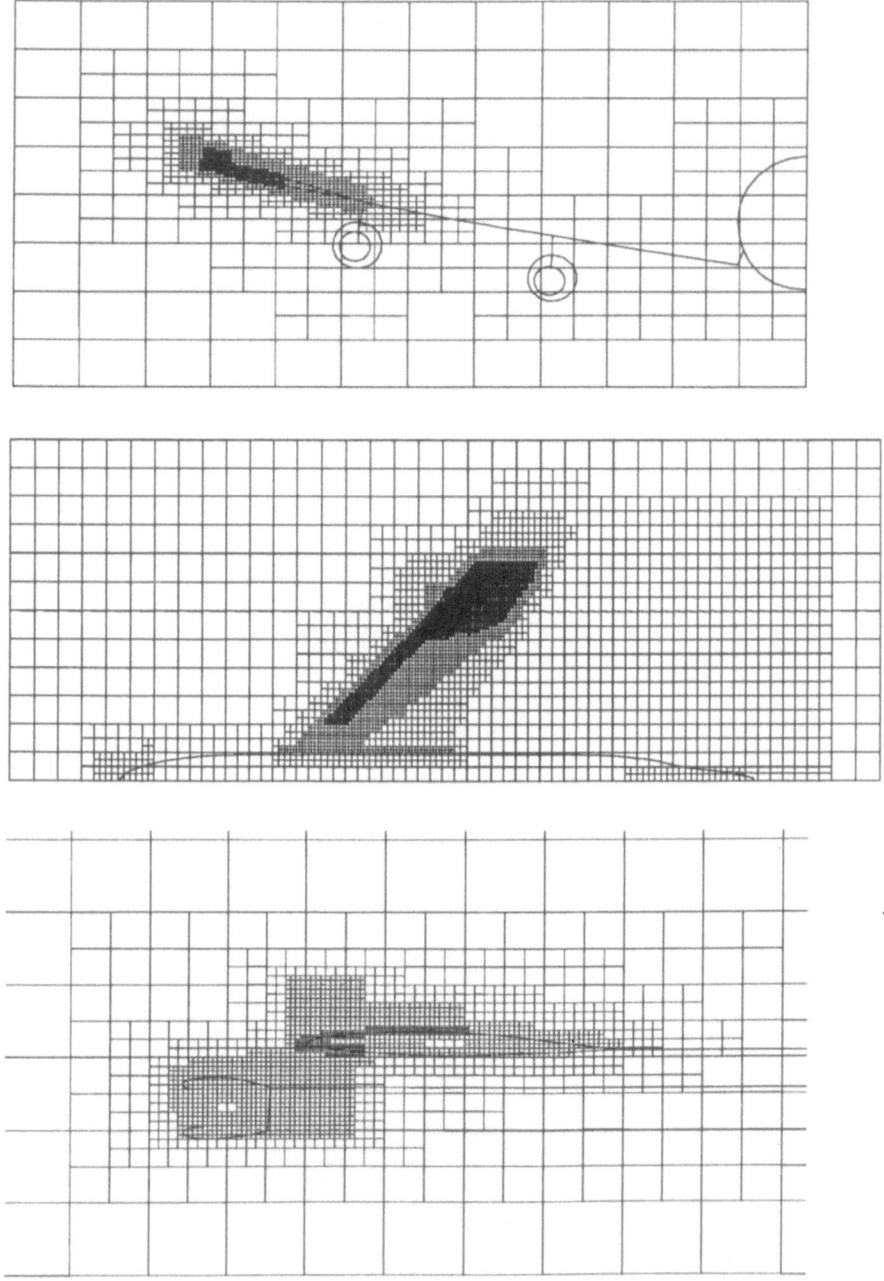

Figure 3. Computational Grid for 747 Configuration Case

Solution Process

The discrete system of equations produced by the finite element discretization is solved iteratively with preconditioned GMRES convergence acceleration. Our implementation of the GMRES algorithm[17] is very similar to that presented by Wigton, Yu and Young[16, 5, 6]. One recent modification to the procedure has been the addition of a damping strategy similar to that described by Bank and Rose[18].

The basic iterative approach is fairly standard. At the nth step we wish to find $\delta\Phi$ such that:

$$L(\Phi^{(n)} + \delta\Phi) = b \tag{15}$$

If $\delta\Phi$ is small, the equation for the correction is:

$$\bar{L}(\delta\Phi) \approx b - L(\Phi) = R \tag{16}$$

where $\bar{L}$ is the linearization of L about Φ. If L is linear (as in acoustics and electromagnetics applications), $L = \bar{L}$. The iterative solution technique is obtained by choosing some appropriate approximate inverse (a "preconditioner") for $\bar{L}$ so that we may solve directly for $\delta\Phi$. The GMRES algorithm[17] then finds a solution which minimizes the residuals over a finite dimensional subspace of orthogonal search vectors. These search vectors are generated by a Gramm-Schmidt orthogonalization process using R in equation 16 and powers of $\bar{L}$ applied to R.

Our use of preconditioners within the GMRES algorithm is somewhat novel in that we employ diagonal scaling of the original operator in conjunction with preconditioning on both the left and the right. In this process a number of matrix splittings are introduced over the various classes of unknowns in our solution vector. There are variations in how this is done within various applications. For purposes of explanation consider the transonic flow applications, e.g. solutions of the full potential equation. We first examine the source unknowns Q. On one hand the source parameterization is a convenient approach to assure the correct far field behavior in the solution when computations are performed on a finite grid. However, this parameterization also serves as a preconditioner. Its effectiveness is determined by how closely the operator $\bar{L}$ resembles the constant coefficient operator T. Recalling the definition of the sources Q,

$$Q = T\Phi \tag{17}$$

or equivalently,

$$\Phi = T^{-1}Q \tag{18}$$

our iterative algorithm may be written as

$$\bar{L}T^{-1}(\delta Q) = R \tag{19}$$

Since the combination $\bar{L}T^{-1}$ may be written as $I + \Delta\bar{L}T^{-1}$, one might simply approximate $\bar{L}$ by the unit operator in Equation 16 and write

$$\delta Q = R = b - \bar{L}T^{-1}Q \tag{20}$$

To the extent that $\Delta\bar{L}$ is small, this will be an effective iterative solution strategy. In our work we have found that this is sufficient over part of the flow field. However, in regions close to the boundary surfaces, particularly when there are strong discontinuities in the solution, this preconditioner alone is not sufficient to provide rapid convergence.

This situation arises in most applications. For example, in fluid flow, when the boundaries are impermeable, there is typically a jump in both the field and its normal derivative. In electromagnetic scattering, similar discontinuities occur across conducting boundaries and in acoustics, the same situation occurs for hard scattering surfaces. Qualitatively one can understand why simply preconditioning on the right is not very effective for these problems by realizing that the solution is built up from a sequence of fields generated by the summation:

$$\Phi = \sum_n T^{-1}Q^{(n)} \tag{21}$$

For Poisson and Helmholtz kernels, the T^{-1} operator is a "smoothing operator". Fields generated by such summations tend to be rather smooth. Thus it takes many terms in the summation to generate a solution with strong discontinuities. In Fourier transform language, the terms in Equation 21 have large amplitudes at low spatial frequencies, and thus it is difficult to construct a solution with large amplitudes at high spatial frequencies without taking a large number of terms.

To improve the effectiveness of the iterative solution we have added additional preconditioning on the left. Going back into field space for a moment, the operator L in Equation 19 is a sparse operator on a three dimensional volume grid. Decomposition of the corresponding matrix over the whole domain is a large computational task on grids of even modest size (100,000 boxes). Furthermore, closure conditions (particularly for the Helmholtz problem) destroy the sparsity of the matrix. These facts discourage the consideration of a left preconditioner.

However, the regions where the discontinuities typically take place are confined to the neighborhood of a two dimensional surface embedded in that volume. If the operation of the sparse matrix preconditioner is confined to those regions which contain the discontinuities, a considerable reduction in matrix size is obtained. In addition, because of the nature of the regions (i.e., a 2D surface embedded in 3D), a sparse matrix solver with nested dissection ordering[19] (to reduce the fill occurring in the decomposition process) is particularly attractive.

Thus for our left preconditioner, we consider only a subset of the complete three dimensional field for our sparse matrix solver which we call the "reduced set". For transonic flow problems the reduced set consists of points near the boundary, points in regions of refinement of the global Cartesian grid and points in volumes of supersonic flow. In the case of the Helmholtz applications, the reduced set consists of only those points near the boundary.

The set of equations is an open system which must be closed in order to have a unique solution. We close the system by adding in neighboring unknowns in the outer periphery of the reduced set and define a closure equation for these closure unknowns. The particular choice of closure equation is dependent on the application. For transonic flow we use Dirichlet and Neumann conditions. For Helmholtz-related problems we define a discrete (local) radiation condition as a closure condition.

To further reduce fill during the factorization of the sparse matrix, we employ a drop tolerance in the factorization process. This is described in detail in a recent report[20]. During the decomposition process, as the LU factorization is being generated, entries in the decomposition are compared against the current values of the corresponding column and row diagonals. If the ratio of the entry to both diagonals is smaller than a tolerance, the entry is set to zero. In our transonic flow applications this results in a matrix decomposition which is only about twice the size of the original sparse matrix. The number of iterations required to converge to a given error level is increased when a drop tolerance is used, but the cost of each iteration is reduced due to the smaller size of the decomposition. The net effect in typical transonic flow applications is that the number of iterations required doubles, but the cost per iteration is reduced by about a factor of three, thus producing a net reduction in CPU time required to obtain a solution.

The correction given by the sparse solver is to the variable ϕ, a field, rather to the variable Q, a source. On the global grid the field is transformed to a source by applying the constant coefficient operator T to the correction field. In summary, at points in the reduced set, the correction sources are defined by:

$$\delta Q = \nu T N^{-1}(b - LT^{-1}Q) \tag{22}$$

The two preconditioners are combined in the following fashion. In our present applications, the source unknowns are confined to the global Cartesian grid. The unknowns defined at the nodes of the locally refined grid are kept in field space. The boundary basis function Ψ unknowns are also kept in field space. In addition, for lifting transonic flow problems, doublet surface parameters are defined at leading edges of wake surfaces. These doublet parameters are included in the reduced set of the sparse matrix solver and remain as doublet parameters in the iteration equation. There are then a total of five classes of unknowns in a given transonic flow problem. They are: $Q^{(1)}$, the source unknowns at global grid points which are not in the reduced set (this includes global grid unknowns at closure points); $Q^{(2)}$, the source unknowns at global grid points in the reduced set; $\Phi^{(3)}$, the values of the velocity potential at points on locally refined grids; $\Psi^{(4)}$, the values of velocity potential in the boundary basis functions; and finally, $\mu^{(5)}$, the doublet strengths at leading edges of wake networks.

Using these five classes of unknowns, we can summarize the fundamental iterative equation for TRANAIR. First, define the vector of unknowns, $\mathcal{X}$, as a vector over these five classes of unknowns:

$$\mathcal{X} = \begin{pmatrix} Q^{(1)} \\ Q^{(2)} \\ \Phi^{(3)} \\ \Psi^{(4)} \\ \mu^{(5)} \end{pmatrix} \tag{23}$$

with the correction vector $\delta\mathcal{X}$ being defined in a similar manner. Then the correction equation is:

$$\delta\mathcal{X} = \nu \mathcal{T} \mathcal{N}^{-1}(b - L\mathcal{T}^{-1}\mathcal{X}) \tag{24}$$

The operators $\mathcal{T}$ and $\mathcal{N}$ are defined as:

$$\mathcal{T} = \begin{pmatrix} T_{(1)(1)} & T_{(1)(2)} & 0 & 0 & 0 \\ T_{(2)(1)} & T_{(2)(2)} & 0 & 0 & 0 \\ 0 & 0 & I & 0 & 0 \\ 0 & 0 & 0 & I & 0 \\ 0 & 0 & 0 & 0 & I \end{pmatrix} \tag{25}$$

$$\mathcal{N} = \begin{pmatrix} I & 0 & 0 & 0 & 0 \\ 0 & N_{(2)(2)} & N_{(2)(3)} & N_{(2)(4)} & N_{(2)(5)} \\ 0 & N_{(3)(2)} & N_{(3)(3)} & N_{(3)(4)} & N_{(3)(5)} \\ 0 & N_{(4)(2)} & N_{(4)(3)} & N_{(4)(4)} & N_{(4)(5)} \\ 0 & N_{(5)(2)} & N_{(5)(3)} & N_{(5)(4)} & N_{(5)(5)} \end{pmatrix} \tag{26}$$

For dimensional reasons, some scaling of the source unknowns relative to the field unknowns is required. Note that from dimensional analysis alone, a source differs from a field quantity by a factor of the inverse length squared. Source scale factors are defined so that the GMRES convergence history is independent of what physical coordinates are used to define a configuration (e.g., meters or inches).

The effectiveness of these preconditioners is illustrated in Figures 4 through 7. Figure 4 shows the execution time required to perform the decomposition of the sparse matrix preconditioner N^{-1} for a variety of problem sizes and applications. For a linear problem this is an overhead cost that is incurred only once. For a nonlinear problem, several (typically three to five) decompositions are performed in the course of obtaining a solution to a problem. The use of the drop tolerance reduces the computational cost of decomposing the matrix by a factor of five to ten. Figure 5 illustrates the cost for a single

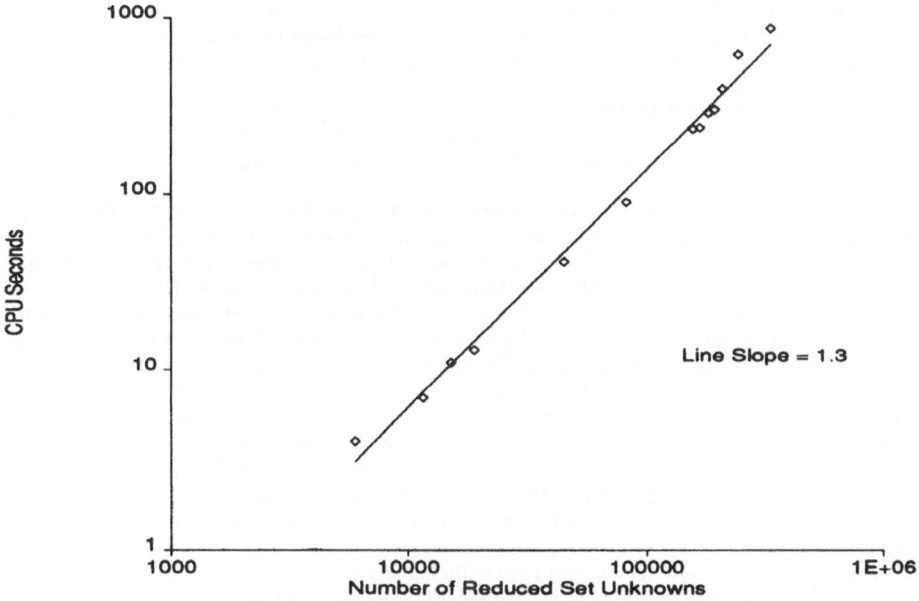

Figure 4. CPU Cost for Decomposition of Sparse Matrix

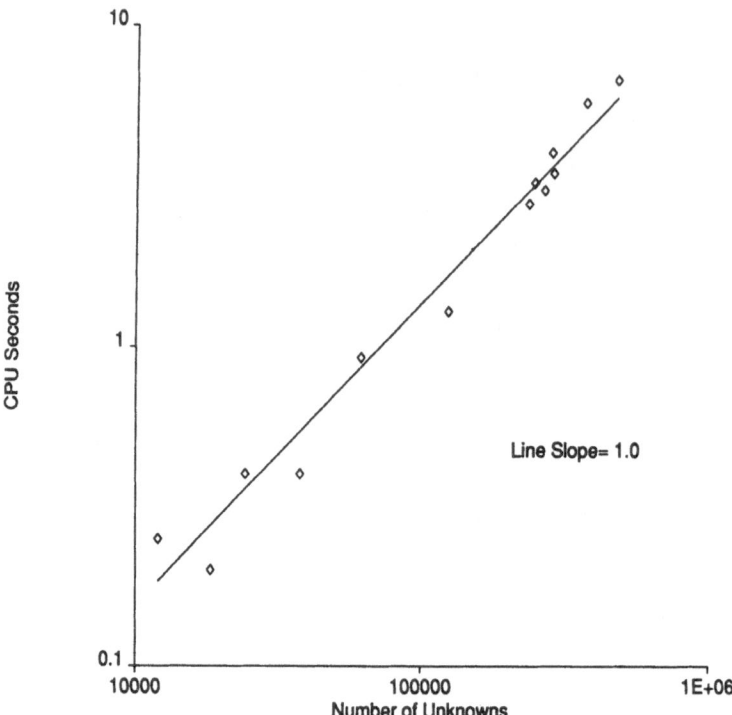

Figure 5. CPU Cost for a Single Iteration

iteration (the cost of evaluating Equation 24), again for a variety of applications and problem sizes. Note here that use of a drop tolerance saves about a factor of two in overall execution time. Figure 6 characterizes the convergence behavior of a typical linear problem (and the behavior of the linear solution portion of nonlinear problems). Typically nine digits of error reduction occurs in ten to twenty iterations. Finally Figure 7 shows that there is an optimal choice of drop tolerance in that, as the drop tolerance is increased, the quality of the preconditioner decreases (more iterations are required to achieve convergence), but the cost of an iteration decreases. We have found in transonic flow applications that a moderate range of drop tolerances (from 1×10^{-3} to 1×10^{-4}) provides an adequate balance in these trends for a wide variety of configurations. Selecting an optimal drop tolerance is not a critical factor in most applications, since use of the drop tolerance is primarily to reduce storage requirements (on the SSD) for large problems. The drop tolerance has not yet been extensively used in applications other than fluid flow.

Because of the effectiveness of our preconditioners, linear problems can be solved quite rapidly. We depend upon "nonlinear GMRES"[16] to solve the nonlinear full potential problem. Our nonlinear GMRES is based on Newton's method. For subsonic problems, Newton's method converges quickly and robustly. However, for problems with supersonic flow and shocks, Newton's method can sometimes diverge. This seems to be due to the fact that a linear model is only accurate very locally in this case. Physically the situation corresponds to a condition in which a steep shock develops in the wrong place. Once such a condition occurs, Newton's method (or any method based on a local model) has difficulty. Only a few points can be moved through the shock in a stable manner by any local method. Since, in a large 3D problem, there could be tens of thousands of points that must pass through the shock to arrive at the correct solution, local optimization methods are of limited usefulness.

There are many methods that have been used to damp Newton's method to prevent such blowups (for example, Fletcher[21], Bank and Rose[18] and Burkhart and Young[22]). We tried a number of such methods and found that they all led to stagnation of convergence for more difficult transonic flow problems.

One effective cure for this difficulty is to partially solve a series of "viscous" problems to obtain a good initial guess. Note that in order to obtain solutions to the full potential equation which satisfy the correct entropy condition, an "artificial" numerical dissipation is introduced[3]. In TRANAIR we introduce an upwinded density whenever the local Mach number exceeds a specified value (the "cutoff Mach number"). To provide the additional stability and robustness in the nonlinear solution process, we increase the amount of this dissipation initially in the solution process, and reduce it as the solution converges. These "viscous" problems are obtained from the original discrete problem by lowering the cutoff Mach number for the density biasing and multiplying the switching function by an appropriate constant. Each successive problem has a higher cut off Mach number and a smaller constant for the switching function until the original problem is reached. This has the effect of obtaining solutions initially with very smeared shocks that are, however, centered in the right place. As the viscosity is reduced the shocks steepen and remain in the correct position.

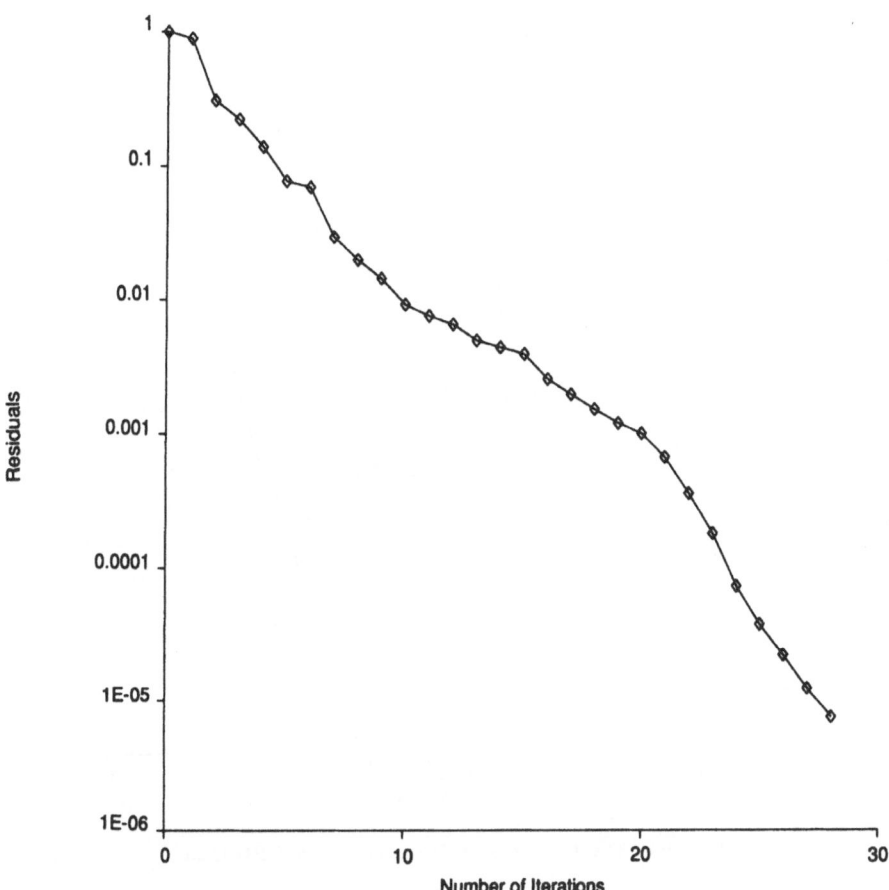

Figure 6. Convergence Properties of Linear Solutions

This strategy has proved very robust in TRANAIR. The number of viscous problems we have used varies from 0 for small problems or ones with very weak shocks to 4 for the most challenging problems. For each viscous problem the residual is reduced by about one digit. This generally takes two to three Newton steps per viscous problem. In addition, Jacobians are sometimes required for the viscous problems. These are more expensive than the Jacobian for the original problem (since the cutoff Mach number is reduced, there are more points in the reduced set).

A much more efficient method of finding a good initial guess for Newton's method is to use grid sequencing in which the problem is solved on finer and finer grids until the final desired grid is reached. The coarse grid problems are much less expensive than the one on the final grid and the problem on the final grid can be converged very rapidly. In addition, Jacobians on the final grid do not need to be computed for the viscous problems, a very expensive procedure. We estimate that grid sequencing could reduce the cost of TRANAIR by a factor of 2 to 3. It is natural to implement this strategy as the first step in solution adaptive grid refinement.

It is still desirable to have some form of damping to prevent divergence in case not enough viscosity damping has been used. The solution could then be restarted with more viscosity to assure convergence. For this purpose, we have chosen the damping of Bank and Rose[18]. This damping is a simple procedure for controlling the step size based on current and previous behavior of the residual. It is very simple and easy to implement and seems to perform as well in most situations as other more complex algorithms[18]. In fact, our nonlinear GMRES subroutine with this damping is less than 150 lines of FORTRAN.

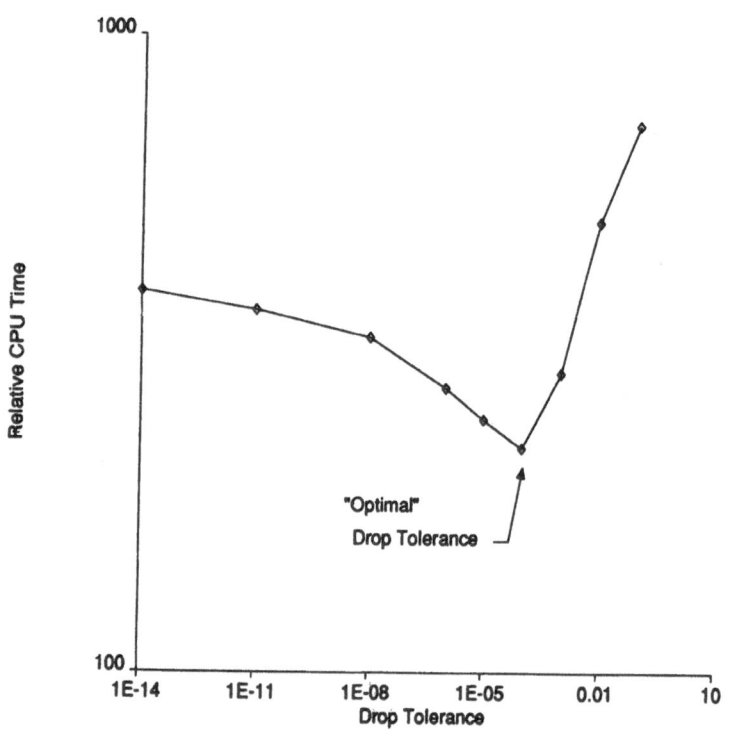

Figure 7. Effects of Drop Tolerance on Total CPU Cost

RESULTS

The TRANAIR methodology has been applied to a number of engineering problems in the areas of fluid flow and wave scattering. In this section we provide some examples of these engineering problems, emphasizing the manner in which different aspects of the numerical technology have contributed to the final solutions.

Linear and Full Potential Flow

The TRANAIR program solves both the linear and nonlinear full potential equation describing transonic fluid flow about lifting configurations. At the present time this is the most developed of all of our applications. Figures 8 and 9 illustrate the power of the hierarchical grid refinement to accurately capture short length scale effects in the solution. These figures show the pressure distribution on an ONERA M6 wing under linear flow conditions. TRANAIR results on a uniform Cartesian grid are compared with the solution from a traditional boundary element method in Figure 8. The uniform grid clearly is not sufficiently fine to accurately represent the leading edge pressure peak. Figure 9 compares a TRANAIR solution on a locally refined grid (based on a coarser Cartesian grid) with the same traditional boundary element solution. By using local grid refinement, sufficient resolution has been obtained to accurately represent the leading edge pressure peak. The total problem size for TRANAIR in Figure 9 is about the same size as in Figure 8. Figure 10 compares a traditional boundary element solution on an F-16 configuration (freestream Mach number, $M_\infty = 0.6$ and angle of attack, $\alpha = 4°$) with a TRANAIR solution on the same configuration. This computational model contains about 3000 surface panels and requires about 3600 CPU seconds of execution time on a single-CPU CRAY X-MP when the traditional boundary element technique is used. The TRANAIR solution on this case required about 1600 seconds of execution time.

In addition to solving for incompressible (linear) flow problems, TRANAIR can also solve for compressible transonic flows. Figure 11 compares wind tunnel measurements of flow about an F-16 configuration (Mach=0.9, $\alpha = 4°$) with TRANAIR computations of the pressure distribution on the wings. The shock strength and position and the leading edge pressure peak are all accurately captured by the calculations.

The versatility offered by TRANAIR's geometry representation system and the advantages of the loose coupling between the geometry definition and the computational grid is illustrated by figure 12, which compares transonic solutions on a bare F-16 configuration with solutions on a configuration containing tip missiles and under-wing fuel tanks. Traditional field methods require a significant expenditure of resources just to generate a computational grid about the more complex configuration. With the TRANAIR approach to modeling, only the surface geometry of the added portions of the configuration need to be generated. The evidence of supersonic flow in the neighborhood of the wing and tank pylon junction attests to the need for transonic flow analysis. While a few traditional surface-fitted grid field methods have been used to compute flows about the bare F-16 configuration, we know of no other methods which have been able to compute the flow about the more complex (and more realistic) F-16 configuration with wing tanks and tip missiles.

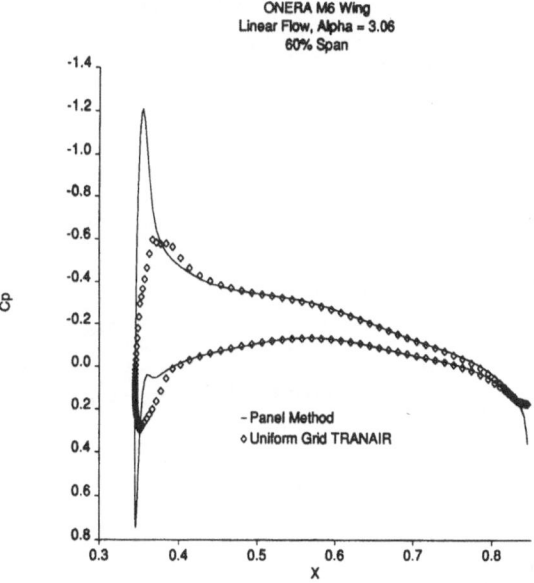

Figure 8. Uniform Grid: Linear TRANAIR Versus Panel Method for ONERA M6 Wing

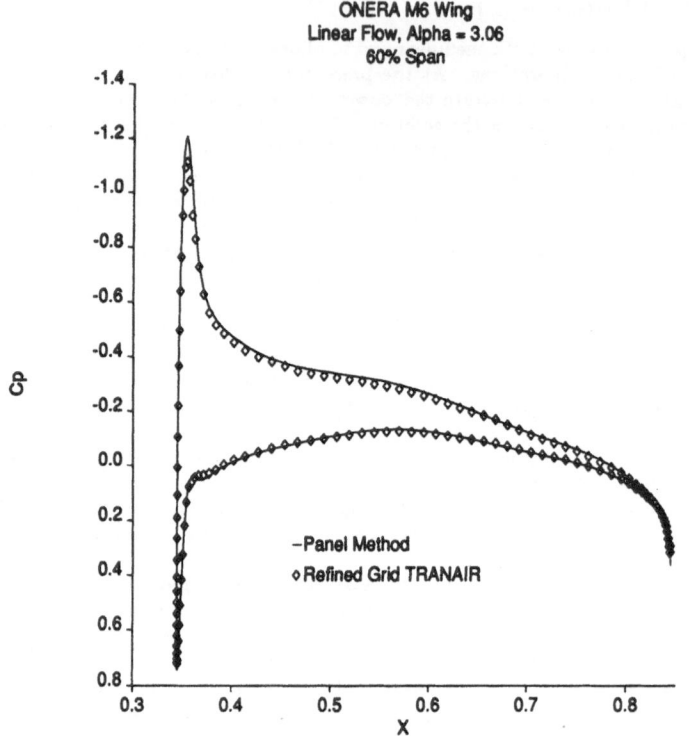

Figure 9. Locally Refined Grid: Linear TRANAIR Versus Panel Method for ONERA M6 Wing

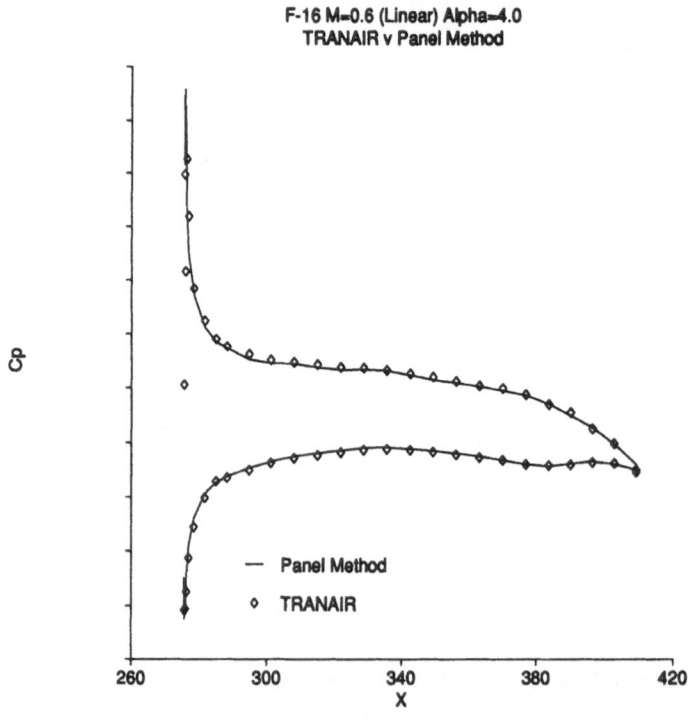

Figure 10. TRANAIR Versus Linear Panel Solution for F-16 Configuration

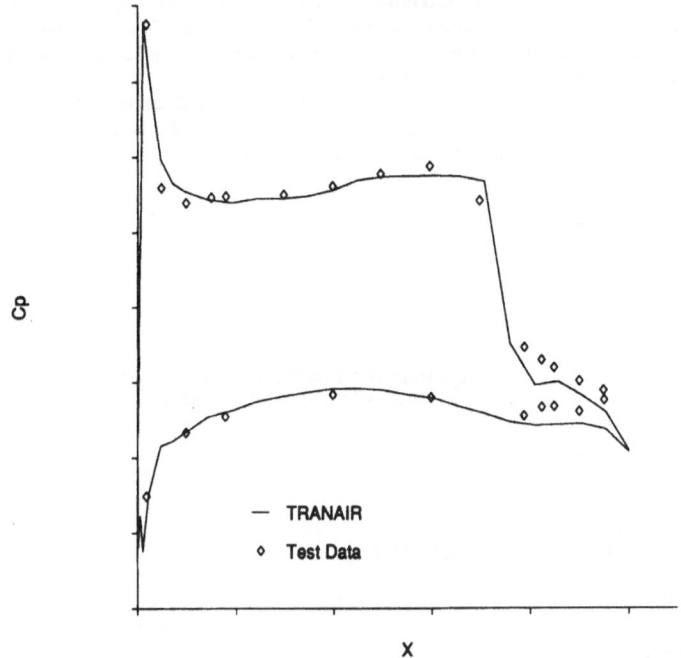

F-16 M=0.9 Alpha=4.0
TRANAIR v Test Data
71% Span
Cp
TRANAIR
Test Data
X

Figure 11. Nonlinear Transonic TRANAIR Solution for F-16 Configuration

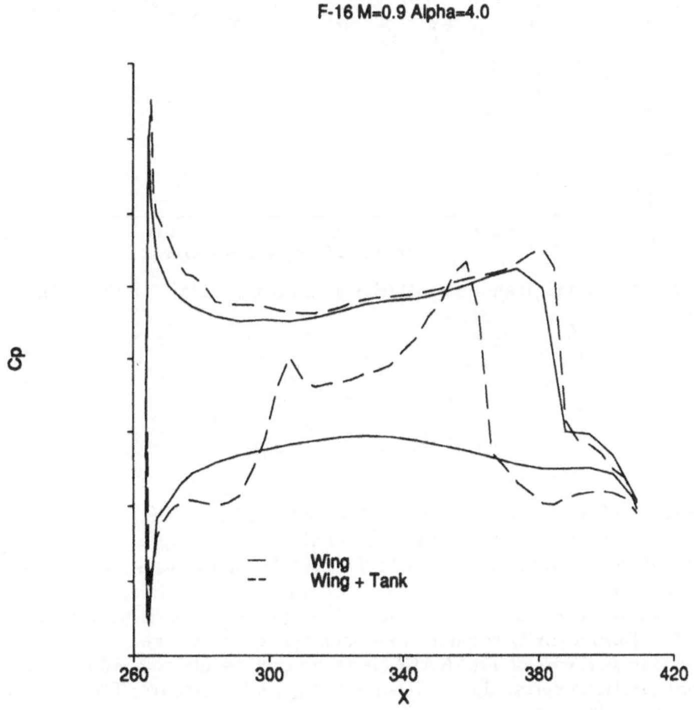

F-16 M=0.9 Alpha=4.0
Cp
Wing
Wing + Tank
260
300
340
380
420
X

Figure 12. Transonic Effects of Under Wing Tank on Wing Pressures, F-16 Configuration

Maxwell's Equations

The same numerical technology used in transonic flow has been applied to solutions of Maxwell's equations, describing the scattering of electromagnetic fields from material objects. These applications to date have been restricted to calculations on a uniform Cartesian grid. The same surface panel definition of configuration geometries has been used. In fact it is possible to use the same aerodynamical panel model for both fluid flow and electromagnetic scattering problems.

Figure 13 compares EM-TRANAIR solutions for scattering of a plane wave from a spherical conductor at a frequency corresponding to a wavenumber-radius product, $ka = 16$. A convergence behavior study was performed on this case showing that the far field signature (the radar cross section, RCS) is

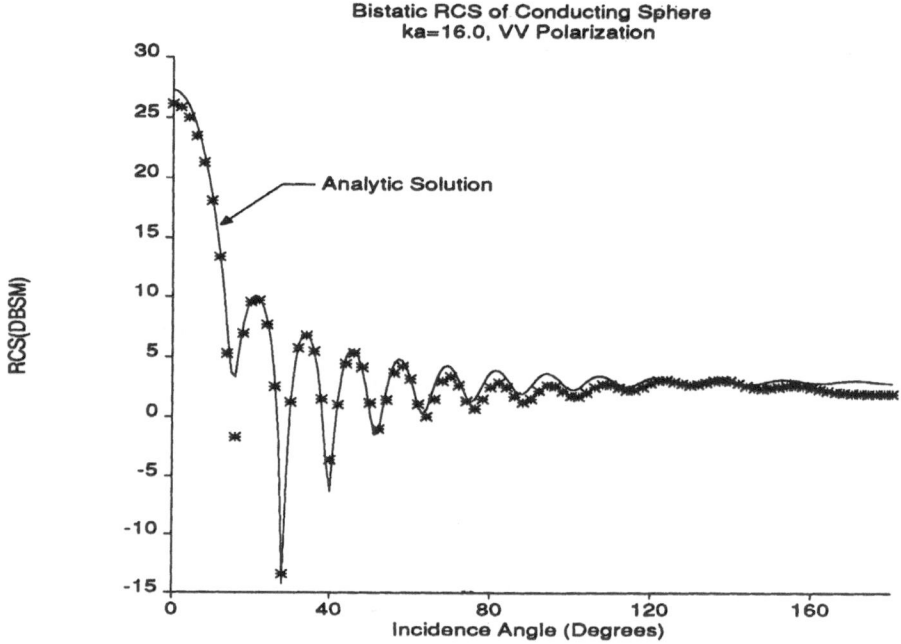

Figure 13. Electromagnetic Scattering From A Conducting Sphere, $ka = 16$

predicted with near-quadratic accuracy as a function of the grid spacing. (Note that for wave equation solutions it is necessary to have a minimum of six points per wavelength.)

A side benefit of the approach used in EM-TRANAIR for the solution of scattering problems is that it is capable of correctly predicting scattering from objects containing a resonant cavity structure. Figure 14 compares EM-TRANAIR predictions with scattering measurement data for scattering of an electromagnetic field from a cavity resonator, an open cylindrical cavity.

The execution time and cost of TRANAIR electromagnetic scattering solutions is dependent on the volume of the configuration expressed in wavelengths. Figure 15 compares TRANAIR costs for a variety of configuration volumes[6].

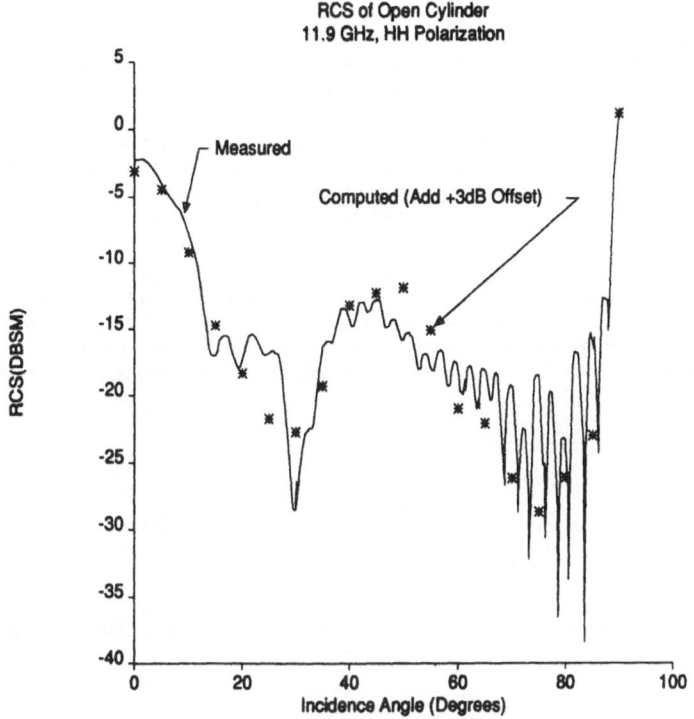

Figure 14. Electromagnetic Scattering from a Resonant Cylindrical Cavity

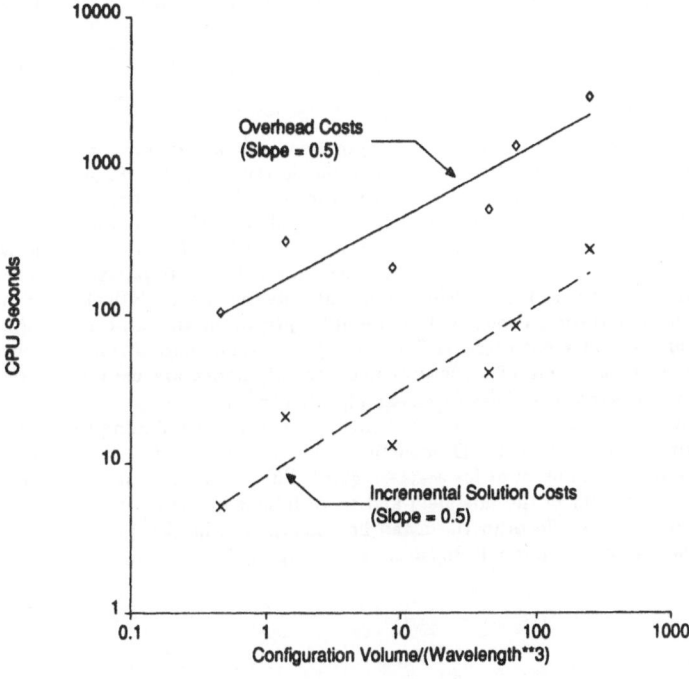

Figure 15. EM-TRANAIR Execution Cost Versus Configuration Volume

Acoustics

The acoustics applications of TRANAIR technology are both the newest and the least developed. The initial capabilities in acoustics came from adapting a 2D electromagnetic scattering program for two dimensional acoustical phenomena.

The validity of the numerical method for solving acoustic scattering problems was demonstrated by comparing the predictions with closed form solutions and test data for scattering of acoustic plane waves and line sources at various frequencies near a cylinder. As shown in Figure 16, excellent agreement was achieved between the numerical predictions and test data. Details of this work have been reported elsewhere[23].

Figure 17 compares the computed scattering of a dipole noise source at the trailing edge of a NACA 0012 airfoil with analytic solutions for such dipole scattering at low and high frequency limits. The numerical calculations agree well with the asymptotic predictions and moreover provide predictions about the intermediate frequency case where no analytic predictions are known.

In order to predict the acoustic transmission through flexible structures, it is necessary to:

- calculate the incident and scattered pressure fields with a scattering code,

- discretize the structure and its interior acoustic volume using a finite element code,

- integrate the pressure field on each structural element to obtain the nodal forces, and

- solve for the coupled structural response and radiated noise due to the above forces.

We have used the 2D acoustics code in conjunction with a structures analysis code to examine the structural response of a cylindrical shell to acoustic stimulations. Figure 18 compares analytic predictions of the structural response functions with the numerical predictions for the coupled structural displacement and the sound pressure level transmitted inside a thin aluminum cylinder. (The values are plotted against a non-dimensional frequency parameter kR, where k is the acoustic wave number and R is the cylinder radius. The value of kR at a typical propeller blade passage frequency is about 7.6, and is also shown in the figure.) Excellent agreement between the analytical and numerical solutions suggests that such computational tools will be useful for predicting the airborne noise transmission into an aircraft, provided the incident pressure field from the exterior sources can be accurately quantified, and provided adequate numbers of structural and acoustic nodes are used to represent the coupled system. Details of this work have been discussed elsewhere[24].

At the present time a 3D acoustics code is being developed by combining the fluid flow and electromagnetic scattering codes. The new 3D acoustics code will solve for harmonic perturbations about a mean transonic flow and will account for scattering and diffraction phenomena due to the configuration boundary, the mean velocity of the flow field and to variations in the local density and speed of sound (induced by compressibility effects in the mean flow field). It will also provide local surface pressure information on the configuration in a form suitable for coupling to one or more structural analysis codes.

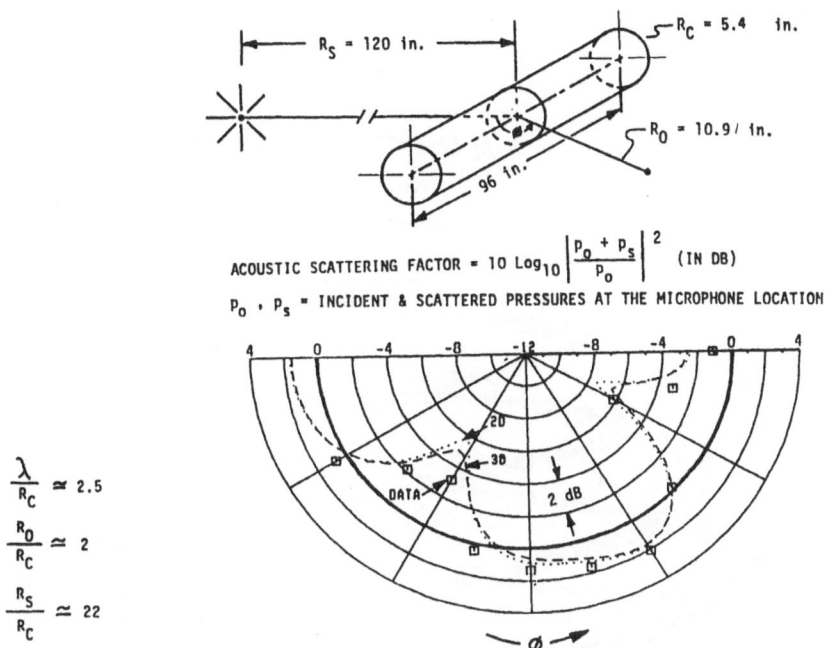

ACOUSTIC SCATTERING FACTOR = $10 \log_{10} \left| \dfrac{P_0 + P_s}{P_0} \right|^2$ (IN DB)

P_0, P_s = INCIDENT & SCATTERED PRESSURES AT THE MICROPHONE LOCATION

$$\frac{\lambda}{R_C} \simeq 2.5$$

$$\frac{R_0}{R_C} \simeq 2$$

$$\frac{R_S}{R_C} \simeq 22$$

NOTE: IN THE 2D PREDICTION, THE SOURCE IS ASSUMED TO BE A LINE SOURCE PARALLEL TO THE AXIS OF THE CYLINDER, WHICH IS ASSUMED TO BE INFINITELY LONG.

Figure 16. Acoustical Scattering Factor Prediction Versus Test Data (1000 Hz, $kR_c = 2.5$)

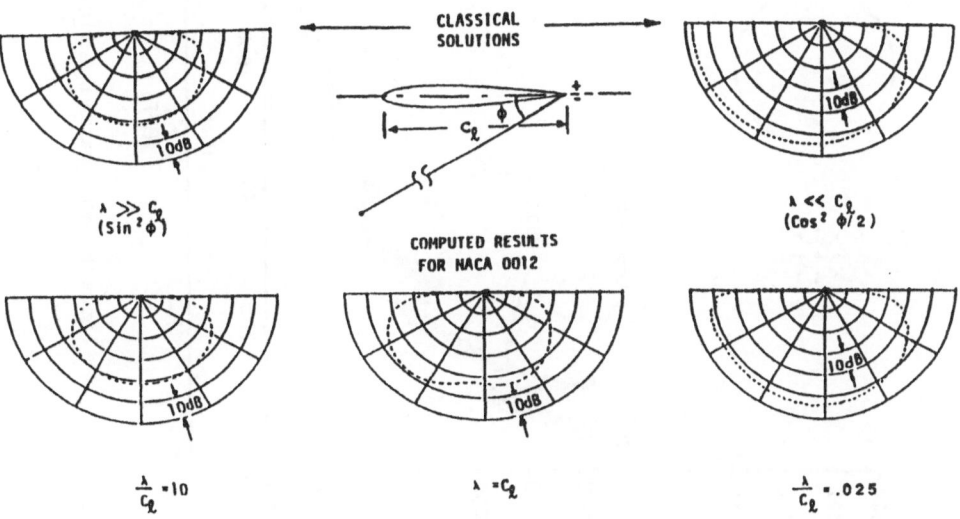

Figure 17. Acoustical Scattering of Dipole Noise Source at Trailing Edge of Wing

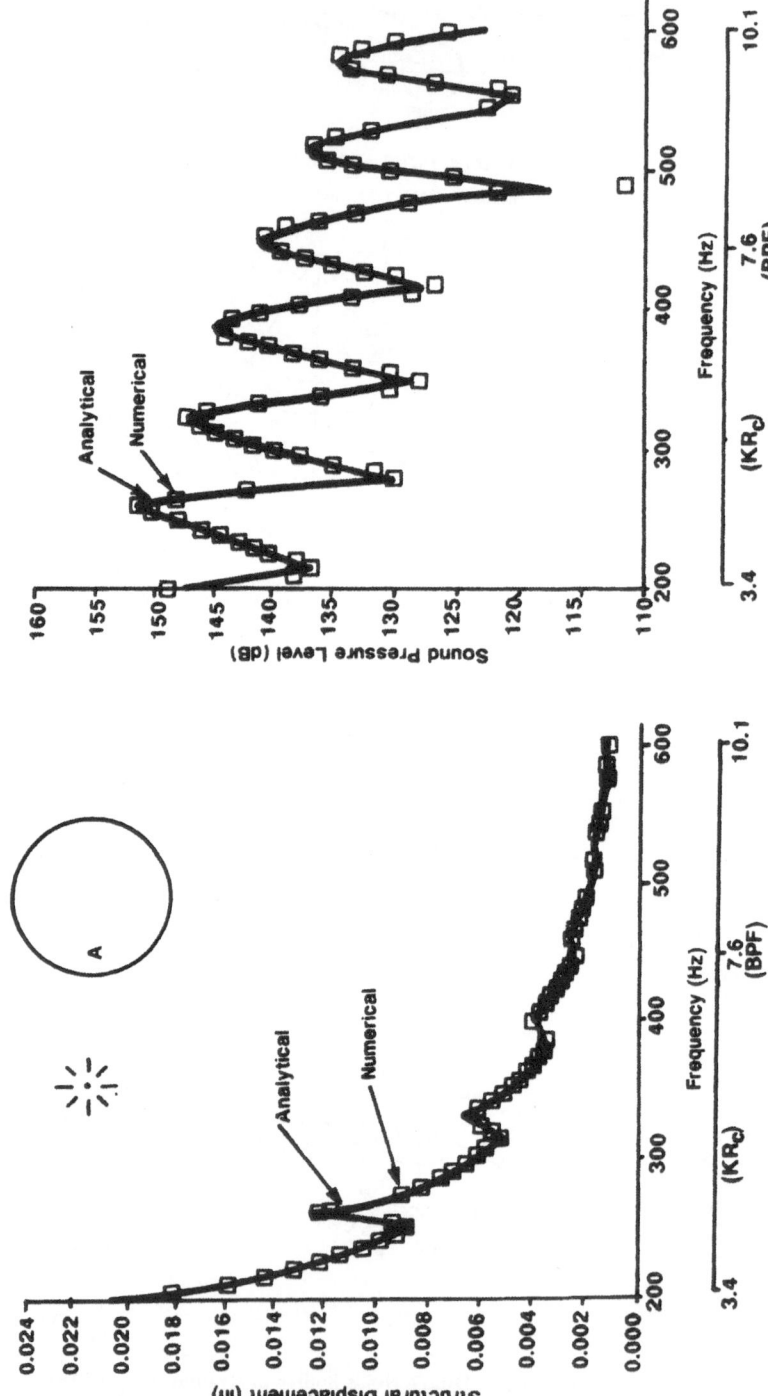

Figure 18. Cylindrical Shell: Analytic and Computed Structural Response and Transmitted Noise

RESEARCH ISSUES

The development of TRANAIR and its related programs is an ongoing project. Based on our experiences to date, we have identified some areas for future development. In this section we first review the current performance of the TRANAIR codes on CRAY X-MP architectures. This forms a baseline of capabilities on current computer systems and defines appropriate hardware platforms for the codes. Then we consider implications of newly emerging hardware on these performance characteristics. New machines are just becoming available which provide similarities in architecture to that of the Cray X-MP but which offer a sufficient number of CPUs to make parallel processing a sensible strategy. We have designed the algorithms in TRANAIR with an eye towards these systems and have some general ideas about the kinds of performance advantages that could be obtained on such systems. Finally, there are other areas of engineering technology which may benefit from the application of these algorithmic ideas. We briefly review some candidates.

Performance of TRANAIR Codes

Performance limits on problem size in the TRANAIR codes are derived from experiences with users of aerodynamic analysis tools. For this community an acceptable but large and expensive solution may require between one and three hours of execution time on a CRAY X-MP class machine. In addition, turnaround is expected within twenty-four hours in a batch processing environment with many users. TRANAIR was designed with these requirements and perceptions in mind and our definition of a "large" problem reflects this situation.

The central memory requirements of TRANAIR codes are based on the central memory available on our development machines: 4 Megawords or less. In addition all TRANAIR development has been done in a single tasking mode (i.e., for execution on one processor). TRANAIR has been developed as an out-of-core code. We have found that this has little impact on execution efficiency and cost as long as sufficient space is available on the CRAY X-MP's Solid-state Storage Device (SSD) for a given problem. I/O wait times for even the largest of jobs remain quite small (less than 200 seconds for jobs with execution times on the order of 4000 CPU seconds or more).

In our experiences to date, SSD usage of TRANAIR is the fundmental limiting resource which determines the problem size limits of the code. Solution of large problems (on the order of 300,000

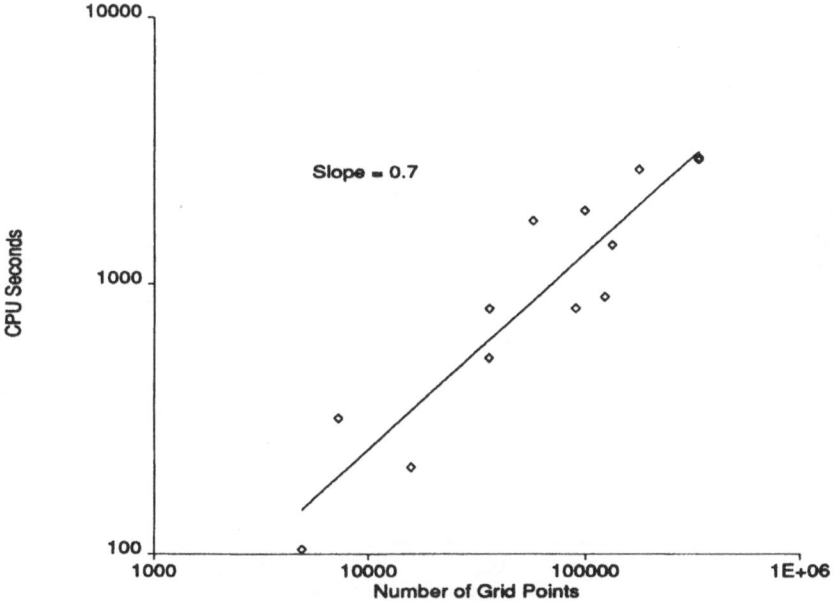

Figure 19. Overhead Costs for EM-TRANAIR Versus Problem Size

unknowns) can require over 100 MW of storage on an SSD. Execution times for these problems remain under two CPU hours for transonic, nonlinear flows and well under one CPU hour for linear potential flow problems.

We characterize the solution times for use of the TRANAIR system of codes by two parameters. The first is an overhead cost (in CPU seconds) representing the cost of setting up the problem, generating the locally refined mesh (where applicable) and computing the finite element operators (for those unknowns in the neighborhood of the boundary). The second is the cost of iteratively solving the system of equations for a single right hand side (single incidence condition and Mach number or frequency). The importance of each of these types of costs is somewhat application dependent.

Figures 19 and 20 present the overhead and incremental costs associated with the first release of the EM-TRANAIR code. This code solves a linear system of equations. Typical applications require solutions for many incidence conditions (many right hand sides), and so the overhead cost is less of an issue than the incremental cost. The unknown in these problems is a 3D vector field defined on a Cartesian grid. From the figures it is evident that solutions on a grid of over 300,000 grid points representing nearly 1,000,000 components of the vector field can be obtained with overhead costs around 2000 CPU seconds. The incremental costs of obtaining a solution for a new right hand side are less than 300 CPU seconds for these problems. Taking a 10,000 CPU second limit as maximum allowable execution time, EM-TRANAIR permits about 30 right hand side solutions for this problem size within the time limit. In physical terms this problem size is characterized by a configuration volume of about 900 cubic wavelengths. Note also that both the overhead cost and incremental costs are approximately linear functions of the problem size. In fact, asymptotically, one portion of the incremental cost is of order $N \log N$, but it does not contribute significantly to the iterative solution cost for most of the problems in the set of data used to establish the curve. As mentioned previously, the EM-TRANAIR

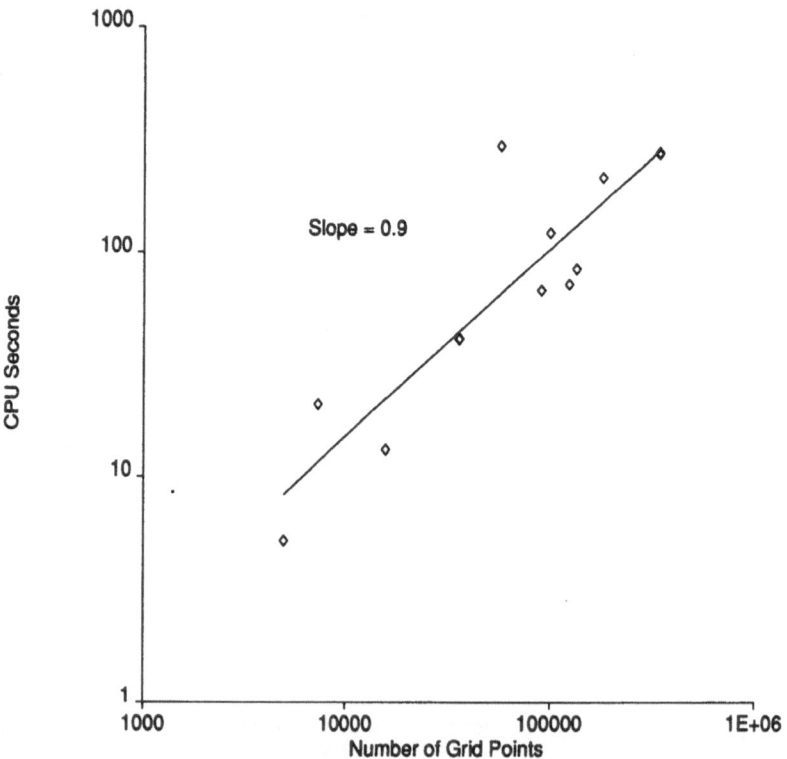

Figure 20. Incremental Costs for EM-TRANAIR Versus Problem Size

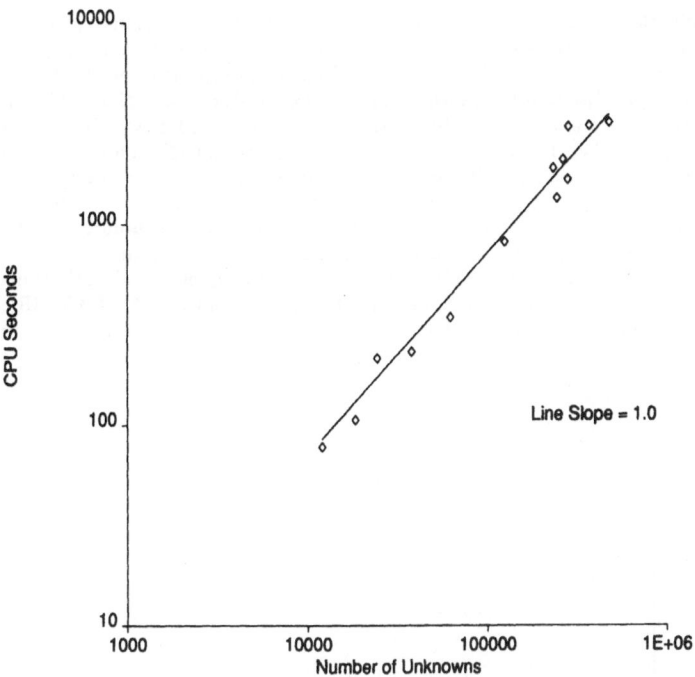

Figure 21. Overhead Costs for TRANAIR Versus Problem Size

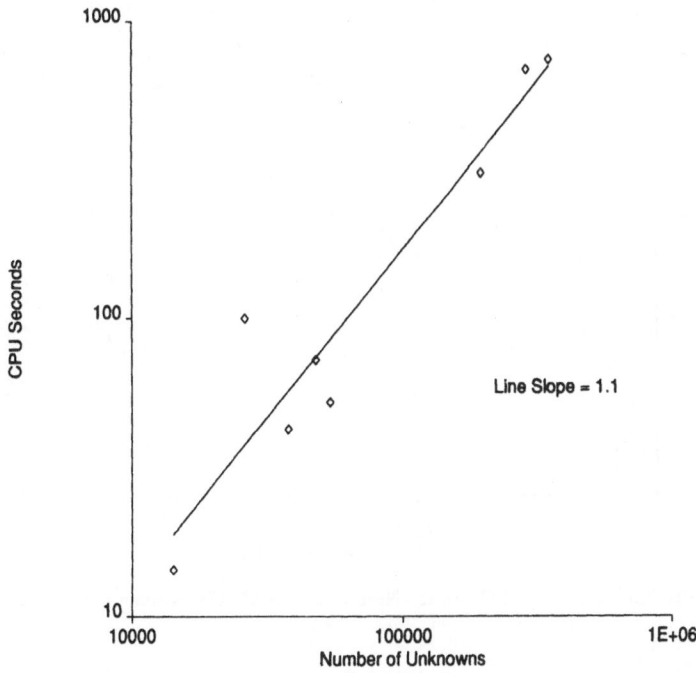

Figure 22. Incremental Costs for Linear TRANAIR Versus Problem Size

code is based on earlier implementations of our technical approach. Considering the current state of the TRANAIR technology in transonic flow, significant speed-ups in execution and significant reductions in SSD resource requirements are possible and will be implemented in the near future.

Figures 21 through 23 present the overhead and incremental costs associated with the current version of the TRANAIR full potential code. Since this code solves a PDE which may be either linear or nonlinear, the incremental costs are presented in two figures: 22 and 23. Figure 22 shows the incremental solution cost incurred for solving a linear system. Figure 23 shows the incremental cost associated with solving a nonlinear system. The costs associated with the solution of a nonlinear equation have somewhat greater variability than those for the linear system in that the number of cycles of the nonlinear GMRES algorithm depend upon the character of each problem. Thus some configurations only require five linearizations and two Jacobian calculations to achieve convergence, while others may require twenty linearizations and eight Jacobian calculations. Note that just as in the EM-TRANAIR case, the solution costs for both linear and nonlinear problems increase approximately linearly as a function of problem size.

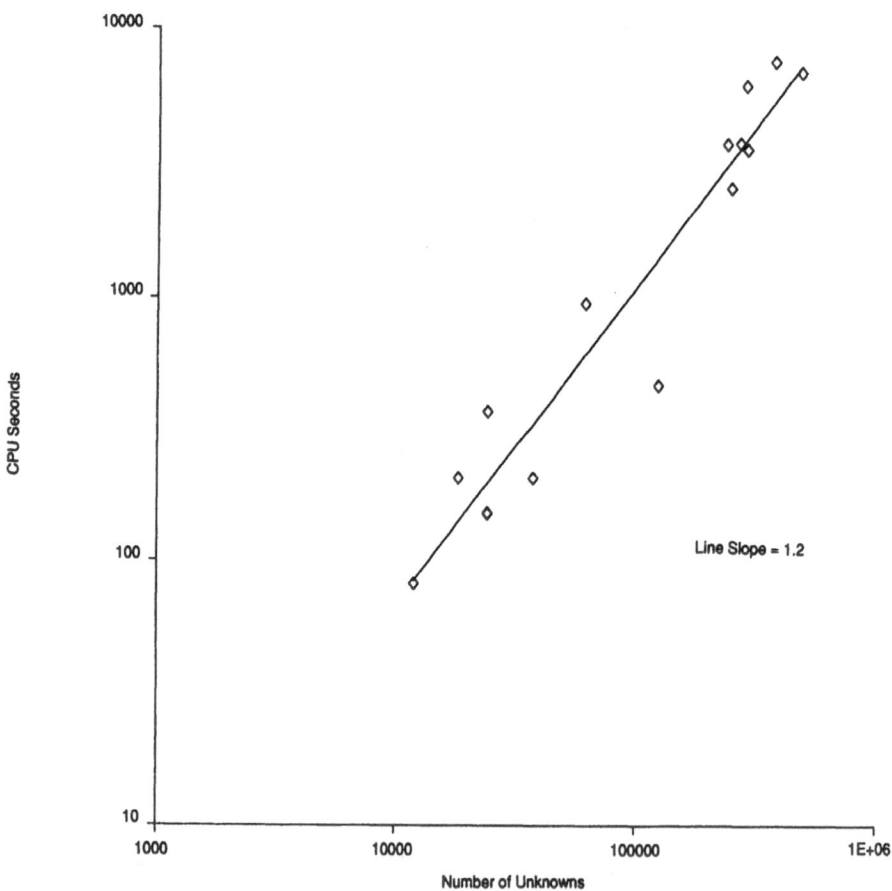

Figure 23. Incremental Costs for Nonlinear TRANAIR Versus Problem Size

Parallel Computing Aspects of The Technology

It is natural to consider how the technology which makes up the TRANAIR codes might be extended to permit the solution of even larger problems, particularly since new, more powerful computer architectures are becoming available. The codes have been developed on a CRAY X-MP architecture and make heavy use of an SSD for out-of-core storage. As discussed previously, the main limit to problem size comes from insufficient SSD storage capacity. Thus, the easiest way to extend the capabilities of the code would be to add an SSD of greater capacity to an existing CRAY X-MP system. Indeed, within the past year Cray Research has announced the availability of SSDs in sizes up to 512 MW, thus quadrupling the capacity of the SSD space presently available on our development systems. In addition, Cray Research has developed operating system support for use of the SSD as a large cache buffer to disk storage. This could prove quite effective in removing the problem size limits due to a shortage of SSD space, particularly since the basic I/O operations which the TRANAIR codes perform involve transfers of large arrays (on the order of 100,000 words) mostly in a fairly sequential fashion.

The second direction in which to push the capacity to solve larger problems is to increase the solution speed. The code is already highly vectorized. The fluid dynamics version of the code has already been through several cycles of efficiency modifications. There is always some capacity for further improvement, but the major portion of the "easy" modifications have been done. However, with the advent of the new CRAY Y-MP architecture, a new horizon for enhancing code performance has emerged. The CRAY Y-MP system has up to eight CPUs which can be employed in a multitasking (macrotasking and microtasking) manner to make both parallelism and vectorization available to improve performance. We have made initial investigations as to how TRANAIR might exploit five or more CPUs in a CRAY Y-MP system. These investigations are conceptual only, but they do indicate that, given five CPUs, one can reasonably expect a reduction of a factor of four in the incremental solution cost, and possibly a similar reduction in the overhead cost. This conclusion is based upon the following observations concerning portions of the TRANAIR codes.

Consider the basic iteration solution sequence in TRANAIR. The first operation is the exterior Poisson or Helmholtz solver, represented by the T^{-1} operator. This involves an FFT convolution performed using the James algorithm[13]. At present, the slowest part of this operation is the plane by plane convolution of the Green's function with sources on the computational boundary (the middle phase of the James algorithm). There are similar plane by plane operations in the first and last phases of the algorithm. It is quite straightforward to modify the code to perform the plane by plane operations over multiple planes using multiple CPU's in either a macrotasking call or through microtasking directives controlling the compilation of the loop over planes of data. There is complete data independence in this part of the algorithm.

The second major step in the iterative sequence is the evaluation of the operator L. In TRANAIR the code is structured so that the data at nodal points is retained in a large array in central memory. Data associated with box centroids (and region centroids) is blocked and stored out of core on the SSD. The operator evaluation consists of first looping over blocks of boxes and then computing velocities and densities for each box in the block from the current value of potential in the nodal array. Then the nodal array is initialized to zero and is used to accumulate the residuals at each nodal point by again looping over blocks of boxes and throwing contributions out to the eight nodes that are influenced by each box. Inside the loops over blocks of boxes, a vector loop over each box is executed in the present version of the code. Again, considering a system which offers multiple CPUs, it is straightforward to allocate the processing of multiple blocks over multiple CPUs. The evaluation of block-based quantities from nodal values of potential contains complete data independency. The evaluation of the residuals has some potential conflicts in that different CPUs might attempt to update the same nodal value "simultaneously" within different vector loops. It is certainly possible, at the cost of some additional overhead, to sort the blocks of boxes so that such data dependencies can be eliminated. At least half of the cost of evaluating the operator can be performed in a parallel-vector manner.

The next step is the forward and backward substitution operations of applying the decomposition of the sparse matrix N to the residuals. With standard ordering of unknowns in the x, y and z coordinate directions, forward and backward substitution is recursive and cannot be efficiently implemented on parallel-vector computers. This situation is partially remedied with orderings which decompose N so that the forward and backward substitutions for the resulting submatrices are independent and can be carried out in parallel[25, 26]. The nested dissection ordering of N used in TRANAIR is on such ordering.

The final step in the processing of an iteration is the evaluation of the operator T at global grid points. At present this is performed in a loop over x, y and z lines in the grid, with outer loops over z

and y and an inner vector loop over x. This structure parallelizes over multiple CPUs by distributing the outer loops over multiple CPUs.

The other major operation in the nonlinear iterative sequence is the decomposition of the sparse matrix N. The basic processing sequence in this operation consists of choosing a single block of the matrix and performing Gaussian elimination of other blocks against the chosen one. This process can be parallelized by having multiple CPUs eliminate multiple blocks simultaneously against a single block with synchronization occurring when choosing the next single block.

Most of the overhead cost associated with the code occurs in computing the operator coefficients for cells which contain a portion of the boundary. This part of the code is presently structured as a loop over cells containing a portion of the boundary. The calculations within the loop are quite lengthy and involve many subroutine calls, but have complete data independence. Thus it is very natural to speed up these computations by macrotasking the loop cycles across multiple CPUs.

There will always be some scalar code with data dependencies which prevent distribution of calculations across multiple CPUs in the TRANAIR codes. However, as outlined in this discussion significant reductions in the more costly portions of the code can be obtained with computer systems containing sufficiently many vector CPUs. This will permit us to solve problems involving 1 to 2 million unknowns in a nominal "parallel hour" of CPU time (one hour of residency time with five CPUs available to the job during residency).

One unfortunate aspect to the sizing and architecture of our preferred development machine, the CRAY Y-MP, is that the SSD size will probably again prove to be the limiting factor in that it is only a factor of four larger than the SSD on our present CRAY X-MP systems. Given the current excessively high price of dynamic RAM, it is not likely that large SSDs will be available at affordable prices within the next year. Our choice of SSD size for a CRAY Y-MP architecture would be about eight times the present maximum size. (This preference is based on a perception that on a two CPU CRAY X-MP, a 512 MW SSD would permit two large TRANAIR jobs to run concurrently. However, it is possible that the Cray Research operating system enhancements which permit the use of the SSD as a cache-to-disk will provide sufficient performance capabilities to remove our perception of a need for additional SSD storage.) With four times the number of CPUs, faster clock cycles and reports of overall execution speedups of a factor of two in CRAY Y-MP systems over CRAY X-MP performance, it is quite reasonable to project a capability to use eight times the disk space on problems eight times larger, all while maintaining machine residency times comparable to that presently experienced in a CRAY X-MP environment.

New Applications of TRANAIR

TRANAIR's approach to the solution of second order PDEs has been demonstrated in two areas of engineering technology: fluid flow and electromagnetic scattering. We plan to extend our work to acoustic scattering. Our acoustics code can be merged with the transonic fluid flow capabilities to permit analysis of harmonic perturbations about a mean transonic flow field. In this mode of analysis, the transonic flow field about a configuration would be computed to obtain local sound speed and density variations. These may be used to describe sound propagation through the mean flow field. This kind of analysis may also have application to unsteady aerodynamic flows.

We also plan to extend our TRANAIR analysis scheme to more complex physics in fluid flow. The main focus is to provide some capabilities for describing viscous effects. Initially this will be accomplished through the addition of boundary layer equations to the set of fluid flow equations. These will be solved simultaneously with the outer full potential equations using the nonlinear GMRES solver. The preconditioner for the boundary layer equations will be the usual marching solver employed in typical boundary layer schemes. It is expected that by using GMRES as a driver, a more stable iterative scheme will emerge even in the case of local upstream flows in the boundary layer.

Another coupled system we plan to examine in the next year is an aero-elastic analysis capability formed by combining equations describing a structural model of an aircraft with the fluid flow system of equations. As in the boundary layer scheme, the structural equations and fluid equations will be solved simultaneously within the same GMRES iteration. Our out-of-core sparse matrix solver will be used as a preconditioner for the structural model. This approach will employ the surface geometry definition to define the coupling between the fluid dynamics problem and the internal structural model. That is, the surface patches (panels) will provide local surface pressure information which will be coupled into the internal structural model. This permits a large measure of independence between the definition of the structural grid and the grid on which the fluid flow equations are solved.

In a more speculative vein, we are curious as to how well the overall strategy that has emerged from our development might apply to fourth order PDE's describing structural problems alone. That is, it would be interesting to see how well one might be able to define a "TRANAIR structural model" This would consist of a locally refined Cartesian grid enclosing the volume around a structure, a set of two dimensional surfaces embedded in that volume which mark the boundary between regions of different (structural) material properties and a list of the structural material properties in each region. The entire volume would be discretized and the whole system would be solved using appropriate preconditioners. Such a scheme would liberate structures problems from the requirement of generating a computational finite element mesh. Users would only be required to specify the bounding surfaces between regions of differing materials.

CONCLUSION

We have demonstrated a new approach to the solution of second order partial differential equations which can accurately describe fluid flow and scattering phenomena about complex geometrical configurations with a fairly minimal definition of the configuration. Some powerful iterative solution techniques have been devised to solve a discretized system of either linear or nonlinear equations. It has been shown that with supercomputer systems like the CRAY X-MP large engineering problems of practical interest can be solved in reasonable execution times (one to three CPU hours). The technical approach should be easy to adapt to the CRAY Y-MP computer systems. On such systems, considerably larger engineering problems will become feasible. The technical approach seems general enough to have application in other engineering problems and we will address these in future work.

Acknowledgments

This work was supported by a variety of contracts and Boeing internal funding sources. In particular we would like to acknowledge support by NASA/Ames Research Center through contracts NAS2-11851 and NAS2-12513, by the U. S. Air Force Wright Aeronautical Laboratories through contract F33615-83-C-3031, and by Boeing Military Airplane and Boeing Commercial Airplane Internal Research and Development funds.

REFERENCES

[1] Johnson, F. T. "A General Panel Method for the Analysis and Design of Arbitrary Configurations in Subsonic Flows", NASA CR-3079, (1979).

[2] Axelson O. and Barker, V. A. Finite Element Solutions of Boundary Value Problems: Theory and Computation, Academic Press, (1984).

[3] Rubbert, P. E., Bussoletti, J. E., Johnson, F. T., Sidwell, K. W., Rowe, W. S., Samant, S. S., SenGupta, G., Weatherill, W. H., Burkhart, R. H., Everson, B. L., Young, D. P. and Woo, A. C. "A New Approach to the Solution of the Boundary Value Problems Involving Complex Configurations", Symposium on Future Directions in Computational Mechanics, Anaheim, Ca, December 10-13, (1986).

[4] Samant S. S., Bussoletti, J. E., Johnson, F. T., Burkhart, R. H., Everson, B. L., Melvin, R. G., Young, D. P., Erickson, L. L, Madson, M.D. and Woo, A. C., "TRANAIR: A Computer Code for Transonic Analyses of Arbitrary configurations", AIAA Paper 87-0034, (1987).

[5] "TRANAIR Computer Code (Theory Document)", NASA Contractor's Report, (unpublished), August (1987).

[6] "EM-TRANAIR: A Computer Program for the Solution of Maxwell's Equations in Three Dimensions: Volume 1, Theory Manual", Technical Report AFFDL-TR-87-3082, September (1988).

[7] Young, D. P., Melvin, R. G., Bieterman, M. B., Johnson, F. T., Samant, S. S. and Bussoletti, J. E, "A Locally Refined Rectangular Grid Finite Element Method", Boeing Computer Services Scientific Computing and Analysis Technicl Report, SCA-TR-108, 1988.

[8] Goodsell, A. M., Madson, M. D., and Melton, J. E., "TRANAIR and Euler Computations of a Generic Fighter Including Comparisons With Experimental Data", AIAA Paper 89-0263, January (1989).

[9] Bojarski, N. N., "K-space Formulation of the Electromagnetic Scattering Problem", Technical Report AFWAL-TR-71-75, (1971).

[10] Mittra, R. and Ko, W. L., "Transform Approach to Electromagnetic Scattering", Proceedings of the IEEE, Vol. 67, No. 11, November (1979).

[11] Buneman, O., "Analytic Inversion of the Five-Point Poisson Operator", Journal of Computational Physics, Vol. 8, pp. 500-505, (1971).

[12] Hockney, R. W., "The Potential Calculation and Some Applications", Methods in Computational Physics, Vol 9, Academic Press, Inc., New York and London, (1969).

[13] James, R. A., "The Solution of Poisson's Equation for Isolated Source Distributions", Journal of Computational Physics, Vol 25, pp 71-93, (1977).

[14] Samet, H., "The Quadtree and Related hierarchical Data-Structures", Computing Surveys, Vol 16, No.2, (1984).

[15] Samant, S. S., Bussoletti, J. E., Johnson, F. T., Melvin, R. G. and Young, D. P., "Transonic Analysis of Arbitrary Configurations using Locally Refined Grids", in Proceedings of the 11th International Conference on Numerical Methods in CFD, Williamsburg, Va., June (1988).

[16] Wigton, L. B., Yu, N. J., and Young, D. P., "GMRES Acceleration of Computational Fluid Dynamics Codes," AIAA Paper 85-1494, July (1985).

[17] Saad, Y. and Schultz, M. H., "GMRES: A Generalized Minimal RESidual Algorithm for Solving Nonsymmetric Linear Systems", Research Report YALEU/DCS/RR-254, August (1983).

[18] Bank R. E., and Rose, D. J., "Global Approximate Newton Methods", Numer. Math. 37, pp. 279-295, (1981).

[19] George, A. and Liu, J. W., Computer Solution of Large Sparse Positive Definite Systems, Prentice-Hall, Englewood Cliffs, New Jersey, (1981).

[20] Bussoletti, J. E., Young, D. P., Johnson, F. T., Melvin, R. G., Samant, S. S., Wigton, L. B., "Application of Sparse Matrix Solvers as Effective Preconditioners", Boeing Computer Services Engineering and Scientific Services Technical Report, ETA-TR-99, August (1988).

[21] Fletcher, R., Practical Methods of Optimization, Volume 1, John Wiley and Sons, (1980).

[22] Burkhart, R. H. and Young, D. P., "GMRES Acceleration and Optimization Codes", Boeing Computer Services Technical Report ETA-TR-88, (1988).

[23] SenGupta, G., "Computation of Aeroacoustic Scattering Effects", AIAA Paper 87-2669 October (1987).

[24] SenGupta, G. and Weatherill, W. H, "Numerical Prediction of Airborne Noise Transmission into a Fuselage" AIAA Paper 87-2736 Presented at the AIAA 11th Aeroacoustics Conference, October (1987).

[25] Dongarra, J. J. and Johnson, L. "Solving Banded Systems on a Parallel Processor", Parallel Computing, Vol. 5, pp. 219-246, (1987).

[26] Duff, I. S., Gould, N. I. M., Lescreinier, M., and Reid, J. K., "The Multifrontal Method in a Parallel Environment", Rept. CSS 211, Computer Science and Systems Division, AERE Harwell, (1987).

COMPONENT MODE ITERATION TECHNIQUE APPLIED TO

MODAL CALCULATION FOR FLUID/STRUCTURE INTERACTION

Jean-Guy Beliveau

CEME Department, Votey Engineering Bldg.
The University of Vermont
Burlington, VT 05405

ABSTRACT

Numerical methods to solve the dynamic interaction of
fluid/structure interaction fall into three general areas; boundary
element techniques, element-by-element solvers, and reduction
algorithms. Amongst the reduction methods there are two general
approaches: subspace iteration involves the reduction to a smaller
eigenvalue problem by assuming Ritz mode shapes whereas component mode
synthesis techniques discretize the structure/fluid system into smaller
substructures, such that a much smaller global system is solved at the
end. The reduced number of eigenvalues and eigenvectors thus calculated
are approximations to the lowest natural frequencies and mode shapes of
vibration.

The component mode iteration technique combines advantages of
subspace iteraction with component mode synthesis. Whereas classical
modal synthesis uses substructure modes of vibration which are
calculated once and for all for each of the substructures, the component
mode iteration technique changes the fixed interface normal substructure
modes at each iteration to satisfy the equations of structural dynamics.
For the fluid/structure interaction formulation involving displacements
in the structure and pressure in the fluid, the equations of
fluid/structure interaction at the substructure level are imposed in the
iterative approach. The methodology of using component mode iteration
for fluid/structure interaction is presented. A simple example is used
to demonstrate its practicality.

INTRODUCTION

The requirements for incorporating the compressibility of the fluid
in acoustic problems of fluid/structure interaction, coupled with the
need to obtain information at fairly high frequencies, necessitates the
use of small elements in numerical modeling procedures leading to
systems having many degrees of freedom. As an example, the requirements
of having the wavelength, λ , discretized into ten (N=10) segments at a

frequency (f=100Hz.) means the element length, ℓ , must be smaller than 60" for sea water with an acoustic wave speed (c=60,000 in/s)

$$\ell = \lambda/N = c/Nf = 60000/10(100) = 60" \tag{1}$$

The higher the frequency the smaller the elements have to be(1).

Thus, the modeling of a surface ship or submerged submarine at sea requires matrices of large dimension. The relatively small energy dissipation in such problems also necessitates large geometry as well as special transmitting boundaries at the limits of the grid.

Although large capacity supercomputers which operate at very high speeds are increasingly available, efficient numerical techniques are required to deal with these large problems in order to maintain the analytical costs at acceptable levels.

The three common techniques used towards this end are boundary elements, element-by-element iterative solvers and reduction techniques. These are all discussed in this symposium. This paper deals with the latter. Amongst the reduction schemes, there are methods which reduce the overall problem to a smaller scale eigenvalue problem, the eigenvalues and eigenvectors of which are representative of the lower frequency range of the original problem. The modern Lanczos algorithm, subspace iteration, and a modified Rayleigh quotient all fall into this scheme of things. Alternatively, substructuring techniques break down the system into small components, whose dynamic characteristics are integrated together to get a reduced order equivalent matrix system whose dynamic characteristics are representaive of the whole structure, again in the low frequency regime.

The component mode iteration technique incorporates beneficial aspects of both substructuring and reduction techniques, in that substructure dynamic information is altered at each iteration in order to satisfy the equations of dynamics of the substructure(2). This is not the case for classical substructuring. Thus, the requirement for accuracy of the substructure is not so critical for component mode iteration. Linear material behavior in the structure is assumed and the fluid is limited to the acoustic regime. Damping is not considered here. Although frequency response is of interest, only eigenvalue solutions are discussed here. A small example is used to demonstrate the method. Steady state vibration, fluid/structure interaction and the component mode iteration technique are first discussed.

STEADY STATE VIBRATION

Steady state excitation, f , and response, x , at the angular frequency ω may be represented as

$$f = Fe^{i\omega t} + F*e^{-i\omega t} \tag{2}$$

$$x = Xe^{i\omega t} + X*e^{-i\omega t} \tag{3}$$

where * represents the complex conjugate. The steady state response to the equations of structural dynamics with no damping, for M and K, the mass and stiffness matrices, respectively.

$$Mx + Kx = f \tag{4}$$

is given by the frequency response matrix, H

X=HF (5)

The impedance matrix, Z , is the inverse of the frequency response matrix, H

$$Z=K-\omega^2 M+i\omega C \tag{6}$$

$$Z=H^{-1} \tag{7}$$

in which $^{-1}$ superscript represents the matrix inverse and "i" is the imaginary symbol $i=\sqrt{-1}$.

Normal modes, ϕ , for ' , representing the transpose, reduces both the mass and stiffness matrices to diagonal form

$$\phi'M\phi=I \tag{8}$$

$$\phi'K\phi=\Omega^2 \tag{9}$$

in which I is the identity matrix, and Ω is the undamped natural frequency in radians per second. The frequency response matrix H may be represented in terms of modal quantities, ϕ and Ω

$$H=\phi[\Omega^2-\omega^2 I]^{-1}\phi' \tag{10}$$

Due to the nature of the frequency response matrix H , the response in a particular frequency range is well approximated by a small number of terms in Eqn. 10,

$$H \approx \sum_{k=1}^{m} \phi_k\phi_k'/(\Omega_k^2-\omega^2) \tag{11}$$

In fact near the resonances the corresponding term completely dominates the response when damping is left out. "m" is thus much less than "n" dimension of the original matrices.

FLUID/STRUCTURE INTERACTION

The degree of freedom in classical finite elements are displacements at the nodes. For a general three-dimensional cube with eight nodes there would be twenty four degrees of freedom per element before integration with the other elements. This can be incorporated into a three-dimensional finite-element formulation in an approximate manner by assuming an almost incompressible material for the fluid. ($\nu=.49999$, k=345600 psi and $\rho=.000096\#-s^2/in^4$ for sea water) with a zero shear modulus.

$$\sigma=E\epsilon \tag{12}$$

$$E=k \begin{bmatrix} 1 & 1 & 1 & 0 & 0 & 0 \\ 1 & 1 & 1 & 0 & 0 & 0 \\ 1 & 1 & 1 & 0 & 0 & 0 \\ 0 & 0 & 0 & 0 & 0 & 0 \\ 0 & 0 & 0 & 0 & 0 & 0 \end{bmatrix} \tag{13}$$

in which the standard sequence of stress $\sigma_x \sigma_y \sigma_z$, $\tau_{xy}\tau_{yz}\tau_{zx}$ and strain ϵ_x, ϵ_y, ϵ_z, $\gamma_{xy}\gamma_{yz}\gamma_{zx}$, is used. This approach leads to many degrees of freedom in the fluid and has often led to numerical difficulties(3).

Pressure Analogy

The pressure analogy considers the pressure at the node as the unknown rather than the three displacements, thus reducing the number of degrees of freedom considerably. The equilibrium equation for pressure, "p" , for "ρ" the fluid mass density, is

$$\nabla p + \rho \ddot{q} = 0 \qquad (14)$$

and the compressibility equation is

$$p = -k \nabla \cdot q \qquad (15)$$

k is the bulk modulus, ∇ is the gradient operator, and q is the displacement. Eqns. 14 and 15 may be combined to yield upon two time derivatives of Eqn. 15

$$\ddot{p}/k = \nabla \cdot (\nabla p / \rho) \qquad (16)$$

For homogeneous fluid in one dimension this yields the well known wave equation having a corresponding wave speed $c = \sqrt{k/\rho}$

$$\partial^2 p / \partial t^2 = (k/\rho) \; \partial^2 p / \partial x^2 \qquad (17)$$

Eqn. 16 may be recast as the classical equation for the motion of solids with only one independent displacement, u , with both v and w identically zero.

$$R \; \partial^2 u / \partial t^2 = \partial \sigma_x / \partial x + \partial \tau_{xy} / \partial y + \partial \tau_{zx} / \partial z \qquad (18)$$

in which density, ρ , and bulk stiffness, k , play inverse roles to their usual meaning

$$R = 1/k \qquad (19)$$
$$\kappa = 1/\rho \qquad (20)$$
$$u = p \qquad (21)$$
$$\sigma_x = \kappa \; \partial p / \partial x \qquad (22)$$
$$\tau_{xy} = \kappa \; \partial p / \partial y \qquad (23)$$
$$\tau_{zx} = \kappa \; \partial p / \partial z \qquad (24)$$

in which the stress-strain law, Eqn. 12 has ($\nu = 5000$, $G = 1/\rho$) or

$$E = \kappa \begin{bmatrix} 1 & -1 & -1 & 0 & 0 & 0 \\ -1 & 1 & -1 & 0 & 0 & 0 \\ -1 & -1 & 1 & 0 & 0 & 0 \\ 0 & 0 & 0 & 1 & 0 & 0 \\ 0 & 0 & 0 & 0 & 1 & 0 \\ 0 & 0 & 0 & 0 & 0 & 1 \end{bmatrix} \qquad (25)$$

The pressure analogy implies apppropriate boundary conditions. Similarly at the fluid/structure interface the equaitons are not symmetric(4).

$$\begin{bmatrix} M_s & 0 \\ A & M_f \end{bmatrix} \begin{Bmatrix} \ddot{x}_s \\ \ddot{x}_f \end{Bmatrix} + \begin{bmatrix} K_s & -A' \\ 0 & K_f \end{bmatrix} \begin{Bmatrix} x_s \\ x_f \end{Bmatrix} = \begin{Bmatrix} f_s \\ f_f \end{Bmatrix} \qquad (26)$$

In which the subscript "s" and "f" refer to structure and fluid respectively, with x_s being displacements in the solid and x_f , pressures in the fluid. f_s and f_s are loads on the structure and in the fluid respectively and A is a matrix of areas at the interface degrees of freedom between the structure and the fluid.

Symmetric Formulation for Steady State

For steady state motion, there results upon reduction of Eqn. 26 for steady state sinusoidal motion at the frequency, ω.

$$[K_s - \omega^2 M_s - \omega^2 A'(K_f - \omega^2 M_f)^{-1} A]\{X_s\}$$

$$= \{F_s\} + A'[K_f - \omega^2 M_f]^{-1}\{F_f\} \qquad (27)$$

This may further be represented by modal information of the fluid

$$[K_s - \omega^2 M_s - \omega^2 A'\chi(k - \omega^2 m)^{-1}\chi'A]\{X_s\}$$

$$= \{F_s\} + A'\chi(k - \omega^2 m)^{-1}\chi\{F_f\} \qquad (28)$$

in which k and m are diagonal matrices with χ being fluid modes, of pressure in the acoustic field.

$$k = \chi'K_f\chi \qquad (29)$$

$$m = \chi'M_f\chi \qquad (30)$$

The fluid effect may also be represented approximately by the first "q" modes

$$\chi(k - \omega^2 m)\chi' \approx P = \sum_{i=1}^{q} \chi_i\chi_i^! / (k_i - \omega^2 m_i) \qquad (31)$$

and the overall fluid/structure system may be represented in symmetric form

$$(K - \omega^2 M)X = F \qquad (32)$$

in which

$$K = K_s \qquad (33)$$

$$M = M_s - A'PA \qquad (34)$$

$$F = F_s + A'PF_f \qquad (35)$$

Using this approach, the structural stiffness matrix is left intact and only the mass matrix is affected by the interaction of fluid and structure. It now is, however, a function of the excitation frequency.

COMPONENT MODE INTERATION

The basic notions of this approach have been presented elsewhere(5). The extension to fluid/structure interaction problems is straightforward and is the focus of the following discussion.

For analytical, rather than experimental modal analysis, it is convenient to use fixed interface normal modes. Thus, at the interface of the fluid and structure and at separation of fluid domains, this is represented by no loads in the fluid based on the pressure analogy

$$\partial p/\partial n=0 \tag{36}$$

in which n is the normal to the surface.

Component Mode Synthesis

Substructuring, also known as component mode synthesis, may be considered as a transformation, α , of the structural degrees of freedom, "x" , by a reduced set "u". At the substructure level this is given by

$$x=\alpha u \tag{37}$$

in which the interior degrees of freedom are represented as fixed interface normal modes ϕ and static constraint modes, ψ , corresponding to unit static displacements at the interface degrees of freedom(6).

Thus the structural degrees of freedom are

$$X_s = \left\{ \begin{matrix} X_e \\ X_i \end{matrix} \right\} \tag{38}$$

$$u = \left\{ \begin{matrix} U_e \\ U_m \end{matrix} \right\} \tag{39}$$

$$\alpha = \begin{bmatrix} I & 0 \\ \psi & \phi \end{bmatrix} \tag{40}$$

for "i" and "e" representing interior and exterior degrees of freedom. ϕ diagonalizes the portion of the mass and stiffness matrices associated with the interior degrees of freedom

$$M_s = \begin{bmatrix} M_{ee} & M_{ei} \\ M_{ie} & M_{ii} \end{bmatrix} \tag{41}$$

$$K_s = \begin{bmatrix} k_{ee} k_{ei} \\ K_{ie} K_{ii} \end{bmatrix} \tag{42}$$

$$\phi' K_{ii} \phi = \Omega^2 \tag{43}$$

$$\phi' M_{ii} \phi = I \tag{44}$$

$$\psi = -K_{ii}^{-1} K_{ie} \tag{45}$$

A global transformation imposes compatibility of the boundary degrees of freedom between substructures

$$u=\beta v \tag{46}$$

resulting in a final reduced symmetric matrix equation

$$\overline{M}\ddot{v}+\overline{K}v=\overline{f} \tag{47}$$

in which

$$\bar{K} = \beta'\alpha' K\alpha\beta \qquad (48)$$

$$\bar{M} = \beta'\alpha' M\alpha\beta \qquad (49)$$

$$\bar{f} = \beta'\alpha' f \qquad (50)$$

The eigenvectors, V, of the reduced matrix equation for the structure diagonalize the M and K matrices

$$V'\bar{M}V = I \qquad (51)$$

$$V'\bar{K}V = \Lambda \qquad (52)$$

in which Λ is a diagonal matrix having as elements the square of the natural frequencies of vibration of the total structure.

Free Vibrations

For a substructure, the equations of steady state vibration at a number of frequencies is given by

$$KX - MX\Lambda = 0 \qquad (53)$$

in which X are the respective amplitudes of motion at Λ which are the squares of the resonant frequencies. Eqn. 53 can be reordered into primary "e" and secondary "i" degrees of freedom resulting in the equation

$$K_{ii}X_i = M_{ii}X_i\Lambda + M_{ie}X_e\Lambda - K_{ie}X_e \qquad (54)$$

For fluid/structure interaction

$$K = K_s = \begin{bmatrix} K_{ee} K_{ei} \\ K_{ie} K_{ii} \end{bmatrix} \qquad (55)$$

$$M = M_s - A'PA = \begin{bmatrix} M_{ee} M_{ei} \\ M_{ie} M_{ii} \end{bmatrix} \qquad (56)$$

In order for the internal degrees of freedom to satisfy the equations of dynamics for free vibrations, Eqn. 54, must be satisfied. An update to X_i is then easily obtained by a decomposition of the stiffness matrix K_{ii} associated with the interior degrees of freedom, K_{ii}, the mode shapes at the primary degrees of freedom, X_e, the eigenvalues, Λ, and initial estimates to the elements of the eigenvectors associated with the interior degrees of freedom, X_i.

This iterative equation, Eqn. 54, has been shown to converge rapidly(5). This is because the X and Λ correspond to the global eigenvectors and eigenvalues of the whole structure and not those of the substructure. Λ is usually much smaller than Ω^2, the square of the natural frequencies of the substructure with fixed interface degrees of freedom.

Algorithm

Letting X be the modal vectors of the substructure associated with the global eigenvectors, V, these are related by a transformation matrix, T

$$X = T\beta V \tag{57}$$

β chooses the appropriate degrees of freedom of the substructure from the global coordinates, both boundary and modal. From Eqns. 37 and 46

$$T = \alpha\beta \tag{58}$$

For a given substructure let "ne" be the number of primary degrees of freedom, "nm" represent the number of substructure modes, and "nt" be the number of mode shapes of the whole structure which are calculated, corresponding to the order of the transformed matrices K and M in Eqn. 48 and 49. The dimension for these three matrices are then $X(ne+ni, nt)$, $T(ne+ni, ne+nm)$, $V(ne+nm, nt)$.

At each iteration in the process, the transformation matrix T is adjusted to satisfy Eqn. 54 resulting in a local transformation having the form of α in Eqn. 40.

$$T = XZ \tag{59}$$

This is done by choosing

$$Z = \begin{bmatrix} X_{ee}^{-1} & -X_{ee}^{-1}X_{er} \\ 0 & I \end{bmatrix} \tag{61}$$

such that for X partitioned as
$$X = \begin{bmatrix} X_{ee}X_{er} \\ X_{ie}X_{ir} \end{bmatrix} \tag{62}$$

there results for r representing the remainder (nr=nt-ne)

$$T = \begin{bmatrix} I & 0 \\ X_{ie}X_{ee}^{-1} & X_{ir}-X_{ie}X_{ee}^{-1}X_{er} \end{bmatrix} \tag{63}$$

The dimensions of Z to be consistent with the previous relations are $Z=(nt, ne+nm)$

Considerations for situations where the number of interface degrees of freedom "ne" is larger than the number of modes of the whole structure as well as pivoting for improved conditioning are discussed in Bennighof(5). Also because the modes are iterated, there is no need to obtain substructure modes. Ritz modes, based on subspace iterations approaches can be used as initial estimates. The following steps are then required to adopt the component mode iteration technique to fluid/structure iteraction.

1. First, a study of the fluid field by itself is performed. This can utilize the pressure analogy, and fixed iterface normal modes

between sub-domains in the fluid, corresponding to zero loads at the interface nodes with both the structure and other fluid domains. Of course the fluid modes could be iterated as well.

2. Obtain "m" limited number of fluid modes and determine a convenient frequency of excitation ω to obtain P , k , and m of Eqn. 29-31.

3. Decompose the structure into substructures and calculate the static constraint modes and either fixed interface normal nodes or Ritz modes for each substructure as initial transformation.

4. Obtain the initial global matrices.

5. Solve the corresponding eigenvalue problem. If convergence occurs, stop, otherwise, continue.

6. At the substructure level, carry out Eqn. 54, to obtain modified substructure modes which satisfy the free vibration constraints on the secondary degrees of freedom, using the mode shapes and natural frequencies from Step 4 on the right hand side.

7. Obtain the new transformation matrix from Eqn. 63 and the corresponding global stiffness and stiffness matrices and return to Step 5.

EXAMPLE

As an example of the procedure, consider the fluid/structure interaction of a shear type structure with a diagonal mass matrix and a tridiagonal stiffness matrix (k=5x10⁸, m=1, n=9).

$$M = m I \tag{64}$$

$$K = k \begin{bmatrix} 2 & -1 & & & \\ -1 & 2 & -1 & & \\ & -1 & 2 & . & . \\ & & . & . & 2 & -1 \\ & & & & -1 & 1 \end{bmatrix} \tag{65}$$

The natural frequencies of vibration for this structure with no fluid structure interaction are given by (2)

$$f_i(Hz)=\sqrt{k/m} \, \sin \, [\pi(2i-1)/(2n+1)/2]/\pi \tag{66}$$

The fluid is assumed to act on an area A=100 at each degree of freedom. The fluid is partitioned into two subdomains, and the structure has three substructures. The fluid modes are used and one mode for each substructure plus the two boundary modes. Thus, the global system is of order five instead of nine.

Each subdomain of the fluid is then analyzed with fixed interface, corresponding in the pressure analogy to zero loads at the interface with the structure and the adjoining fluid. A modal synthesis of the fluid yields pressure modes. These are then substituted into Eqn. 31-34 for each of the three substructures which have fixed interface normal modes.

The eigenvalues of Eqn. 32 are then calculated. Results for the three lowest resonant frequencies without and with fluid/structure interaction are given in the Table after six iterations.

CONCLUSION

Combination of the pressure analogy with component modal synthesis of the pressure field in the fluid and component mode iteration in the substructures yields a technique for fluid/structure interaction which utilizes aspects of subspace iteration and modal synthesis.

In this paper the concepts for implementing the method are presented. Although, only small models have been analyzed thus far, it appears that the method will work for large dimensional matrices as well. The efficiency depends on the number of iterations. The extension from eigenvalue analysis to steady state response is straightforward.

ACKNOWLEDGMENTS

The author wishes to thank the Summer Faculty Program of ASEE/NAVY for fellowships granted in the summers of 1986, 1987, and 1988 in which this work was performed. Members of the Applied Mechanics Division of the Naval Underwater Systems Center in New London, CT were helpful in motivating and in collaborating on this work. In particular, I wish to thank D. Cox, A. Carlson and M. Tuchio for their support and M. Drew who typed this article.

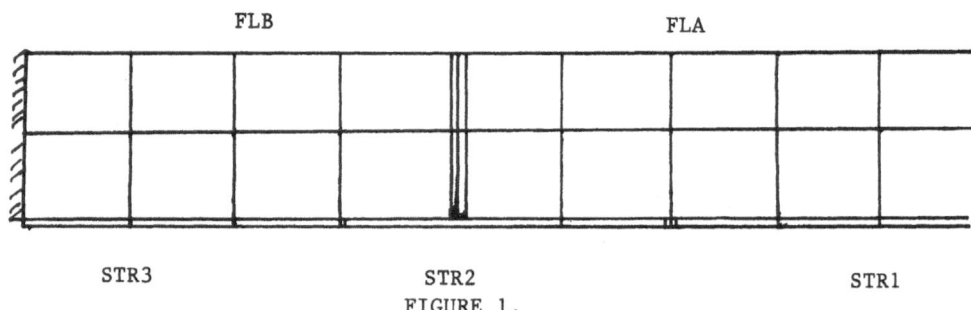

FIGURE 1.

TABLE FREQUENCY (Hz)

	FLA	FLB	FLUID	F/S	STRUCTURE	STR1	STR2	STR3
1.	830	832	385	584	588	2724	3559	1584
2.	859	880	394	1739	1747	5033	6164	4438
3.	925	988	421	2900	2859	6576	--	6413

REFERENCES

(1) Beliveau, J-G. "Modal Techniques for Steady State Response of Three-Dimensional Fluid/Structure Interaction", Technical Report for NUSC, 1986.

(2) Beliveau, J-G, "The Component Mode Iteration Technique", Technical Report for NUSC, 1987.

(3) Kalinowski, A.J. and Nebelung, C.W. "Media-Structure Interaction Employing Frequency Dependent Mesh Sizes with the Finite Element Method", Shock and Vibration Bulletin, Vol. 51, 1981, pp. 173-193.

(4) MacNeal, R., "Acoustics and Fluid-Structure Coupling", Section 7.3, MSC/NASTRAN Handbook for Dynamic Analysis, Version 63, M.A. Gockel, Editor, 1983, MacNeal-Schwendler Corporation.

(5) Bennighof, J.K. "Component Mode Iteration for Frequency Calculations", Journal of American Institute of Astronautics and Astronautics, Vol. 25, No. 7, 1987, pp. 996-1002.

(6) Craig, R. R., Jr., "Structural Dynamics: An Introduction to Computer Methods", John Wiley and Sons, 1981.

THE COUPLING OF FINITE ELEMENTS AND

BOUNDARY ELEMENTS FOR STRUCTURAL ACOUSTICS

Gordon C. Everstine

Applied Mathematics Division (184)
David Taylor Research Center
Bethesda, Maryland 20084

ABSTRACT

Computational structural acoustics is concerned with the numerical prediction of the far-field sound pressure field radiated or scattered by a harmonically-excited, submerged, arbitrary, 3-D elastic structure. This paper discusses a large-scale computational capability which addresses this problem by coupling a finite element model of the structure with a boundary element model of the surrounding fluid. The problems of interest are large enough that many hours of supercomputer time are often needed to obtain solutions. In the solution procedure, the surface fluid pressures and velocities are first calculated by coupling a finite element model of the structure with a discretized form of the Helmholtz surface integral equation for the exterior fluid. Far-field radiated pressures are then calculated from the surface solution using the Helmholtz exterior integral equation. The numerical capability is very general, highly automated, and requires no independent specification of the fluid mesh. To solve very large problems, a new out-of-core block equation solver was written for the fully-populated, nonsymmetric, complex systems of equations which arise. The use of NASTRAN as the structural analyzer permits a variety of graphical displays of results, including the computer animation of the dynamic response, an essential tool for interpreting the results of large-scale calculations. The overall approach is illustrated using known analytic solutions for submerged spherical shells subjected to both incident pressure and non-uniform mechanical loads.

INTRODUCTION

Two fundamental problems in structural acoustics are (1) the calculation of the far-field acoustic pressure field radiated by a general, submerged, three-dimensional, elastic structure subjected to internal time-harmonic loads and (2) the calculation of the far-field acoustic pressure scattered by an elastic structure subjected to an incident time-harmonic wave train. These problems are usually solved by combining a finite element model of the structure with a fluid loading computed using either finite element[1-3] or boundary element[4-10] techniques.

Although both approaches are computationally expensive for large structural models, the fluid finite element approach is burdened with the additional complications caused by the approximate radiation boundary condition at the outer fluid boundary, the requirements on mesh size and extent (sometimes leading to frequency-dependent meshes[3]), and the difficulty of generating the fluid mesh.

In contrast, the boundary element or boundary integral equation (BIE) approach for generating the fluid loading is mathematically exact (except for surface discretization error) and requires little or no additional modeling effort to convert an existing model of a dry structure for use in submerged analyses.

Although several other coupled finite element/boundary element capabilities have been developed previously[4,7,9] to solve the fluid-structure interaction problem, none was developed for a widely-used, general-purpose structural analysis code such as NASTRAN.[11,12] Here we describe a large-scale computational capability known as NASHUA which couples a NASTRAN finite element model of a structure with a fluid loading calculated with a discretized form of the Helmholtz surface integral equation.

This implementation differs from the previous approaches in several important ways. First, the use of NASTRAN allows considerable generality in the structural model, including the capability to treat internal fluid volumes.[13,14] In addition, the acoustic analysis of a structure can be integrated with other dynamic and stability analyses, and many of the pre- and postprocessors developed for use with NASTRAN (e.g., computer animations[15]) can be used. Moreover, the reduction in structural stiffness caused by hydrostatic pressure can be easily accounted for using the geometric stiffness matrix used in elastic stability analyses.

Second, no independent fluid mesh need be specified. The NASHUA approach locates fluid grid points at the wet structural points (rather that at structural element centroids), and obtains normals and areas by applying a unit, outwardly-directed static pressure load on the wet surface. Since a first-order fluid approximation is used (to allow convenient coupling with the first-order shell elements often used by the finite element practitioners), explicit integration over the surface elements is avoided by using Chertock's approach[16] to compute the influence of a fluid point on itself (the "self" terms in the fluid matrices).

Third, a new out-of-core block equation solver[17] was written to solve the large, complex, fully-populated, nonsymmetric system of algebraic equations which arises. As a result, very large problems (with tens of thousands of structural degrees of freedom) can be solved.

In general, the finite element program (NASTRAN) is used to generate the structure's stiffness, mass, and damping matrices and to perform various matrix manipulations. Other programs are used to generate the fluid matrices, perform the field calculations, and display the results.

This paper will summarize the theoretical approach used, illustrate the application of the approach to two problems with classical series solutions (acoustic radiation and scattering from spherical shells), and discuss some of the computational aspects of such problems.

THEORETICAL APPROACH

The Structure

Consider an arbitrary, submerged, three-dimensional, elastic structure subjected to either internal time-harmonic loads or an external time-harmonic incident pressure wave train. If the structure is modeled with finite elements, the resulting matrix equation of motion for the structural degrees of freedom (DOF) can be written as

$$Zv = F - GAp, \tag{1}$$

where Z = structural impedance matrix (dimension s x s),
   v = complex amplitude of the velocity vector for all structural DOF (wet and dry) in terms of the coordinate systems selected by the user (s x r),
   F = complex amplitude of the vector of mechanical forces applied to the structure (s x r),
   G = rectangular transformation matrix of direction cosines to transform a vector of outward normal forces at the wet points to a vector of forces at all points in the coordinate systems selected by the user (s x f),
   A = diagonal area matrix for the wet surface (f x f), and
   p = complex amplitude of total pressures (incident + scattered) applied at the wet grid points (f x r).

In this equation, the time dependence $\exp(i\omega t)$ has been suppressed, where $i = \sqrt{-1}$, and ω is the circular frequency of excitation. In the above dimensions, s denotes the total number of independent structural DOF (wet and dry), f denotes the number of fluid DOF (the number of wet points), and r denotes the number of load cases. If first-order finite elements are used for the surface discretization of the structure, surface areas, normals, and the transformation matrix G can be obtained from the calculation of the load vector resulting from an outwardly-directed static unit pressure load on the structure's wet surface.

In Eq. 1, the structural impedance matrix Z, the matrix which converts velocity to force, is given by

$$Z = (-\omega^2 M + i\omega B + K)/i\omega, \tag{2}$$

where M, B, and K are the structural mass, viscous damping, and stiffness matrices, respectively. For structures with a nonzero loss factor, K is complex.

The Exterior Fluid

The total fluid pressure p satisfies the Helmholtz differential equation

$$\nabla^2 p + k^2 p = 0, \tag{3}$$

where $k = \omega/c$ is the acoustic wave number, and c is the speed of sound in the fluid. Equivalently, for smooth surfaces, p is the solution of the Helmholtz integral equation[7,18]

$$\int_S p(\underline{x})(\partial D(r)/\partial n)dS - \int_S q(\underline{x})D(r)dS = \begin{cases} p(\underline{x}')/2 - p_I, & \underline{x}' \text{ on } S \\ p(\underline{x}'), & \underline{x}' \text{ in } E \end{cases} \tag{4}$$

where S and E denote surface and exterior fluid points, respectively, p_I is the incident free-field pressure, r is the distance from $\underline{x}$ to $\underline{x}'$ (Fig. 1),

D is the Green's function

$$D(r) = e^{-ikr}/4\pi r, \tag{5}$$

$$q = \partial p/\partial n = -i\omega\rho v_n, \tag{6}$$

ρ is the mass density of the fluid, and v_n is the outward normal component of velocity on S. As shown in Fig. 1, $\underline{x}$ in Eq. 4 is the position vector for a typical point P_j on the surface S, $\underline{x}'$ is the position vector for the point P_i which may be either on the surface or in the exterior field E, the vector $\underline{r} = \underline{x}' - \underline{x}$, and $\underline{n}$ is the unit outward normal at P_j. We denote the lengths of the vectors $\underline{x}$, $\underline{x}'$, and $\underline{r}$ by x, x', and r, respectively. The normal derivative of the Green's function D appearing in Eq. 4 can be evaluated as

$$\partial D(r)/\partial n = (e^{-ikr}/4\pi r)\ (ik + 1/r)\ \cos\beta, \tag{7}$$

where β is defined as the angle between the normal $\underline{n}$ and the vector $\underline{r}$, as shown in Fig. 1.

The substitution of Eqs. 5-7 into the surface equation (4) yields

$$p(\underline{x}')/2 - \int_S p(\underline{x})\ (e^{-ikr}/4\pi r)\ (ik + 1/r)\ \cos\beta\ dS$$
$$= i\omega\rho \int_S v_n(\underline{x})\ (e^{-ikr}/4\pi r)dS + p_I, \tag{8}$$

where $\underline{x}'$ is on S. This integral equation relates the total pressure p and normal velocity v_n on S. If Eq. 8 is discretized for numerical computation[19,20] (the details of which will be omitted here), we obtain the matrix equation

$$E p = C v_n + p_I \tag{9}$$

on S. These discretizations yield fluid matrices E and C which are fully-populated, complex, nonsymmetric, frequency-dependent matrices of dimension f x f.

In addition, the matrices E and C are singular at the discrete frequencies of the resonances of the corresponding interior acoustic cavity with Dirichlet (zero pressure) boundary conditions.[5] Such frequencies (also referred to as the "characteristic," "critical," or "forbidden" frequencies of the problem) can be avoided by using the Schenck[5] or Burton and

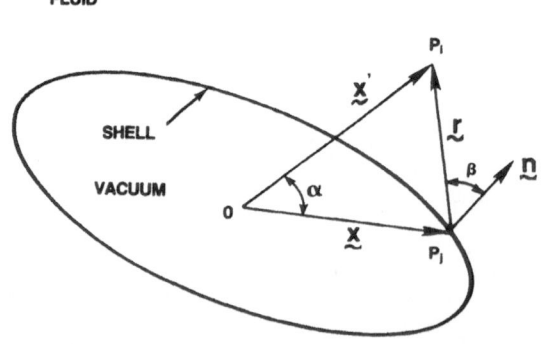

Fig. 1. Notation for Helmholtz integral equation.

Miller[8-10] approaches, neither of which has yet been implemented in this procedure. As a result, only low frequency acoustic radiation and scattering can be handled.

The Coupled System

The normal velocities v_n in Eq. 9 are related to the total velocities v by the same rectangular transformation matrix G:

$$v_n = G^T v, \tag{10}$$

where the superscript T denotes the matrix transpose. If velocities v and v_n are eliminated from Eqs. 1, 9, and 10, the resulting equation for the coupled fluid-structure system is

$$H p = Q + p_I, \tag{11}$$

where

$$H = E + C G^T Z^{-1} G A \tag{12}$$

and

$$Q = C G^T Z^{-1} F. \tag{13}$$

Since H and Q depend on geometry, material properties, and frequency, Eq. 11 can be solved to yield the total surface pressures p. Since the two right-hand side terms in Eq. 11 correspond to mechanical and incident loadings, respectively, only one of the two terms would ordinarily be present for a given case. We note that H exhibits the same critical frequency singularities as E and C.[19]

The vector v of velocities at all structural DOF may then be recovered by solving Eq. 1 for v:

$$v = Z^{-1} F - Z^{-1} G A p. \tag{14}$$

Surface normal velocities v_n may be recovered by substituting this solution for v into Eq. 10.

The Field Solution

With the solution for the total pressures and velocities on the surface, the exterior Helmholtz integral equation, Eq. 4, can be integrated to obtain the radiated (or scattered) pressure at any desired location $\underline{x}'$ in the exterior field. We first substitute Eqs. 5-7 into the second of Eq. 4 to obtain a form suitable for numerical integration:

$$p(\underline{x}') = \int_S [i\omega\rho v_n(\underline{x}) + (ik + 1/r)p(\underline{x}) \cos \beta] (e^{-ikr}/4\pi r)dS, \tag{15}$$

where all symbols have the definitions used previously, and $\underline{x}'$ is in the exterior field. Thus, with the total pressure p and normal velocity v_n on the surface S, the radiated or scattered pressure at $\underline{x}'$ can be determined by numerical quadrature using Eq. 15. Alternatively, to compute the far-field pressure at range x', the following asymptotic form[19] of Eq. 15 could be used:

$$p(\underline{x}') = (ike^{-ikx'}/4\pi x') \int_S [\rho c v_n(\underline{x}) + p(\underline{x}) \cos \beta]e^{ikx \cos \alpha} dS, \tag{16}$$

where, in the far-field, β is the angle between the normal $\underline{n}$ and the direction of interest $\underline{x}'$, and α is the angle between $\underline{x}$ and $\underline{x}'$.

With the surface and far-field solutions, a variety of other quantities of interest can be computed, including average and root-mean-square surface velocities and pressures, surface acoustic intensity, radiated power, and radiation efficiency.[19] For example, the acoustic intensity at a point on the surface is the product of the pressure there and the component of normal velocity which is in phase with the pressure:

$$I = \text{Re}(pv_n^*), \qquad (17)$$

where the asterisk denotes the complex conjugate. The total power radiated is the surface integral of acoustic intensity.

Summary of Theoretical Approach

The solution procedure uses the finite element program (NASTRAN) to generate the matrices K, M, B, and F and to generate sufficient geometry information so that the matrices E, C, G, A, and p_I can be computed by a separate program. Then, given all matrices on the right-hand sides of Eqs. 12 and 13, standard matrix operations are used to compute H and Q. Eq. 11 is then solved for the total surface pressures p using E.A. Schroeder's block solver OCSOLVE.[17] Next, the surface normal velocities v_n and the vector v of velocities at all structural DOF are recovered to complete the surface solution. Then, with this solution for the total pressures and velocities on the surface, the asymptotic (far-field) form of the Helmholtz exterior integral equation is used to compute far-field radiated pressures.

NUMERICAL EXAMPLES

In this section, two problems with known series solutions are solved to illustrate the application of the structural-acoustic capability described.

Radiation From Sector-Driven Spherical Shell

Consider first the problem of a thin spherical shell with a uniform pressure drive over a sector, as shown in Fig. 2. We solved numerically the problem with the following characteristics:[21]

a = 5 m	shell radius
h = 0.15 m	shell thickness

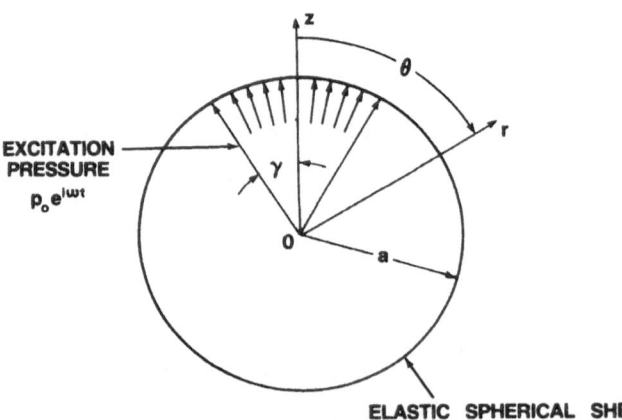

Fig. 2. Submerged elastic spherical shell driven over sector.

$$E = 2.07 \times 10^{11} \text{ Pa} \qquad \text{Young's modulus}$$
$$\nu = 0.3 \qquad \text{Poisson's ratio}$$
$$\rho_s = 7669 \text{ kg/m}^3 \qquad \text{shell density}$$
$$\eta = 0 \qquad \text{shell loss factor}$$
$$\rho = 1000 \text{ kg/m}^3 \qquad \text{fluid density}$$
$$c = 1524 \text{ m/s} \qquad \text{fluid speed of sound}$$
$$\gamma = 36 \text{ degrees} \qquad \text{polar angle of load}$$

One octant of the shell was modeled with triangular membrane/bending finite elements, as shown in Fig. 3. With 20 elements along each edge of the domain, the model has 231 wet points and 1263 structural DOF. However, with a one-octant model of the sphere, the numerical solution of this problem requires running both symmetric and anti-symmetric parts of the problem, providing a good check on the ability to combine symmetry cases.

The benchmark solution to which the numerical results are compared is a series solution which was developed by Henderson[22] based on equations in the Junger and Feit book.[23] The results of this comparison are shown in Table 1 for four different nondimensional drive frequencies ka, where a is the radius of the sphere. None of the drive frequencies is near a critical frequency, the first 13 of which are located at at ka = π, 4.49, 5.76, 2π, 6.99, 7.73, 8.18, 9.10, 9.36, 3π, 10.4, 10.5, and 10.9.[24] For each drive frequency ka, the normalized far-field pressure $|p_r r/p_0 a|$ is listed for each colatitude angle θ, where p_r is the far-field pressure at distance r from the origin, and p_0 is the internally-applied pressure. The far-field pressure pattern for one of these frequencies (ka = 5) is also shown in Fig. 4. The numerical solution agrees very well with the exact (i.e., converged series) solution, even at ka = 2, which coincides with a resonance of the submerged shell.

Scattering From a Spherical Shell

Here we illustrate the solution of scattering problems by solving the problem of the submerged thin spherical shell subjected to an incident time-harmonic planar wave train, as shown in Fig. 5. The solution of this problem exhibits rotational symmetry about the spherical axis parallel to the direction of wave propagation. The benchmark solution to which the

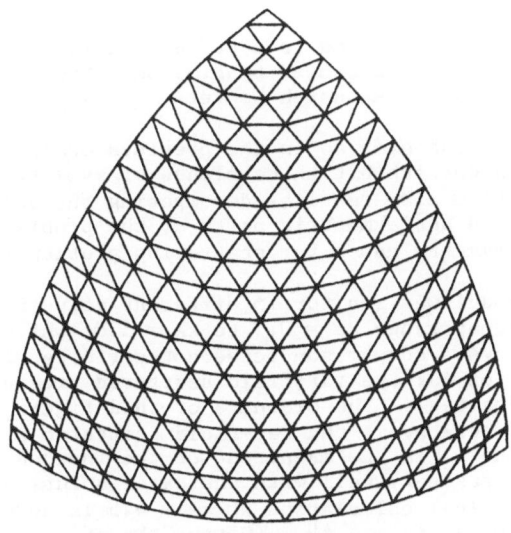

Fig. 3. Finite element model of one octant of spherical shell.

Table 1. Comparison of computed solution with converged series solution for sector-driven spherical shell.

ka	Colatitude θ (degrees)	Normalized Far-Field Pressure, $\lvert p_r r/p_o a \rvert$		
		Computed	Exact	% Error
0.5	0	0.0514	0.0514	0.0
	30	0.0445	0.0445	0.0
	60	0.0257	0.0258	0.4
	90	0.0035	0.0035	0.0
	120	0.0258	0.0259	0.4
	150	0.0446	0.0446	0.0
	180	0.0515	0.0515	0.0
1.0	0	0.0887	0.0889	0.2
	30	0.0744	0.0745	0.1
	60	0.0434	0.0434	0.0
	90	0.0235	0.0237	0.8
	120	0.0448	0.0448	0.0
	150	0.0784	0.0786	0.3
	180	0.0939	0.0942	0.3
2.0	0	1.183	1.163	1.7
	30	0.278	0.276	0.7
	60	0.667	0.666	0.2
	90	0.131	0.128	2.3
	120	0.721	0.716	0.7
	150	0.757	0.695	8.9
	180	1.977	1.860	6.3
5.0	0	0.510	0.512	0.4
	30	0.292	0.292	0.0
	60	0.020	0.017	17.6
	90	0.100	0.097	3.1
	120	0.161	0.160	0.6
	150	0.169	0.163	3.7
	180	0.177	0.170	4.1

numerical results will be compared is a series solution published in the Junger and Feit book[23] and programmed by F.M. Henderson, a recently retired employee of the David Taylor Research Center.

The shell used is the same as in the previous example. Since the incident loading does not exhibit three planes of symmetry, the numerical solution of this problem also requires decomposing the solution and the load into both symmetric and antisymmetric parts of the problem, providing a check on combining symmetry cases for scattering problems.

The numerical model was run for 15 different drive frequencies in the nondimensional frequency range ka = 0.5 to ka = 5.0, where a is the shell radius. Figure 6 shows a comparison between the finite element calculation and the series solution for the far-field scattered pressure in the forward direction (θ = 180 degrees). The ordinate of this figure is the normalized pressure $\lvert p_r r/p_o a \rvert$, where p_r is the far-field scattered pressure at distance r from the origin, and p_o is the magnitude of the incident pressure. The computed solution agrees very well with the exact solution. We note that, as expected, the numerical calculation at ka = 4.5 is adversely affected by the critical frequency at ka = 4.49. (Often, the effect of a critical frequency is very severe.[19])

For several of the excitation frequencies, we also tabulate in Table 2 the far-field scattered pressure patterns. Again the agreement between the calculations and the series solution is excellent, even at ka = 1.6, which is near a resonant peak. At the sharper resonant peaks, the results would be much more sensitive to small changes in frequency.

COMPUTATIONAL ASPECTS

The numerical procedure outlined above is computationally intensive and could require many hours of supercomputer time for large problems. The principal problem-dependent parameters on which the computer time and memory depend are the number of independent structural DOF (denoted as 's' in the discussion following Eq. 1), the number f of fluid DOF (the number of wet points on the fluid-structure interface), and the average wavefront W_{avg} of the structural stiffness matrix. A related parameter is the root-mean-square matrix wavefront W_{rms}, which is slightly larger than W_{avg} but will be used interchangeably with W_{avg} to simplify this discussion.

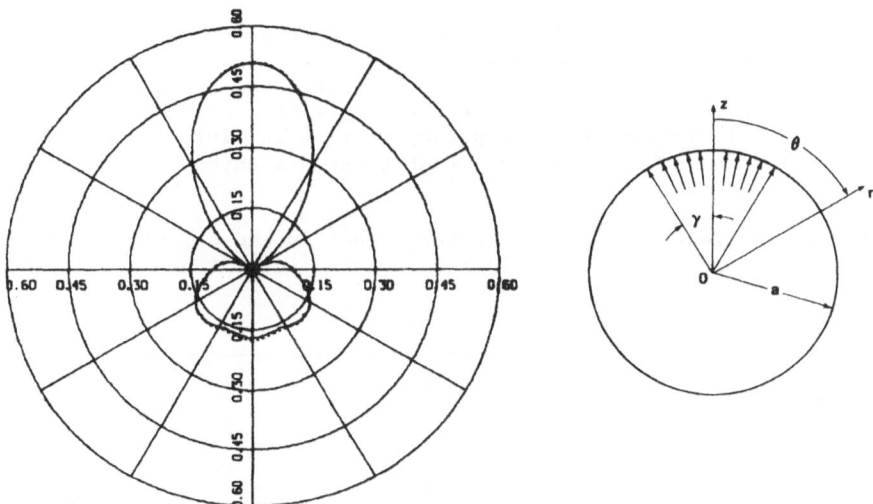

Fig. 4. Far-field radiated pressure $|p_r r/p_o a|$ at ka = 5 for sector-driven spherical shell (solid is exact, dashed is computed).

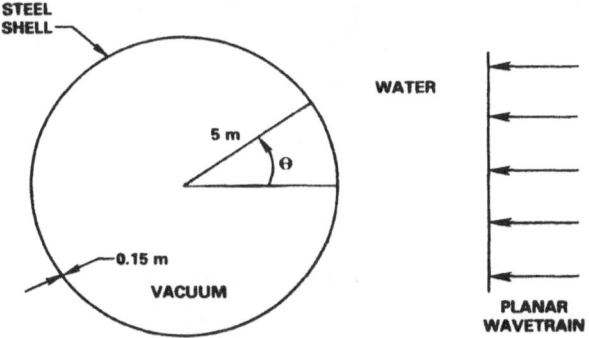

Fig. 5. Plane wave scattering from an elastic spherical shell.

Three steps in the procedure have the potential to require large amounts of computer time:

1. the triangular factorization of H in Eq. 11, for which the classical operation count is $f^3/3$ complex operations (since H is fully-populated, complex, and nonsymmetric),

2. the triangular factorization of Z (which is performed only once) in Eqs. 12-14, for which the operation count is $sW_{rms}^2/2$ complex operations, and

3. the forward/backward substitution (FBS) required in Eq. 12, in which the matrix $Z^{-1}GA$ is computed given the triangular factors of Z and the matrix product GA. The operation count for this FBS is $2sfW_{avg}$ complex operations. (Although the cost of FBS operations is usually insignificant in equation solving, this FBS step is significant because of the large number (f) of columns of GA.)

We note that the operation counts cited above are useful only as a general indication of the relative computer effort required since, on a supercomputer, all these steps would be vectorized to the extent possible.

Although asymptotically (i.e., for very large problems), the factorization of H might be the most time-consuming operation, in practice, the presence of internal structure increases the order s and wavefront W_{avg} of the structural matrices, so that the FBS operation cited often requires the most time. To illustrate this point, we list in Table 3 the problem size statistics for several problems (not those presented in the preceding section).

Another interesting numerical issue arises in the calculation of surface acoustic intensity (Eq. 17) and radiated power (the surface integral

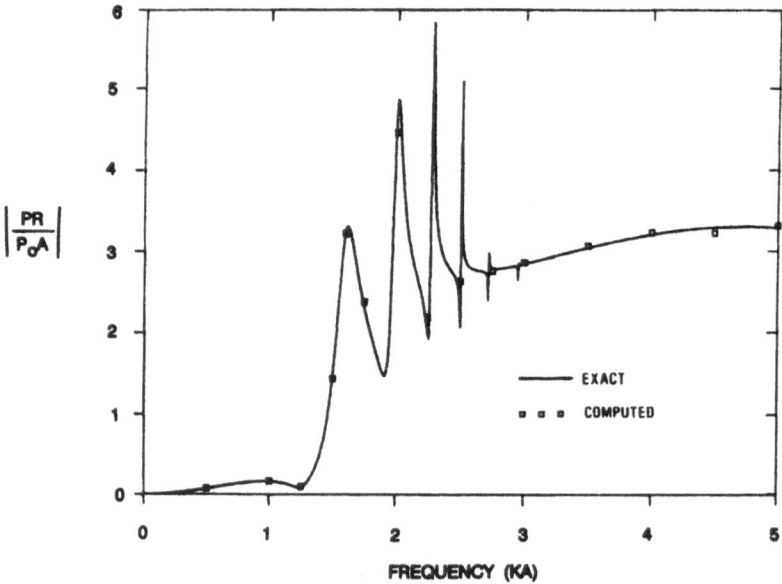

Fig. 6. Normalized far-field pressure $|p_r r/p_o a|$ scattered in the forward direction (θ = 180 degrees) by a spherical shell; solid curve is converged converged series solution, and square boxes are numerical solution.

Table 2. Comparison of computed solution with converged series solution for scattering from spherical shell; $\theta = 0$ corresponds to the back-scattered direction.

ka	Angle θ (degrees)	Normalized Far-Field Pressure, $\lvert p_r r/p_o a\rvert$		
		Computed	Exact	% Error
0.5	0	0.0083	0.0081	2.5
	30	0.0144	0.0143	0.7
	60	0.0299	0.0299	0.0
	90	0.0481	0.0481	0.0
	120	0.0626	0.0626	0.0
	150	0.0708	0.0708	0.0
	180	0.0734	0.0733	0.1
1.0	0	0.0892	0.0903	1.2
	30	0.0382	0.0389	1.8
	60	0.0886	0.0886	0.0
	90	0.1926	0.1930	0.2
	120	0.2208	0.2210	0.1
	150	0.1893	0.1887	0.3
	180	0.1662	0.1652	0.6
1.6	0	3.146	3.149	0.1
	30	1.993	1.995	0.1
	60	0.325	0.320	1.6
	90	1.515	1.498	1.1
	120	0.561	0.540	3.9
	150	2.058	2.092	1.6
	180	3.213	3.245	1.0
4.0	0	0.166	0.160	3.8
	30	0.105	0.101	4.0
	60	0.068	0.069	1.4
	90	0.274	0.269	1.9
	120	0.562	0.554	1.4
	150	1.769	1.757	0.7
	180	3.228	3.205	0.7

of intensity). Since for low frequencies the pressure p and the normal velocity v_n have nearly orthogonal phase (i.e., the fluid behaves like an added mass), the intensity calculation (and hence the power integral) can be sensitive to small errors in p and v_n on the surface. To circumvent this numerical problem, the radiated power can alternatively be computed by integrating the acoustic intensity over the far-field sphere, where pressure and velocity are in phase (since $p = \rho c v_n$). The calculation of far-field pressure is not sensitive to small errors in the surface solution, since it is calculated using the exterior Helmholtz integral equation, Eq. 16, which has the effect of smoothing out small errors in the surface solution.

DISCUSSION

A very general computational capability has been described for predicting the acoustic sound pressure field radiated or scattered by arbitrary three-dimensional elastic structures subjected to time-harmonic loads. Sufficient automation is provided so that, for many structures of practical interest, an existing finite element structural model can be adapted for acoustic analysis within a few hours.

Table 3. Number of complex operations for three problems.

Prob.	No. of Structural DOF (s)	No. of Fluid DOF (f)	Average Wavefront (W_{avg})	Decomp. of H $(f^3/3)$	Decomp. of Z $(\approx sW_{avg}^2/2)$	FBS $Z^{-1}GA$ $(2sfW_{avg})$
1	2973	496	129	4.1×10^7	2.5×10^7	3.8×10^8
2	12980	971	317	3.1×10^8	6.5×10^8	8.0×10^9
3	16158	971	411	3.1×10^8	1.4×10^9	1.3×10^{10}

One of the major benefits of having the acoustic analysis built around a general-purpose finite element program is the ability to integrate the acoustic analysis of a structure with other dynamic and stability analyses. Thus the same finite element model can be used for modal analysis, frequency response analysis, linear shock analysis, and underwater acoustic analysis. In addition, many of the pre- and postprocessors developed for use with the finite element package become available for acoustics as well.

The two main limitations of the approach described are (1) that the symmetries inherent in axisymmetric structures (a special case of the more general 3-D problem) are not being exploited, and (2) that operation below the fundamental critical frequency of the Helmholtz integral equation is required. As indicated earlier, this latter limitation can be removed by using either the Schenck[5] or Burton and Miller[8,9] approaches.

REFERENCES

1. G.C. Everstine, "A Symmetric Potential Formulation for Fluid-Structure Interaction," J. Sound and Vibration 79(1), 157-160 (1981).
2. G.C. Everstine, "Structural-Acoustic Finite Element Analysis, with Application to Scattering," Proc. 6th Invitational Symposium on the Unification of Finite Elements, Finite Differences, and Calculus of Variations, ed. by H. Kardestuncer, Univ. of Connecticut, 101-122 (1982).
3. A.J. Kalinowski and C.W. Nebelung, "Media-Structure Interaction Computations Employing Frequency-Dependent Mesh Sizes with the Finite Element Method," The Shock and Vibration Bulletin 51(1), 173-193 (1981).
4. L.H. Chen and D.G. Schweikert, "Sound Radiation from an Arbitrary Body," J. Acoust. Soc. Amer. 35(10), 1626-1632 (1963).
5. H.A. Schenck, "Improved Integral Formulation for Acoustic Radiation Problems," J. Acoust. Soc. Amer. 44(1), 41-58 (1968).
6. F.M. Henderson, "A Structure-Fluid Interaction Capability for the NASA Structural Analysis (NASTRAN) Computer Program," Report 3962, David Taylor Naval Ship R&D Center, Bethesda, Maryland (1972).
7. D.T. Wilton, "Acoustic Radiation and Scattering From Elastic Structures," Int. J. Num. Meth. in Engrg. 13, 123-138 (1978).
8. A.J. Burton and G.F. Miller, "The Application of Integral Equation Methods to the Numerical Solution of Some Exterior Boundary-Value Problems," Proc. Roy. Soc. Lond. A 323, 201-210 (1971).
9. I.C. Mathews, "A Symmetric Boundary Integral-Finite Element Approach for 3-D Fluid Structure Interaction," in Advances in Fluid-Structure Interaction - 1984, edited by G.C. Everstine and M.K. Au-Yang (American Society of Mechanical Engineers, New York, 1984), PVP-Vol. 78 and AMD-Vol. 64, pp. 39-48.

10. I.C. Mathews, "Numerical Techniques for Three-Dimensional Steady-State Fluid-Structure Interaction," J. Acoust. Soc. Amer. 79(5), 1317-1325 (1986).
11. "NASTRAN User's Manual," NASA SP-222(08), Computer Software Management and Information Center (COSMIC), University of Georgia, Athens, Georgia (1986).
12. "MSC/NASTRAN User's Manual," The MacNeal-Schwendler Corporation, Los Angeles, California (1985).
13. G.C. Everstine, "Structural Analogies for Scalar Field Problems," Int. J. Num. Meth. in Engrg. 17(3), 471-476 (1981).
14. G.C. Everstine, "Dynamic Analysis of Fluid-Filled Piping Systems Using Finite Element Techniques," ASME Trans. J. Press. Vessel Tech. 108, 57-61 (1986).
15. R.R. Lipman, "Computer Animation of Modal and Transient Vibrations," Fifteenth NASTRAN Users' Colloquium, NASA CP-2481, National Aeronautics and Space Administration, Washington, DC, 88-97 (1987).
16. G. Chertock, "Integral Equation Methods in Sound Radiation and Scattering from Arbitrary Surfaces," Report 3538, David Taylor Naval Ship Research and Development Center, Bethesda, Maryland (1971).
17. E.A. Schroeder, "A New Block Solver for Large, Full, Unsymmetric, Complex Systems of Linear Algebraic Equations," Report 88/003, David Taylor Research Center, Bethesda, Maryland (1988).
18. H. Lamb, Hydrodynamics, sixth edition, Dover Publications, New York (1945).
19. G.C. Everstine, F.M. Henderson, E.A. Schroeder, and R.R. Lipman, "A General Low Frequency Acoustic Radiation Capability for NASTRAN," Fourteenth NASTRAN Users' Colloquium, NASA CP-2419, National Aeronautics and Space Administration, Washington, DC, 293-310 (1986).
20. G.C. Everstine, F.M. Henderson, and L.S. Schuetz, "Coupled NASTRAN/ Boundary Element Formulation for Acoustic Scattering," Fifteenth NASTRAN Users' Colloquium, NASA CP-2481, National Aeronautics and Space Administration, Washington, DC, 250-265 (1987).
21. H. Huang, and Y.F. Wang, "Asymptotic Fluid-Structure Interaction Theories for Acoustic Radiation Prediction," J. Acoust. Soc. Amer. 77(4), 1389-1394 (1985).
22. F.M. Henderson, "RADSPHERE -- A Computer Program for Calculating the Steady-State, Axially Symmetric, Forced Response and Radiation Field of a Submerged Spherical Shell," Report 87/031, David Taylor Naval Ship Research and Development Center, Bethesda, Maryland (1987).
23. M.C. Junger and D. Feit, Sound, Structures, and Their Interaction, second edition, The MIT Press, Cambridge, Massachusetts (1986).
24. H. Huang, "Helmholtz Integral Equations for Fluid-Structure Interaction," in Advances in Fluid-Structure Interaction - 1984, edited by G.C. Everstine and M.K. Au-Yang (American Society of Mechanical Engineers, New York, 1984), PVP-Vol. 78 and AMD-Vol. 64, pp. 19-38.

VECTOR CONCURRENT COMPUTATIONS OF

EXPLICIT FINITE ELEMENT PROCESSES

Shohei Nakazawa

MARC Analysis Research Corporation
260 Sheridan Avenue, Suite 309, Palo Alto, California 94306

INTRODUCTION

This paper reports observations made through an on-going programming project of a finite element package for viscous, incompressible fluid flow calculations. Attempting to utilize the performance of modern computer architecture to its full extent, experiments are carried out using a prototype software on a machine with the vector processors running concurrently. As the hardware platform, an Alliant FX/8 computer with 4 CE (Computational Elements) is used to measure the performance of the prototype software. The main objective of this exercise is to construct a *fast* and *robust* software for a wide range of fluid mechanics applications. The current version of the prototype code uses an implicit finite element solution for the steady incompressible flows. The possibility of using the explicit finite element processes is being investigated to meet this requirement in particular for the solution of large numerical models in three dimensions which are beyond the limit of implicit finite element processes.

The definition of *super large problems* has been growing in size, perhaps, more rapidly than the evolution of computing machineries. The machines have gotten faster and equipped with a lot more memory, but they have always been *not* sufficient for the solution of largest class of problems imaginable in the computational mechanics. Attempts has been made to extend the limit of hardware by introducing better algorithms and programming techniques, which are faster and less memory intensive for the solution of a given numerical problem. Also the software architecture has been reviewed and modified to stretch the size limit of computer programs. Hardware has also been designed to stretch its limit of the performance, not necessarily with new and improved devices, but implementing a more sophisticated architecture.

With application to science and engineering, the prediction of growth rate for the need of computing power, speculated linearly from the present state-of-the-affairs, tends to underestimate the reality. As we have experienced in the solid mechanics in the past three decades, the progress mode in the computational mechanics had provided better insight to the behavior of the physical system. This, in turn, motivated the advancement of theoretical and experimental mechanics, which then increased the needs for computations much more rapidly than anticipated. The same symptom has started to appear in the area of computational fluid mechanics and electromagnetics. Figure 1 schematically illustrates the growth cycle of the size and complexity of computational mechanics. To design and develop the software capable of incorporating ever increasing demand in such applications, the programming aspects as well as the algorithmic aspects need critical review and improvement to take advantage of the hardware available today and in future.

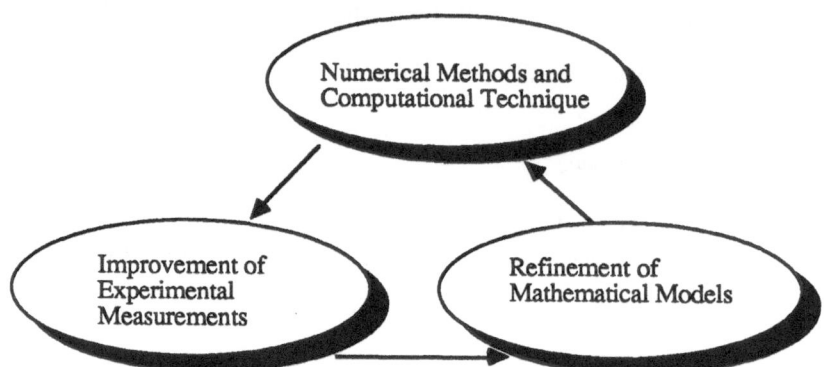

Figure 1. Growth Cycle of Computational Mechanics and Engineering

When the numerical technology for a certain class of difficult problems in engineering and scientifice has matured and become available for applications, we try to solve *quantitatively* more complicated problems using such a well-established method. The extension has however never been straightforward. By tackling large numerical problems using the state-of-the-art technology, we have kept hitting the limit of machine speed and memory space. Also computer system behaved strangely: sometimes due to the hardware/operating system problems when it was pushed to the limit of its performance; often because of the *bugs* which had not been acting up for small-size problems. New algorithms, software architecture and programming techniques (as well as the extra debugging exercise) have been needed to make the breakthroughs for the solution of large problems using the existing numerical techniques.

In the scientific community (among computational mechanicians and numerical analysts), traditionally more emphasis in research has been placed on the formulation and algorithmic development than the programming issues. Amount of literature published on the software architectures and programming techniques is far less substantial than those on the numerical methods themselves. This is because the programming in computational mechanics has not yet been *scientized*; e.g., the formalism and protocol to present observations made through computing experiment is not yet fully established.

The *art and craft of finite element programming* has been discussed sporadically in an *ad hoc* fashion in technical journals despite its significant practical importance. A formal procedure to review the quality of coding has not yet been elstablished in the computational mechanics community. To compare performance of two algorithms by means of CPU time required to solve a benchmark problem, the quality of code often plays more crucial role than those algorithms in question. Simple-minded experimental results may be misleading. Speculations on the performance of algorithms based upon the operation counts without proper coding and experimenting can hardly be trusted (unless there is, at least, two order of magnitude difference clearly visible). The author's recent experience indicates that the performance of an algorithm can differ even by an order of magnitude in CPU time depending on how programs are written. There have not been many computational mechanics articles addressing this point. One of a very few recent exceptions is a operation count reported for various shell elements implemented in DYNA (Hallquist, Benson, 1986B) in conjunction with the detailed performance figures.

The fundamental issues regarding the algorithms and programming are discussed in a recent essay by Fetzer (1988) in which the programming is regarded as a *fallible social process* rather than a strict logical deduction. The program is regarded as a different artifact from the mathematical theory or algorithm it attempts to realize. The theme in Fetzer's article is to seek possibilities and to establish procedures to verify the computer programs. In the field of computational mechanics, the realization of potential numerical performance is one of the most important issues in the program verification. The concerns expressed in this article seems to be quite relevant to the observations made on the programming in computational mechanics and its execution on the real computer systems.

Modern finite element codes, such as NYKE (Hallquist, 1984), DYNA (Hallquist, Benson, 1986A) and PRONTO (Taylor, Flanagan, 1987), achieves their computational efficiency partly due to carefully crafted *Fortran* codes specifically for a certain type of supercomputers such as CRAY. However the computational performance of these codes are often superior performers even on a small machines, such as VAX and PRIME, to the well-established commercial program packages which are originally designed decades ago based upon the architecture acceptable at the beginning of their lives. According to the remark made by Flanagan and Taylor (1987), this speed is gained by the reduced amount of page faults thanks to the well-designed data structure.

In this paper, we discuss programming aspects of finite element computations. An attempt is made to clarify significance of the experimental observations made through the aforementioned programming project. Attention is focused on the techniques used in an explicit finite element program for large scale fluid flow calculations in an commercial environment. We shall also review the past experience of exploiting (or, more precisely attempts to exploit) the full performance of the machines available at the time when the program was written. Some of the techniques (useful then on an old IBM mainframe) are still valid for the modern computing machinaries.

Regardless of the algorithms used, there are utilities and coding rules unique to speeding up the finite element data manipulations, some of which are *invisible* in the algorithmic description of the computational processes. Experimental observations indicate that a significant portion of actual computing time is spent on such operations. The reduction of this computational overhead is on of the most critical issue here. The data structure suitable for unrolling the nested DO-loops in the element assembly is discussed in detail as an example of this argument. Also we look at the trade-offs between the memory usage and the CPU load.

This paper uses fragments of Fortran as a device to outline the ideas, instead of the formal mathematical/algorithmic representations commonly used in technical publications.

EXPLICIT FINITE ELEMENT PROCESSES

We shall refer a class of finite element solution processes to as the *explicit finite element processes*. Those are the finite element solution algorithms and program architectures which do not involve the assembly and factorization of the global finite element arrays. The use of such processes has been very common for the integration of linear and nonlinear hyperbolic equations in solid and fluid mechanics.

The generic form of explicit finite element processes can be written as an *update process* of a global vector

$$u := u + \tau A^{-1} r(u) \tag{1}$$

where u is the vector of nodal variables with A being the preconditioning matrix which is *not* a fully populated finite element array such as the lumped mass matrix and the element-by-element (EBE) preconditioner. τ is a update parameter such as the time step for transient calculations and the relaxation parameter for the iterative update algorithms. The *residual vector r* is calculate in an element-by-element fashion by

$$r(u) = F - \sum_{e=1}^{n_{EL}} r_e(u_e) \tag{2}$$

implying that the computational procedure consists of an element DO-loop and a global vector assembly operation.

The explicit processes have extensively been used in finite element computations in fluids, particularly for convection dominated problems. DYNA 2D and 3D and PRONTO 2D are examples of elplicit code for solid mechanics applications. The conditional stability of this class of numerical methods limits its applications. For a certain calss of problems such as those dealt with by the above code, the advantage of this method, not requiring the assembly and factorization of global finite element coefficient arrays, makes this approach mora favorable option than the implicit ones despite its limitations.

The advantage of time step size being limited only by the *convection velocity over the element size* makes this approach an attractive option for the compressible gas flows (Peraire, Morgan, Zienkiewicz, 1986). For the transient solution of shallow water wave equations in the coastal engineering applications, the explicit scheme has been used extensively for large numerical simulations (Kawahara, Nakazawa, 1981), of which we shall discuss some of the computational aspect in the later section. In case of incompressible flow problems, in which the pressure wave travels at infinite speed, the straightforward explicit scheme suffers from the severe time step size limitation. A family of implicit-explicit scheme has been used to avoid this difficulty in the engineering flow calculations (Brooks, Hughes, 1982).

The exclusive use of linear finite elements is the *only* practical choice in the engineering flow computations. Despite many brave attempts and claims made by finite element researchers, there has *not* been enough evidence that the higher order finite element basis functions can be *upwinded* in a optimal manner as for the linear elements. This is true for both time dependent and independent problems. The lumped mass coefficient, an useful ingredient for setting up the explicit scheme, is readily available and well tested only for these linear elements. The computational implication of exclusive linear finite element approximations is that the length of element vectors and size of element matrices are short and remain unchanged throughout the analysis. In the Fortran coding, this simplicity helps the loop structure of element array manipulations kept far simpler than the conventional solid mechanics codes with many types of elements. The linear elements are under-integrated to avoid the loop over the quadrature points. The kinematic modes are controlled by the additional stabilization mechanism included in the element formulation. These simplifications make the coding strategy far simpler than the numerically integrated finite element assembly processes.

The manipulation of global arrays symbolically written by Equation (1) can readily be vectorized and parallelized by calling the standard vector add and multiply subroutines available on most of the vector and parallel computers. It is straightforward in Fortran programming to achieve the peak performance in the simple array manipulations such as this operation.

The formulation of element array requires the data such as coordinates and nodal variables to be pulled out from the global array to the local element storage area (*gather operations*). And the local element arrays are loaded to the registers for the subsequent algebraic manipulations. Then the result is accumulated in the global array (*scatter operations*). It is obvious from the above schematic computational procedure that the element loop for residual recovery is the most complicated and, perhaps, the most time consuming operation in the explicit finite element processes. The data structure and the element array manipulation subsystem should therefore be designed to take full advantage of the computer architecture particularly in this *element assembly* loop.

DATA STRUCTURE AND CONSTRUCTION OF DO-LOOPS

As pointed out in the recent article by Cowell, Thompson (1986), the bottleneck for the computations on vector machines is the memory access. The results reported therein show that by unrolling the DO-loops and, reducing the data load/store operation, improves significantly the performance of the Fortran program running on CRAY.

In the explicit finite element computations, the assembly of element array and its multiplication with *gathered* element variables is the most time consuming part. The data transfer from the global nodal array to the local element variable storage involves $NELVAR$ load operations, with $NELVAR$ being the number of variables per element with reference to the *element connectivity table*. This operation is referred to as the *gather* operation. For the efficient addressing by this connectivity table, instead of commonly used the *nodal connectivity* array which stores the node number in the form

$$LCONNC (NELNOD, NETOTL) \qquad (3A)$$

where $NELNOD$ and $NETOTL$ denote the number of node per element and the total number of elements respectively, the use of the *degree-of-freedom connectivity table*

$$LCONNC (NELVAR, NETOTL) \qquad (3B)$$

reduces the number of address calculations in each element assembly. The Expression (3B), however, increases the storage requirement for this particular integer array by (neldof-1) times. The global nodal variables are stored accordingly. Instead of the conventional expression

$$GLBVAR (NELDOF, NETOTL) \qquad (4A)$$

we explicitly declare the same array in one dimensional manner

$$GLBVAR (NELDOF*NETOTL) \qquad (4B)$$

This storage scheme of global variable array simplifies the data structure. Furthermore this data structure naturally unrolls the loop over the degree-of-freedom which is the operation performed quite frequently in the finite element programs. To reduce and simplify this bookkeeping, we often use a conversion table or a function subprogram to map the element nodal degree-of-freedom to the global array address, which requires additional computations or at least data load and store operations. In this data storage scheme, the address calculation for the nodal degree of freedom needs to be performed only once as a preprocessing of the finite element topology data.

First, let us briefly review the control structure and array manipulations based on the connectivity storage characterized by Expression (3A). In the element assembly loop, the control structure dictated by this data structure is symbolically written as shown in Listing 1.

This convoluted nested loop structure involves a number of address calculations as well as the setting-up operations for the small loops. Note that the length of vectors appears in the inside of element loop is relatively short (typically 8 for two dimensional computations and 24 for three dimensional computations, which is split into 4 by 2 and 3 by 8 two dimensional arrays) and relative hard to take advantage of vector registers. Mizukami (1988) reports that the significant performance improvement is achieved by the modification of element assembly loop to circumvent the disadvantage of such short nested loops. See Listing 2.

Listing 1   Conventional Loop structure

```
      do 1020 ielem=1,netotl
        do 1010 inode=1,nelnod
        nodeid=lconnc(inode,ielem)
          do 1000 idof=1,ndof
          elmvar(idof,inode)=glbvar(idof,nodeid)
 1000        continue
 1010    continue
       /* codes to assemble element arrays */
 1020 continue
```

Listing 2   Inverted Loop Structure

```
        do 1010 inode=1,nelnod
      do 1020 ielem=1,netotl
        nodeid=lconnc(inode,ielem)
          do 1000 idof=1,ndof
          elmvar(idof,inode)=glbvar(idof,nodeid)
 1000        continue
       /* codes to assemble element arrays */
 1020 continue
 1010    continue
```

By unrolling the internal 1020 loop, the vector length now becomes the total number of elements and the performance is improved by over an order of magnitude on Fujitsu vector processor VP50.

By implementing the Expression (3B) for the connectivity array, the element loop can be simplified significantly as shown in Listing 3. Further unrolling the inner 1010 loop would eliminate the loop structure in the element assembly operations which are not long enough to take advantage of the vector facilities available on the modern computing facility. By unrolling all the internal loops including nested element array calculations, the overall element assembly loop can now be vectorized as exercised in DYNA and other modern finite element structural analysis code.

Listing 3   Simplified Loop with D.O.F. Connectivity Array

```
      do 1010 ielem=1,netotl
         ndofid=lconnc(inode,ielem)
            do 1000 idof=1,nelvar
            elmvar(idof)=glbvar(ndofid)
 1000       continue
         /* codes to assemble element arrays */
 1010 continue
```

The exclusive use of linear elements enables us to unroll *all* the inner loops in element assembly processes. The number of nodes per element, number of degree of freedom per node and number of coordinate entries per node, which are used as control variable for inner loops, remain invariant throughout the analysis. Indeed when the number of repetition in the loop is known *a priori* there is no positive reason to construct loop structure which possibly slows down the execution of finite element codes.

The elimination of inner short loops in the element assembly process improves the performance of the finite element code significantly on the scalar machines as well. It is speculated that the floating point hardware cuts down the CPU time spent on the actual matrix assembly operation considerably and, as the result, the fraction of integer and logical operations in the execution of short DO-loops become more significant. Thus the improvement of loop constructs becomes visible even on the scalar machines.

In the gather operation, the large stride of address in the global array impede the performance of the element assembly loops, causing unnecessary page fault on the virtual storage machine and excess updates of the data cache, if the hardware is equipped with such devices. The numbering scheme for the nodes and elements affects considerably the speed of data transfer. A simple renumbering scheme based upon the geometrical configuration of the finite element model often improves the throughput of large finite element problems.

An experimental evidence indicates that the data transfer between the memories cost considerable amount of computing time. Using the implicit code in which the frontal solution is used for the solution of algebraic system of equations (Irons, 1970), effects of parallelization is studied in the data transfer operations. The solver uses the working space allocated in the blank common area to store the factor calculated for each row of the global coefficient matrix. The front matrix is stored also in the lower address of the same working space. After a row is eliminated, the factor residing in the front matrix is copied into the work space. To perform this write operation, the data residing in the register is copied into the cache and then to the main memory. Using the Alliant Fortran compiler option, this copy operation is selectively parallelized whereas the rest of the code is optimized for the vectorization for this experiments. The results indicate that the 20% of the code is running in parallel for this experimental setting. Indeed this observation qualitatively agrees with the investigation reported by Cowell and Thompson (1987).

A guideline to be set from this experimental observation is that the update of data in the space explicitly allocated in the main memory are to be kept minimal. The dynamic variables used locally does not necessarily stored in the main memory. (Of course, it entirely depends on how the compiler allocates space for those variables, but the codes compiled without the static

157

data storage option seem to run faster. This is a speculation, since no systematic experiments have performed yet.) The memory allocation of element work area such as the discrete gradient matrix (B matrix in the structural analysis) and element stiffness matrix would possibly impede the performance of the finite element computation.

It is interesting to note that the frontal solution procedure used in this numerical experiments runs about 4 times faster in the vector mode than the scalar mode using one vector processor on Alliant FX/8. Note that the Fortran code for this operation is streamlined to reduce the computational overhead. In particular, the connectivity data defined by Expression (3B) is crucial to remove some on unvectorizable loops in the frontal solution process. The result indicates over 90% of the operations are vectorized on this particular hardware platform. Even on the scalar computer, this highly tuned frontal solution runs several times faster than the standard textbook code.

AN EXAMPLE OF LARGE SCALE EXPLICIT FINITE ELEMENT COMPUTATIONS

Figure 2 shows a finite element model used for the numerical simulation of tsunami wave in Tokyo bay consisting of 3,366 nodes and 6,327 linear triangle elements resulting in a system on nonlinear equations with 10,098 unknown variables (Kawahara and Nakazawa, 1981). The computation was carried out in 1977 using IBM 370/168, one of the largest and fastest machines available then. The finite element program used for this analysis was fine-tuned for IBM systems and plug compatible machines. The code was also ported to CDC 6600. Despite the significantly different architecture of these machines, the programming techniques used to improve the performance were more or less the same.

The principle followed in the coding was to minimize the reference to the memory and avoid the page fault (on IBM systems) /the access to the extended core memory (on CDC systems). The exclusive use of linear triangle elements allowed us to unroll all the inner loops for the explicit time integration. A storage scheme is designed in this finite element computation to take advantage of *finite difference like* gridwork. Noting that in Eulerian computations the element geometry remains unchanged throughout the analysis, a preprocessor is devised to reduce the storage requirement. Elements with the same geometry are grouped together and element arrays are precalculated for each element group. The grouping operation is a post-processing and can be performed at the same time as the input data is being checked. In the time integration loop, the residual is calculated without directly referring to the connectivity and coordinate arrays given as the input data except for the gather and scatter operations for the field variables. Figure 3 illustrates the data structure for this computation. As seen from this chart, The major portion of the expensive floating point operations is eliminated from the element loop as well as the reference to some of the data arrays.

One of the disadvantage of finite element methods over the finite differences is the speed. This is partly due to the complication of the data structure and the operations required to assemble the element arrays. By removing most of these complicated arithmetic operations unique to the finite element computations, the above data structure improves the competitiveness of the explicit finite element computations against the well established finite difference methods. The extension of this approach with the self-adaptive mesh refinement would boost the performance of finite element computations beyond the other numerical solution technology.

In the computation reported above, the number of element groups are less than 80, and by storing all the variable in the single precision (32 bits/word) on IBM machine, the entire analysis could be squeezed into a mere 512 Kbytes of the working space. On IBM 370/168, about 2,000 elements are processed in the time integration loop per CPU second. Major problem encountered then was the disk space required to store the time history of all the filed variables at every time step.

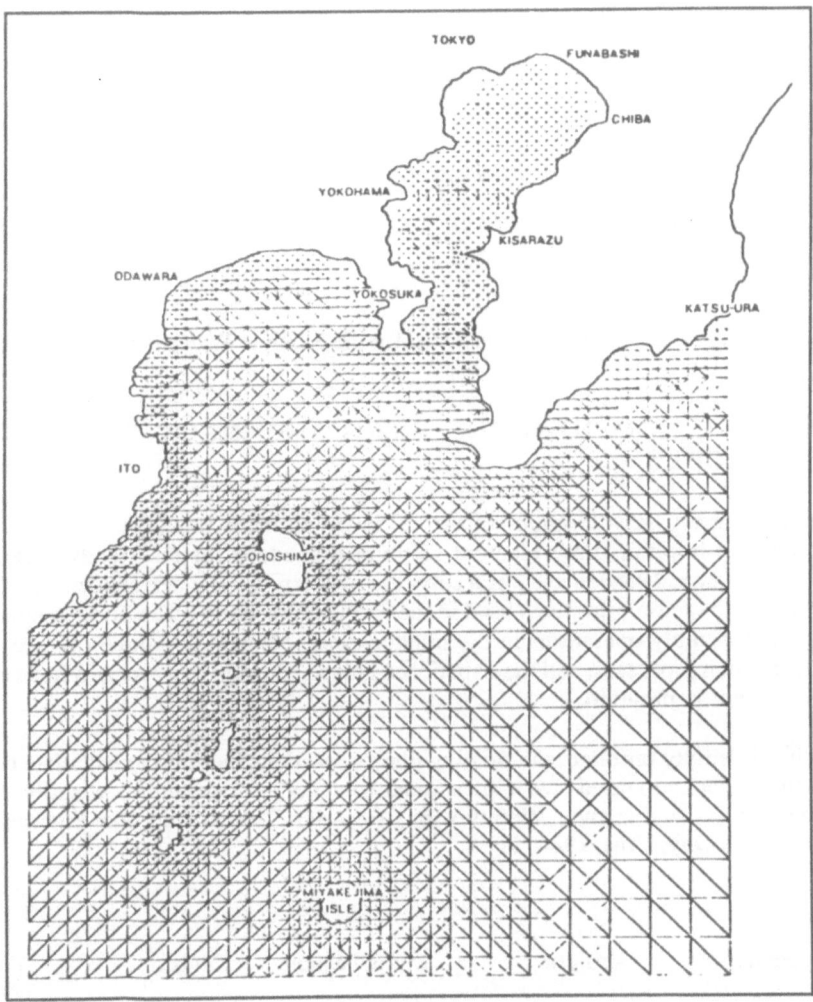

Figure 2. Finite Element Model of the Tokyo Bay (Kawahara and Nakazawa, 1981)

Even with the current technology, the capacity and speed of storage devices get overwhelmed by the processor's number crunching performance. When the computations are carried out, the numerical results are to be stored for post-processing and many other purposes. As we shall discuss in the next section, the speed of Fortran read/write operation to and from the disk drive is indeed one of the major bottlenecks in the modern computing machineries. Also moving the data around in various storage devices in the computer itself is time consuming and need serious look and optimization. Particularly for the large computational mechanics exercise, a fast sequential input/output device with sufficiently large record size would improve the throughput of the number crunching devices. The recent magneto-optics technology and other mass storage ideas needs to be investigated for these applications.

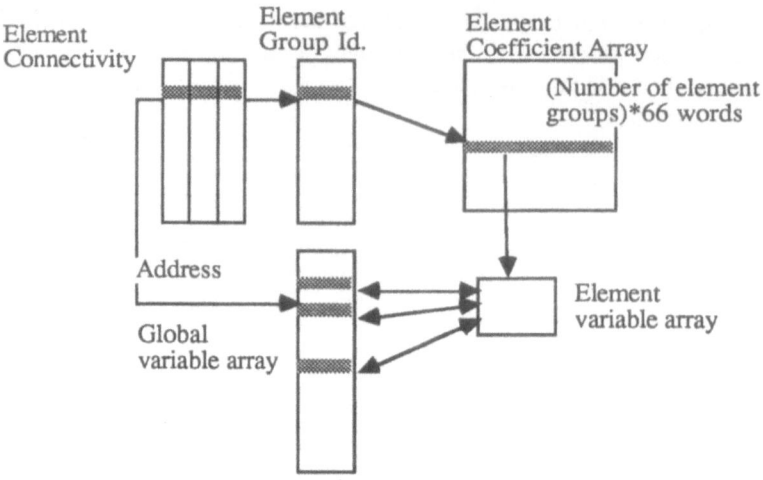

Figure 3. An Addressing scheme Used for the Shallow Water Wave Equations

LARGE SCALE COMPUTATIONS USING THE IMPLICIT FINITE ELEMENT CODES

In the current development effort of finite element package for the fluid flow computations, an implicit finite element solution procedure for the steady Navier-Stokes equations (Nakazawa, 1988) is used as a software platform for the performance evaluation of the element manipulation primitives used in the element assembly. The book-keeping and the element array assembly operations in the implicit finite element computations are similar to those in the explicit processes, and the experience gained is useful to develop a fast and reliable explicit finite element program.

When it is fully unrolled, the floating point operations required to calculate an element array for the under-integrated two dimensional four node quadrilateral with the balancing dissipation (Kelly, Nakazawa, Zienkiewicz, Heinrich, 1980) and the kinematic mode stabilization (Liu, Ong, Uras, 1984) is

Operation	Addition	Multiplication	Division	Function Evaluation
Isoparametric Mapping	34	10	4	
Upwinding	33	24	3	4
Kinematic Mode Stabilization	36	41	2	
Element Formulation	416	960		
Total	519	1035	9	4

On the Alliant FX/8 with one processor, the element assembly loop is executed at a rate of about 700 element/CPU second resulting in about 1.1 million floating operations per second which is about the same performance as the execution speed reported in the performance comparison table compiled by Dongarra (1986) for the linear equation solution with all Fortran code. This indicate that to achieve peak performance of the vector CPU (about 6 MFLOPS), futher stramlining is necessary. A work is still in progress, strictly using the standard Fortan code, to further vectorize the element fomulation on this particular hardware platform.

To assess the effects of miscllaneous computational overhead other than the main finite element solution, a timing study is carried out using the recirculating cavity flow problem consisting of 2,500 element and 2,601 nodes. This is not the largest problem one can handle using the present state of the art technology, but large enough that the accurate timing data are obtainable on this hardware platform.

The prototype code used in this numerical experiment is designed to read a formatted ASCII file as the input date deck. Before starting the execution, a large number of data lines needs to be read and interpreted accordingly. Relative computing time of 0.115 (normalized by the time consumed for matrix assembly and solution) is taken to accomplish this task. Before the finite lement data manipulation, au out-of-core database is built by this prototype finite element code. The time used for setting up a binary direct access file, of which the total number of records is the sum of the numbers of elements and nodes (5101 records in this example) is about 0.334. These figures indicate that the file manipulation takes considerable amount of computing time almost 1/2 of one implicit finite element solution.

Before starting the implicit solution, the elimination table needs to be built for the element connectivity data. The relative computing time of 1.00 is consumed to set up the reverse connectivity table and elimnation sequence for the frontal solution. The modern finite element data manipulation schemes such as the profile solver and element-by-element factorization uses the connectivity table in a similar manner as the frontal solution and this outstanding overhead may become significant as the size of problem grow. Note that the operation discussed here is not optional such as the front matrix size optimization, but a mere table search to find the sequence of forward reduction in the frontal solution process.

In the process of setting up the reverse conncetivity table, a serach to fine the maximum number of element connected to a node is the vital information for book-keeping purposes. A major trade off of the computing time and memory usage is observed in an attempt to cut down computing time for this operation. A naive implementation of this search process is

```
Initialize Max_element_at_node;
Loop over node;
     Initialize Element_at_curent_node
     Loop over element;
          Scan the element connectivity list
          and if entry = current_node
          Update Element_at_current_node;
If Element_at_current_node > Max_element_at_node
Update Max_element_at_node;
```

going through integer IF tests as many as total number of nodes times number of nodes per element times total number of elements. This naive implementation often costs several times more CPU time that the implicit finite element solution. Such significant drawback of integer book-keeping operation was not visible on the computers of previous generation. An order of magnitude of improvement in CPU time is realized by simply adding an integer array

```
        Initialize Array_of_element_at_node;
    Loop over element;
        Loop over node;
            Update entry of array_of_element_at_node;
    Find maximun entry in array_of_element_at_node;
```

This version reduces the loop operation down to only the total entry in the connectivity table with if statement executed on total number of node times. The figure mentioned above is obtained with the modified search procedure. It is highly probable that the modified version of this table search could be vectorized by copying integer arrays into the real array and using the vectorization facility (the vectorization of the naive version makes a very little sense because the loop structure is not optimal, but such an attempt to optimize integer data manipulation has been exercised by many finite element programmers).

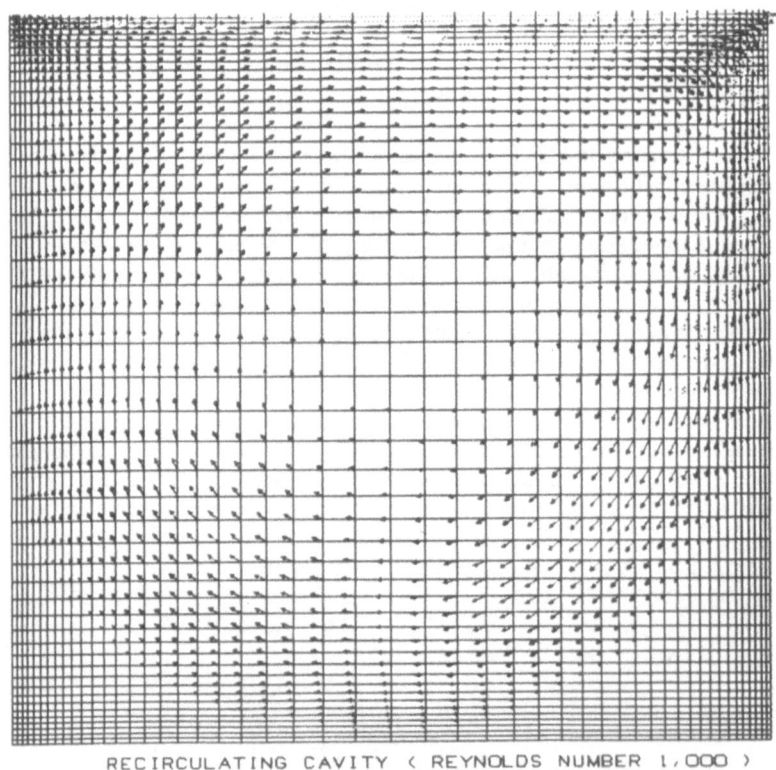

RECIRCULATING CAVITY (REYNOLDS NUMBER 1,000)

Figure 4. Finite Element Mesh and Velocity Field for the Recirculating Cavity Flow
Re=1,000

After the numbers are crunched, the field variable are written onto a formatted file for post processing and visual inspection of the results. This phase of operation costs normalized time 0.17 in the total computation process. The actual wall time for this operation differ siginificantly depending on the usage of the particular disk system to which the result is sotred.

In the explicit finite element computations, the rate at which the numbers are crunched is about 40 times faster (if the resut were stored at every time/iteration step), and proportion of input/output operation in the total computing time grows substantially. This is one of the important areas where the more attention needs to be paid by the finite element programmers.

RECIRCULATING CAVITY (REYNOLDS NUMBER 1,000)

Figure 5. Pressure Field in the Recirculating Cavity Flow (re=1,000)

It is interesting to note that the time used by the post-processing code for the graphic representation of the result as shown in Figures 4 and 5 is comparable or more than the single implicit finite element solution. Considerable amount of computing time is used for the disk file input and data processing in this operation. This observation indicates that, for handling large finite element problems, the data management and the computational procedure for pre- and postprocessing program packages also need to be reviewed and streamlined for the maximum efficiency.

In the production runs of impicit structural analysis codes on the vector concurrent machines, it is often observed that the overhead associated with the file handling and integer book keeping operation overwhelms the actual finite element analysis procedure. It seems necessary to review algorithms used for those operations and streamline the codes for better performance on the modern computers.

CONCLUDING REMARKS

The author's experience indicates that the effective use of the heirachy of strage devices, the scalar and vector registors, high speed cache, main memory and disks is one of the most crucial ingredients in the efficient finite element programming. The storage scheme based on the *nodal-degree-of-freedom* connectivity discussed in this paper seems the most versatile internal data structure for finite element computations to come up with efficient finite element programs providing a natural framework to unroll the element assembly loops.

It is emphasized here again that the programming aspects of finite element computations are equally important to the algorithmic development to fully utilize the capability of modern computing machineries.

A simple example of utility for the integer table search operation indicates that there are rooms in the finite element programs, either implicit or explicit, to cut corners for the beter performance. In particular, it is a nice, if too simple, example to demonstrate the principle that the use of extra memory pays off, i.e., if there is a trade off between memory usage and computing time, it may be wise to go for the more memory intensive option.

ACKNOWLEDGEMENT

The provision of computing facility by MARC Analysis Research Coroporation is gratefully acknowledged. The views, opinions.and methods presented in this paper do not necessarily reflect those of MARC Analysis Research Corporation nor its commercial finite element program packages.

REFERENCES

Brooks,A.N. and T.J.R.Hughes (1982), Streamkine Upwind/Petrov Galerkin Formulation for Convection Dominated Flows with Particular Emphasis on the Incompressible Navier Stokes Equations, *Comp.Meth.Appl.Mech.Eng.*, 32, 199-259.

Cowell,W.R. and C.P.Thompson (1986), Trasnsforming Fortran DO Loops to Improve Performance on Vector Architecture, *ACM Trans.Math.Software*,12, 324-353.

Dongarra,J.J. (1986), *Performance of Various Computers Using Standard Linear Equations Software in Fortran Environment*, Argonne National Laboratory, Methematics and Computer Science Division, Technical Memorandum No.23.

Fetzer,J.H.(1988), Program Verification: The very Idea, *Comm.ACM,* 31, 1048-1063.

Flanagan,D. and L.M.Taylor (1987), Structuring Data for Concurrent Vectorized Processing in a Transient Dynamics Finite Element Program, (ed.A.K.Noor) *Parallel Computations and Thier Impact onnMechanics*, ASME, AMD Vol.86.

Hallquist,J.O. (1984), *NIKE 3D: An Implicit, Finite-Deformation, Finite Element Code for*

Analyzing the Static and Dynamic Response of Three-Dimensional Solids, University of California, Lawrence Livermore National Laboratory, Report UCID-18822.

Hallquist,J.O. and D.J.Benson (1986A), A Comparison of an Implicit and Explicit Implementation of the Hughes-Liu Shell, (eds. T.J.R.Hughes and E.Hinton) *Finite Element Methods for Plate and Shell Structures, Volume 1: Element technology*, Pineridge Press International, Swansea, Chap.16, pp394-431.

Hallquist,J.O. and D.J.Benson (1986B), *DYNA 3D Users Manual (Nonlinear Dynamic Analysis of Structures in Three Dimensions*, University of California, Lawrence Livermore National Laboratory, Report UCID-19592.

Irons,B.M. (1970), A Frontal Solution Program for Finite Element Analysis, *Int..J. num.. Meth.Eng.*, **2**, 5-32.

Kawahara,M. and S.Nakazawa (1981), Finite Element Method for Unsteady Shallow Water Wave Equation, (eds. J.F.Abel, T.Kawai and S.-F.Shen) *Interdisciplinary Finite Element Analysis*, Cornell University, Ithaca.

Kelly,D.W., S.Nakazawa, O.C.Zienkiewicz and J.C.Heinrich (1980), A Note on Upwinding and Anisotropic Balancing Dissipation in Finite Element Approximations to Covection Diffusion Problems, *Int.J.num.Meth.Eng.*, **15**, 1705-1711.

Liu,W.K, J.S.-J.Ong and R.A.Uras (1985), Finite Element Stabilization Matrices - A Unification Approach, *Comp.Meth.Appl.Mech.Eng.*

Mizukami,A. (1988), Private Communication, Nippon Kokan, Applied Technology Research Center.

Peraire,J., K.Morgan and O.C.Zienkiewicz (1986), Convection Dominated Problems, (eds. T.E.Tezduyar and T.J.R.Hughes) *Numerical Methods for Compressible Flows-Finite Difference, Element and Volume Techiniques*, ASME AMD Vol.78, pp129-147.

Robichaud,M.P. and P.A.Tanguy (1987), Finite Element Solution of Three-Dimensional Incompressible Fluid Flow Problems by a Preconditioned Conjugate Residual Metho, *Int.J.num.Meth.Eng.*,**24**, 447-457.

Taylor,L.M. and D.P.Flanagan (1987), *PRONT 2D, A Two DImensional Transient Dynamics Program*, Sandia Report SAND86-0594.UC-32.

TAYLOR'S VORTEX ARRAY: A NEW TEST PROBLEM

FOR NAVIER-STOKES SOLUTION PROCEDURES

Daniel T. Valentine and A. Gaber Mohamed[1]

Mechanical Engineering Department
Clarkson University
Potsdam, N.Y. 13676

ABSTRACT

A direct simulation of the exact, nonparallel flow solution of the Navier-Stokes equations published by G.I. Taylor in 1923 was performed. The solution is a two-dimensional double array of vortices which decay exponentially with time. The unsteady Navier-Stokes equations were solved numerically for a subdomain of four vortices along the boundaries of which the exact solution was specified. The simulation was compared with Taylor's solution to verify a new computational method. A moderate size (1200 lines) program was written to solve a system of coupled, time dependent, nonlinear partial differential equations. The algorithm developed implements a variable time step Crank-Nicolson method with Richardson's extrapolation to discretize the timewise integration. It implements a 4th-order collocation method to solve the linear elliptic problems at the core of the computations. An iterative method is applied to solve for the nonlinearities. The coded algorithm was designed to solve large scale fluid dynamics problems. It was applied to solve the Navier-Stokes equations and, in particular, to simulate the flow in a subdomain of Taylor's vortex array. This simulation was a tests to verify the *new* 4th-order in time, 4th-order in space Navier-Stokes code. This code was executed on an Alliant FX/80 superminicomputer and on the IBM 3090-600E supercomputer. The simulation reproduced the exact solution 'exactly,' *i.e.*, to within the precision of the machine.

INTRODUCTION

The analysis of many problems of incompressible fluid dynamics involves solving a system of unsteady, nonlinear simultaneous partial differential equations. This system of equations includes the Navier-Stokes and continuity equations. In many instances approximate solutions can only be obtained by solving the equations numerically on a digital computer. Because the equations are nonlinear and the behavior of fluids in motion past arbitrarily shaped boundaries is quite complex, the discrete analogues that must be solved require the solution of a large number of simultaneous equations at a large number of time steps. This is particularly so if accurate solutions are sought; hence, as is usually the case, the speed and storage capacity of existing computers limit the accuracy that is realistically obtainable.

The National Research Council (NRC), at the end of 1986, reported that the most pressing

---

[1] Permanent address : Military Technical College, Cairo, Egypt.

need today is for algorithm technology that will lead to practical methods for solving various forms of the Navier-Stokes equations for arbitrary geometries with reliable and predictable accuracy; see Bradley *et al* (1986). In the NRC report, it was also mentioned that there is a need to do direct simulation instead of solving the averaged Navier-Stokes. These needs must be satisfied before any *practical* viscous flow problem can be solved with any degree of reliability.

Orszag and Israeli (1974) pointed out that the principal role of computers in fluid dynamics should be to give physical insight into dynamics. They stated that: "In order to give physical insight, the simulation themselves should be above reproach. All reasonable standards of accuracy should be maintained and the range of applicability of any model should be clearly and explicitly understood before embarking upon a series of lengthly simulation experiments with increasing physical complexity." Hence, high-order computational methods are both desirable and preferred.

The computational algorithm developed in the present investigation was designed to solve large scale fluid dynamics problems. There are two classes of flow problems that motivated the development of this *new* algorithm. One class is modeled by the Reynolds averaged Navier-Stokes equations. These are problems in which time-averaged properties are to be predicted. (When the *new* algorithm is used to solve this class of problems, the time is interpreted as an iteration parameter.) The goal is to develop a predictive capability to solve viscous, incompressible turbulent flows in fluids engineering problems, *e.g.*, to predict the forces acting on afterbodies of bodies-of-revolution moving through incompressible, viscous fluids. The number of simultaneous equations to be solved increases from two for the laminar flow model to at least four for a two-equation turbulence model. This increase in the number of equations in conjunction with the increase in grid density that is required (particularly in the boundary layer regions near solid boundaries) to simulate turbulent flows at moderate to high Reynolds numbers makes these problems large scale problems in computational mechanics.

One of the main problems of turbulence modeling is to ensure that the effects of turbulent diffusion are not masked by numerical diffusion particularly in regions where the flow properties vary rapidly. The 4th-order scheme developed in the present investigation does ensure this. The principal advantage of the *new* algorithm is that the user prescribes the accuracy of the solution sought. The variable time stepping subroutine ensures that excessive computer time is minimized for a prescribed accuracy of solution. In addition, the capabilities of parallel processors are utilized to a great extent.

The other class of problems for which this algorithm was designed to compute is the direct simulation of unsteady incompressible viscous flows. The direct simulations of turbulent flows are indeed super large scale problems. At the NASA Ames Research Center Kim, Moin and Moser (1987) did the first detailed direct simulation of a turbulent channel flow for a Reynolds number of 3300. This Reynolds number is based on the mean centerline velocity and the channel half-width. The number of grid points computed in the x, y, z directions were $192 \times 129 \times 160$, respectively, or about 4×10^6. In their computations all of the essential turbulence scales were resolved on the computational grid and no subgrid model was used. The paper by Kim *et al* illustrates *a fortiori* the limitations of the most powerful computers available today in solving turbulent flows by direct simulation. Their numerical method was based on applying a spectral method in the spatial directions and applying a semi-implicit scheme for time advancement (*i.e.*, Crank-Nicolson for the viscous terms and Adams-Bashforth for the nonlinear terms). Their channel flow computations took about 40 seconds per time step. For the simulation of 10 nondimensional time units it took approximately 250 CPU hours on the NASA Ames CRAY-XMP. The accuracy of their code was examined by numerical experiments. They computed the evolution of small-amplitude oblique waves in a channel. Both decaying (stable) and growing (unstable) waves were tested for appropriately specified, critical flow conditions. They reported that the growth and decay rates were predicted (by their code) to within $10^{-4}\%$ of the values predicted by linear theory; hence, their code was validated.

The computational scheme described in this paper provides an alternative method to solve turbulent flows by direct simulation. It is not clear if the *new* method is faster than the method applied by Kim *et al* (1987); however, in the *new* method the user has direct control over the accuracy. This feature should be advantageous if the required error bound is selected judiciously by the user. The new algorithm is 4th-order in time and 4th-order in space. It uses the unconditionally stable implicit Crank-Nicolson method, which leads to a discrete system of nonlinear elliptic equations. At each time step an iterative scheme is applied to solve the nonlinear system by solving, at each iteration, linear elliptic problems. At each iteration the linear elliptic problems are solved by applying a 4th-order collocation method. Richardson's extrapolation is used at the end of each time step to reduce the error to 4th-order in time. A computer code called NSC (Navier-Stokes Collocation), that uses this computational scheme, was developed by the authors to study the axisymmetric, separated, laminar flow over an inner-radius annular back-step; see Valentine, Mohamed and Hessel (1988). The coded method allows the user to prescribe the maximum allowable error to be accepted in the computations. To verify the coded scheme, Taylor's vortex array, which is an exact solution of a nonparallel flow that was published by G.I. Taylor (1923), was selected as a test problem. This solution is sufficiently complex so that the effect of convective transport across a finite-difference or finite element cell due to a velocity vector that is not perpendicular to the cell face may be investigated. The computational results presented in this paper are from numerical simulations of this test problem.

The principal contributions of the present paper are: (1) An outline of a computational algorithm to solve unsteady, nonlinear simultaneous equations and, in particular, the Navier-Stokes and continuity equations. (2) An application of the computational algorithm to simulate a nonparallel, laminar flow to verify the coded method.

MATHEMATICAL MODEL

The model equations for laminar flow problems are the Navier-Stokes equations. For two-dimensional flows the basic equations describing the motions in Cartesian coordinates are, in dimensionless form:
Conservation of mass (or continuity):

$$\frac{\partial u}{\partial x} + \frac{\partial w}{\partial z} = 0. \tag{1}$$

Momentum equation in the horizontal direction:

$$\frac{\partial u}{\partial t} = -\frac{\partial P}{\partial x} - u\frac{\partial u}{\partial x} - w\frac{\partial u}{\partial z} + \frac{1}{\mathcal{R}}\left(\frac{\partial^2 u}{\partial x^2} + \frac{\partial^2 u}{\partial z^2}\right). \tag{2}$$

Momentum equation in the vertical direction:

$$\frac{\partial w}{\partial t} = -\frac{\partial P}{\partial z} - u\frac{\partial w}{\partial x} - w\frac{\partial w}{\partial z} + \frac{1}{\mathcal{R}}\left(\frac{\partial^2 w}{\partial x^2} + \frac{\partial^2 w}{\partial z^2}\right). \tag{3}$$

In these equations the characteristic velocity is U, the characteristic length is d, and the characteristic time is d/U. The dimensionless pressure is $P \equiv p/(\rho U^2)$ in which ρ is the mass density of the fluid and p is the dimensional pressure. The parameter $\mathcal{R} \equiv Ud/\nu$ is the Reynolds number in which ν is the kinematic viscosity of the fluid.

For two-dimensional flow fields these equations may be simplified as follows. In this case there is only one, finite component of the vorticity vector, *viz*, $\omega\mathbf{j}$. From the definition of the vorticity, which is $\underline{\omega} \equiv \nabla \times \mathbf{u}$, we may write the following:

$$\omega = \frac{\partial u}{\partial z} - \frac{\partial w}{\partial x}. \tag{4}$$

If we take the z-derivative of Equation (2), the x-derivative of Equation (3), and apply the definition of the vorticity given by Equation (4), we may write the following expression for the transport of the vorticity, ω.

$$\frac{\partial \omega}{\partial t} = -u\frac{\partial \omega}{\partial x} - w\frac{\partial \omega}{\partial z} + \frac{1}{\mathcal{R}}\left(\frac{\partial^2 \omega}{\partial x^2} + \frac{\partial^2 \omega}{\partial z^2}\right). \qquad (5)$$

As a consequence of the continuity equation, Equation (1), we may define the stream function, ψ, in terms of the velocity field as follows:

$$u = \frac{\partial \psi}{\partial z}, \quad \text{and} \quad w = -\frac{\partial \psi}{\partial x}. \qquad (6)$$

Substituting (6) into (5), we may write the vorticity transport equation as follows:

$$\frac{\partial \omega}{\partial t} + \frac{\partial \psi}{\partial z}\frac{\partial \omega}{\partial x} - \frac{\partial \psi}{\partial x}\frac{\partial \omega}{\partial z} = \frac{1}{\mathcal{R}}\left(\frac{\partial^2 \omega}{\partial x^2} + \frac{\partial^2 \omega}{\partial z^2}\right). \qquad (7)$$

Substituting (6) into (4), we obtain

$$\omega = \frac{\partial^2 \psi}{\partial x^2} + \frac{\partial^2 \psi}{\partial z^2}, \qquad (8)$$

which is an elliptic differential equation, *viz*, a Poisson equation, that relates the vorticity to the stream function. Equations (7) and (8) are the field equations for the vorticity-stream function model of the two-dimensional, incompressible flow of a Newtonian fluid. These equations form a complete set of two equations in two unknowns, *viz*, ω and ψ.

Once the stream function and vorticity fields are known, we may evaluate other flow properties, *e.g.*, the velocity field by applying the set of Equations (6) and the pressure field by applying the Poisson equation derived by taking the x-derivative of Equation (2) and adding to it the z-derivative of Equation (3). The equation for the pressure is given below. The principal advantage of the vorticity-stream function formulation is that the pressure is eliminated from the prediction procedure for the velocity field. The pressure is subsequently computed by solving the Poisson equation

$$\frac{\partial^2 P}{\partial x^2} + \frac{\partial^2 P}{\partial z^2} = -2\left[\left(\frac{\partial^2 \psi}{\partial x \partial z}\right)^2 - \frac{\partial^2 \psi}{\partial x^2}\frac{\partial^2 \psi}{\partial z^2}\right]. \qquad (9)$$

This completes the set of field equations required to solve for the viscous flow in a prescribed domain with prescribed boundary conditions.

THE TEST PROBLEM

G. I. Taylor in 1923 published the following exact solution of the Navier-Stokes equations written in terms of the stream function and vorticity, *viz*, Equations (7) and (8). It is unique in that it is for a time-dependent, non-parallel flow. Taylor assumed that $\omega = f(\psi)$; hence, the nonlinear terms in Equation (7) cancel. As a special case he selected $\omega = -k\psi$, where k is an arbitrary constant. The example he presented is as follows:

$$\psi = B\cos\pi x \cos\pi z\, e^{-2\pi^2 t}, \qquad (10)$$

where $U \equiv \nu/d$ and $\mathcal{R} \equiv Ud/\nu = 1$. This flow may be interpreted as a double array of vortices which decay exponentially in time. This solution is for the infinite domain, $-\infty < x < \infty$, $-\infty < z < \infty$. Also, the time varies between $0 < t < \infty$. Figure 1 illustrates this flow field.

The test problem is an extension of this problem that confines our attention to a subdomain along the boundaries of which the exact solution is specified. In addition, the initial condition specified within the subdomain is the exact solution at $t = 0$ as given by Equation (10). The computational method is subsequently executed to simulate the decay process of a finite number of adjacent vortices. The quality of the numerical method is judged in terms of how well the numerical predictions match the exact solution as time increases, *i.e.*, for $t > 0$. The mathematical statement of the test problem is given next. How it is applied to help verify a particular computational method is described in a subsequent section.

The boundary-value problem posed requires the solution of Equations (7) and (8) in the domain $-\frac{n_1}{2} < x < \frac{m_1}{2}$, $-\frac{n_2}{2} < z < \frac{m_2}{2}$, where n_1, m_1, n_2 and m_2 are odd, positive integers that are specified by the user depending upon the number of vortex cells desired for the evaluation. If $n_1 = n_2 = 1$ and $m_1 = m_2 = 3$, then four cells are considered; they are illustrated in Figure 2 for $t = 0$. The boundary conditions for this case are as follows:

- $\psi = 0$, $\omega = 0$ for $-1/2 < x < 3/2$, $z = -1/2$.
- $\psi = 0$, $\omega = 0$ for $x = -1/2$, $-1/2 < z < 3/2$.
- $\psi = 0$, $\omega = 0$ for $-1/2 < x < 3/2$, $z = 3/2$.
- $\psi = 0$, $\omega = 0$ for $x = 3/2$, $-1/2 < z < 3/2$.

Note that $\omega = -2\pi^2\psi$. The initial and boundary conditions for ω are found by substituting the initial and boundary conditions for ψ into this expression. The initial condition for ψ is given by Equation (10) by setting $t = 0$. The transient solution for $t > 0$ is sought. $B = 1$ was used for the test case computed in this paper.

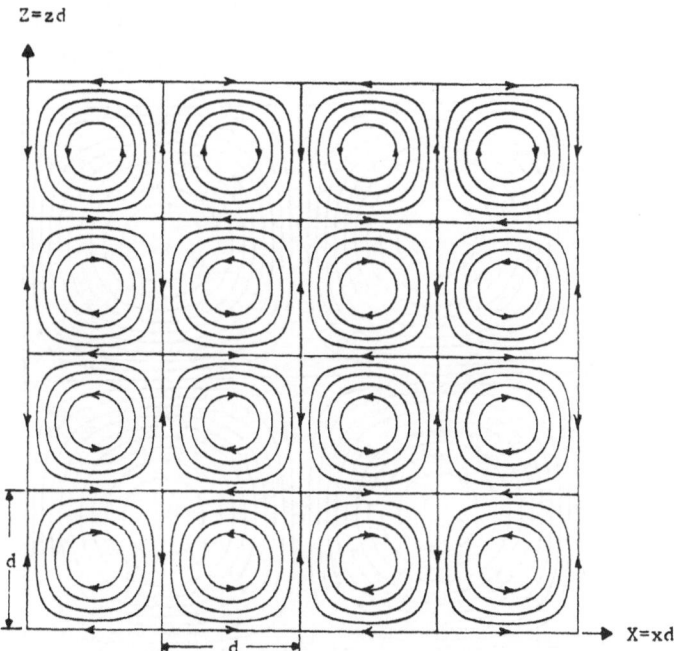

Figure 1. Illustration of Taylor's double array of vortices; $-1 \leq \psi \leq 1$, $\Delta\psi = 0.2$.

COMPUTATIONAL SOLUTION

Equations (7) and (8) are a system of parabolic and elliptic partial differential equations (PDEs). Such systems of PDEs occur in many different engineering applications. A general statement of such a problem and its numerical (or computational) solution are given next. Consider the time dependent two-dimensional nonlinear system of partial differential equations given by:

$$\frac{\partial U}{\partial t} = N_1 U + f_1(x, z, t), \tag{11}$$

$$\frac{\partial V}{\partial t} = N_2 V + f_2(x, z, t), \tag{12}$$

where N_1 is a *nonlinear* elliptic operator, which depends on x, z, t, V, and its derivatives, and N_2 is a *nonlinear* elliptic operator, which depends on x, z, t, U, and its derivatives. In the algorithm presented herein, the stable Crank-Nicolson time discretization is applied; its advantages in comparison to other implicit and explicit methods have been discussed by Isaacson and Keller (1966).

The Crank-Nicolson discretization uses the following approximations to $\partial U / \partial t$ and $\partial V / \partial t$:

$$\frac{\partial U}{\partial t}\Big|_{@(t+\delta t/2)} \approx \frac{U(t + \delta t) - U(t)}{\delta t},$$

$$\frac{\partial V}{\partial t}\Big|_{@(t+\delta t/2)} \approx \frac{V(t + \delta t) - V(t)}{\delta t},$$

Applying these approximations to Equations (11) and (12), this system of equations reduces to the following simultaneous nonlinear elliptic problems:

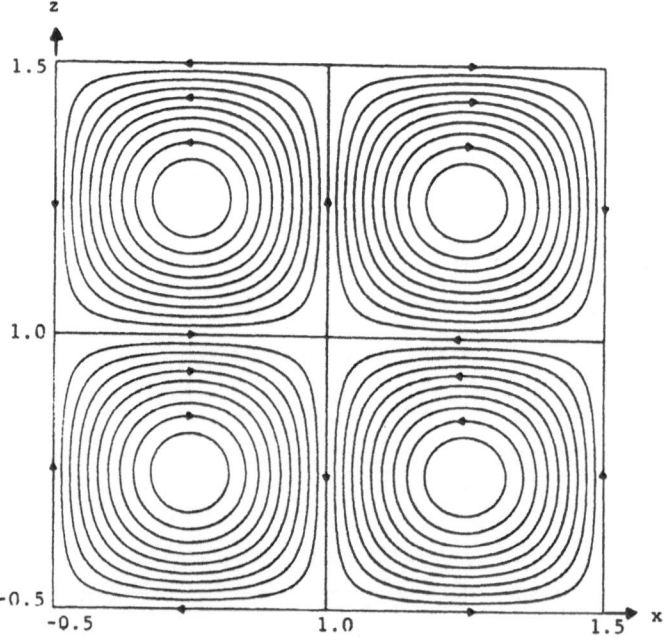

Figure 2. Initial condition for ψ for the test problem; $-1 \leq \psi \leq 1$, $\Delta \psi = 0.1$.

$$
\begin{aligned}
U(x,z,t+\delta t) &= U(x,z,t) + \frac{\delta t}{2}[N_1 U(x,z,t+\delta t) \\
&+ f_1(x,z,t+\delta t) + N_1 U(x,z,t) + f_1(x,z,t)],
\end{aligned}
\tag{13}
$$

$$
\begin{aligned}
V(x,z,t+\delta t) &= V(x,z,t) + \frac{\delta t}{2}[N_2 V(x,z,t+\delta t) \\
&+ f_2(x,z,t+\delta t) + N_2 V(x,z,t) + f_2(x,z,t)].
\end{aligned}
\tag{14}
$$

Equations (13) and (14) are approximate integrations in time that are determined by applying the trapezoidal rule. This discretization in time is unconditionally stable so that time steps δt are not restricted by stability considerations. For each time step one must solve the simultaneous elliptic problems:

$$
\begin{aligned}
N_1 U(x,z,t+\delta t) - \frac{2}{\delta t}U(x,z,t+\delta t) = &- \frac{2}{\delta t}U(x,z,t) - N_1 U(x,z,t) \\
&- f_1(x,z,t) - f_1(x,z,t+\delta t).
\end{aligned}
\tag{15}
$$

$$
\begin{aligned}
N_2 V(x,z,t+\delta t) - \frac{2}{\delta t}V(x,z,t+\delta t) = &- \frac{2}{\delta t}V(x,z,t) - N_2 V(x,z,t) \\
&- f_2(x,z,t) - f_2(x,z,t+\delta t).
\end{aligned}
\tag{16}
$$

We solved this set of simultaneous equations from t to $t + \delta t$ using the following algorithm:

Initialize U and V working arrays.

REPEAT

FOR i=1 TO 2 DO

Solve (15) as a linear elliptic problem for U using current values of V.

Update the current values of U.

OR

Solve (16) as a linear elliptic problem for V using current values of U.

Update the current values of V.

END FOR

UNTIL converged.

This iterative procedure is similar to Gauss-Siedel iteration for a set of linear algebraic equations. As in the linear algebra analogue, the iteration may or may not converge, and the convergence is directly affected by how the equations and unknowns are ordered. The advantage of this procedure is that a single linear elliptic problem is solved at each step; hence, the program could take advantage of the ELLPACK programming environment; see Rice and Boisvert (1985). Note that the terms on the right-hand-sides of Equations (15) and (16) are known and on the left-hand-sides we have linear elliptic operators. At each step, the elliptic equations are solved by applying the 4th-order collocation module (or method) within the ELLPACK programming environment to find $U(x,z,t+\delta t)$ or $V(x,z,t+\delta t)$, respectively. The resulting banded matrices were solved by applying the LINPACK band solver; see Dongarra et al (1966). This procedure can be generalized for k simultaneous nonlinear elliptic equations. The numerical procedure was applied in the present investigation to solve Equations (8) and (7) for ψ (or U) and ω (or V), respectively. The name of the new code is NSC, which stands for Navier-Stokes Collocation.

The collocation method forces the residual ($R = LU - f$, where L is a linear elliptic operator) to be zero at a number of points inside the domain, viz, the interior collocation points, and it forces the residual to be zero at a number of points on the boundary, viz, the boundary collocation points. There is freedom in choosing the collocation points. However, it has been shown by Papatheodorou (1982) that the 'Gauss points' are the best symmetric points for interpolation. The Gauss points were used in the collocation method selected. [For a definition of the Gauss points see the paper by Papatheodorou (1982); see also Rice and Boisvert (1985) and Birkhoff and Lynch (1984).] For a two dimensional domain, if an element is completely inside the domain, four Gauss points are used. The boundary collocation points are placed on the edge of the rectangular partition and then mapped onto the portions of the boundary intersecting the rectangular elements in which the boundaries pass through; this two step mapping procedure is done by applying the method described by Houstis *et al* (1978) and Rice and Boisvert (1985).

The 4th-order collocation method was selected and coded in the ELLPACK language (or programming environment) for the following reasons. The method discretizes a general elliptic operator with general linear boundary conditions on a general two-dimensional domain. The method implements a finite element method that uses Hermite bicubic piecewise polynomial approximations. The coefficients of the approximations are determined to satisfy the elliptic problem exactly at a set of collocation points. Similarly, the boundary conditions are satisfied exactly at a set of collocation points on the boundary. The method solves an elliptic problem if and only if a unique solution exists.

It is usually not economical to keep the time step size constant. The automatic control of it is an important part of the program developed in this investigation. In addition to computing the solution from one time step to the next, the following four things are done to control the error and to select the subsequent time step:

- Estimate the local error. The solution at time $t + \delta t$ is calculated with a time step of δt and $\frac{1}{2}\delta t$. The two estimates at $t + \delta t$ are used to control the local error and to estimate a new step size. The step size is also limited to control the convergence of the iteration.

- Decide whether the computed solution can be accepted, or whether one has to take a shorter step from the previous one. The solution at time $t + \delta t$ is accepted if both

$$\frac{1}{NU_{max}} \sum_{i=1}^{N} |\Delta U_i| \leq \epsilon$$

and

$$\frac{1}{NV_{max}} \sum_{i=1}^{N} |\Delta V_i| \leq \epsilon$$

where N is the number of collocation grid points, ΔU_i and ΔV_i are the differences of U and V between the two solutions at the ith grid point at $t + \delta t$ that are computed by using the time steps δt and $\delta t/2$. The error tolerance, ϵ, is supplied by the user.

- Apply Richardson's extrapolation at the end of each time step to reduce the error in the solution to $O(\delta t^4)$.

- Determine the step size to be used next by applying a time step selection algorithm that is based on the procedure described by Dahlquist *et al* (1974).

The desired accuracy as required by the user is specified by appropriate selections of the grid size, the time step and the error bounds that are imposed on the iteration procedures already described.

To verify the computational procedure numerical simulations of the test problem were executed. A comparison between the exact solution and the numerical simulation produced with the scheme just described is presented and discussed in the next section. The grid size selected for these simulations leads to the computation of 2500 grid points in the subdomain of four vortices of Taylor's vortex array. The number of grid points used in the x, z directions are 50×50, respectively. This is a relatively fine mesh and hence a conservative choice to ensure accuracy. Quantitatively this, the automatically determined time step (which is always less than 0.05), and the error tolerance selected (which is $\epsilon = 10^{-4}$) produced required errors of less than $10^{-5}\%$. Hence, for all practical purposes, the method is expected to produced the exact solution. This is indeed the case as is shown in the next section; hence, the method is verified.

The simulations were computed on the Northeast Parallel Architectures Center (NPAC) FX/80 Alliant superminicomputer at Syracuse University and on the Cornell Theory Center IBM 3090-600E supercomputer at Cornell University. To exploit the parallel architectures of these machines we took advantage of the opportunities for parallel computations that are abundant in the finite element method. The basic premise of finite elements is that the solution is approximated by interpolating functions that have compact support. Thus, calculations like element matrix formulation can be and are (in NSC) performed in parallel. The core calculations in NSC are solutions of systems of linear equations of the form $\mathbf{Ax} = \mathbf{b}$. The direct solver applied allows, not only a high degree of parallelism, but the opportunity to use the so called "blocked" algorithms that can greatly improve the performance of the parallel processors.

Parallelism can be found in the NSC code at several levels. At the lowest level, all vector operations are implemented as calls to Level 1 BLAS routines. Loop level parallelism is used to save execution time in key portions of the code. At a higher level, high performance equation solvers, *e.g.*, the FX/Skyline solvers on the Alliant or the solvers in the ESSL library on the IBM, are used.

Finally, the primary reason for developing the 4th-order code, NSC, is computational reliability. In complex viscous flow problems it is not known *a prior* if there are locations of rapid changes in flow properties within the computational domain. It is well known that low-order (*e.g.*, second-order) schemes will have difficulty resolving steep gradients. Because the method coded in NSC is 4th-order, the *new* scheme has zero second derivative truncation error (or false diffusion). It also has zero third derivative truncation error (or numerical dispersion). One of the methods for suppressing the 'wiggles' (*i.e.*, dispersion or overshoot and undershoot) is to apply a high-order method that eliminates at least the second and third derivative truncation error terms; see, *e.g.*, Valentine (1988). It is, in fact, the implementation of a 4th-order algorithm that enhances the reliability of NSC to simulate accurately Navier-Stokes problems.

RESULTS AND DISCUSSION

In this section the results of the numerical simulation of the test problem are presented and discussed. The principal purpose is to validate the coded algorithm. In addition, the numerical simulation of the breakdown of the flow pattern of the test problem when it is subjected to a small perturbation is described.

Taylor's vortex array is illustrated in Figure 1. The subdomain of four vortices that represents the test problem solved is also identified in this figure. The initial condition for the numerical simulation is shown in Figure 2. (This contour plot is a pictorial illustration of the structure of the flow field. Since the contour program linearly interpolates the solution between the grid points at which it is known, the relatively poor quality of the contour plot does *not* represent the actual quality of the solution. The actual quality of the solution is discussed next.) Relief maps of the stream function are illustrated in Figure 3 for three time steps. Figure 3(a) is the initial condition at $t = 0$, Figure 3(b) is an intermediate time at $t = 0.05$, and Figure 3(c) is close to the steady-state of rest at $t = 0.766$. These maps graphically illustrate the decay of the stream function and, hence, the velocity field with time.

The predicted (or simulated) rate of decay of the energy, *i.e.*, of

$$< E >= \int_{-0.5}^{1.5} \int_{-0.5}^{1.5} \left(u^2 + w^2 \right) dx dz,$$

is exactly equal to the same integral calculated from the exact solution. This decay is illustrated in Figure 4. In Figure 5, the decay of the vorticity at the center of one of the cells is shown. Again excellent agreement between the numerical predictions (or simulation) and the exact solution is observed. Finally, comparisons of the vertical velocity profiles at $z = 1.0$ for $t = 0.05, 0.1$ and 0.15 are illustrated in Figure 6. Again, excellent agreement is observed. Actually, the graphical presentation of results do not do complete justice to the results. In fact, the differences between the exact solution and the predicted solution from the simulation is within a maximum difference of 10^{-9}. This was determined for all times from 0 to 0.94.

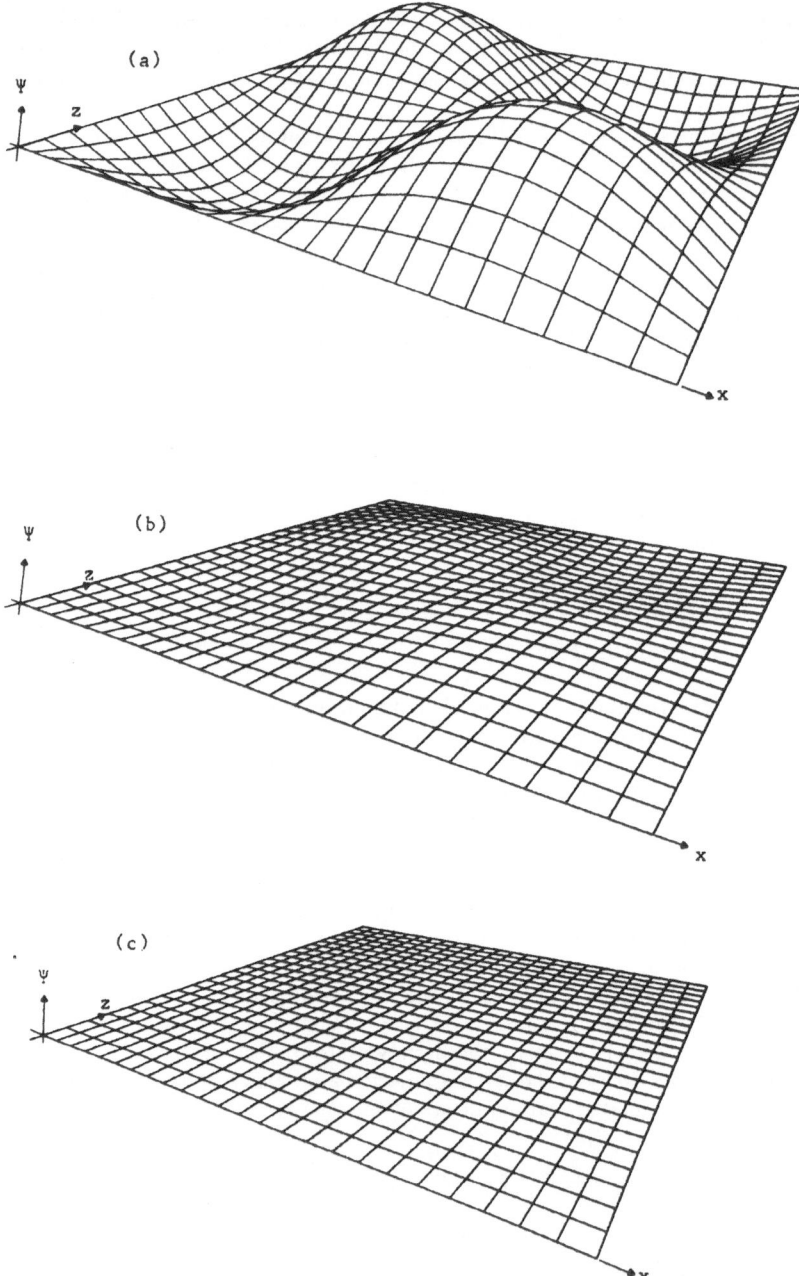

Figure 3. Graph of ψ versus x and z for the test problem at: (a) $t = 0$ ($-1 \leq \psi \leq 1$). (b) $t = 0.05$, and (c) $t = 0.766$.

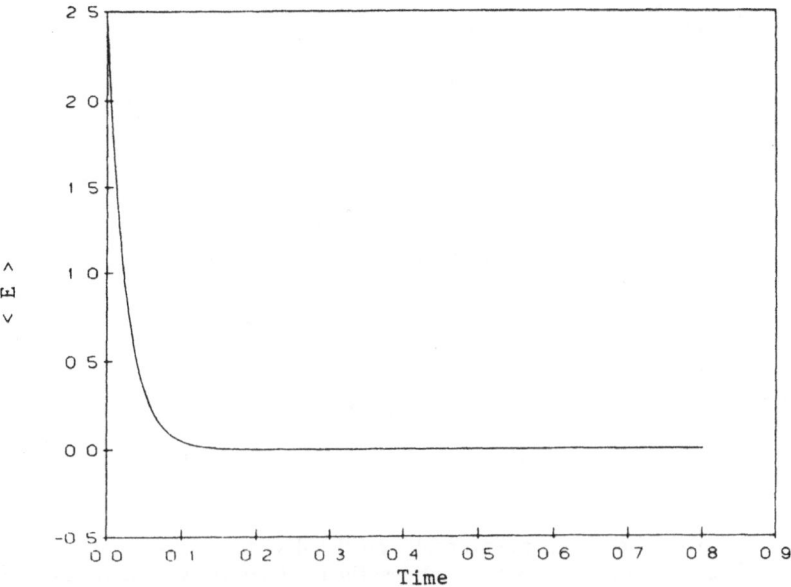

Figure 4. Comparison of the computationally predicted decay of kinetic energy with the exact solution; the two curves are identical.

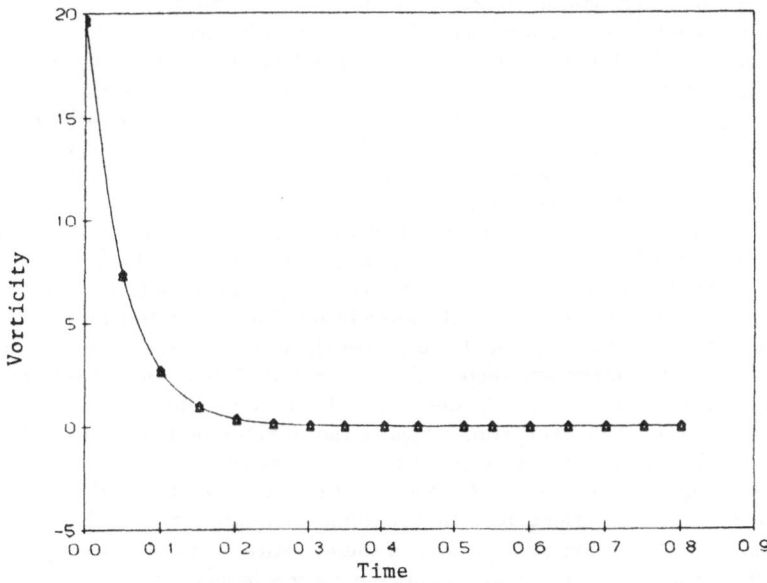

Figure 5. Comparison of the predicted decay of the vorticity at the center of one of the vortices with the exact solution; the two curves are identical.

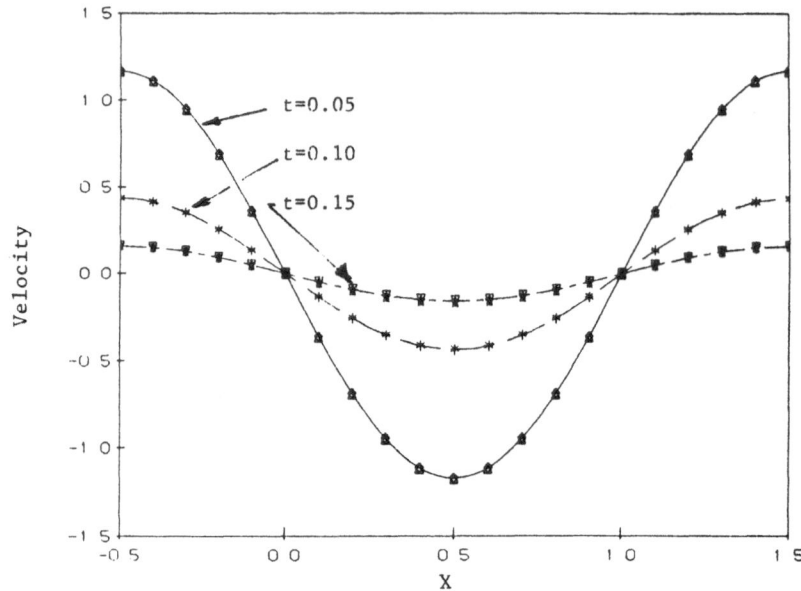

Figure 6. Comparison of the computationally predicted vertical velocity profiles at $z = 1$ for three different times with the exact solutions; the results of the simulation and the exact solutions are identical.

To compute the unsteady flow field over this range of t took 1.7 CPU hours. Over this range, the exact solution is reproduced 'exactly', *i.e.*, to within the precision of the machine.

Recently, there has been a renewed interest in Taylor's vortex array; see the work published by Lin and Tobak (1986, 1987). In fact, it is their work that stimulated the present numerical investigation. Lin and Tobak examined the stability of Taylor's vortex array. It is indeed globally stable. However, the structure of the flow does not necessarily persist when it is perturbed by an infinitesimal perturbation. In fact, it appears that for particular perturbations, the structure of the flow field breaks down. The works of Lin and Tobak (1986,1987) and Lin and Chen (1988) on the linear and nonlinear stability theories of this flow indicate that Taylor's vortex array is expected to breakdown, *i.e.*, the vortices will merge if all but odd-odd mode perturbations are imposed on the initial condition. These theoretical results are examined next in terms of the test problem by imposing a small perturbation on the initial condition and computing the subsequent unsteady solution for the test problem.

Two cases were investigated. In the first case the absolute value of the stream function at the center of each vortex was increased by 3 percent; this is an odd-odd mode type of perturbation. In this case the vortices decayed without merging as expected.

In the second case the perturbation added to the stream function at the center of the two vortices with positive values of ψ is 10^{-4}; hence, the values of the stream function at the center of these vortices were changed from $\psi = 1.0$ to $\psi = 1.0001$ in the initial condition. The perturbation added to the two vortices with negative values of ψ is also 10^{-4}; hence, the values of the stream function at the center of these vortices were changed from $\psi = -1.0$ to $\psi = -0.9999$. This small perturbation caused the vortices to merge as shown in Figure 7. This merging process is precisely what is expected from both the linear and nonlinear perturbation theories of S. P. Lin and his colleagues (1986,1987,1988). In addition, as is expected by the fact that this flow is globally stable, the total energy decays in the identical way as it does for the original four vortices.

CONCLUSION

A 4th-order in time and 4th-order in space finite element method was coded and successfully executed to solve the Navier-Stokes equations for an extension of the viscous decay problem of Taylor's vortex array. The algorithm applied, which was designed to solve simultaneous nonlinear partial differential equations, was described. The solutions to the linear elliptic problems that are at the core of the algorithm were solved by a 4th-order collocation method. The method was coded in the ELLPACK programming language. The code was translated into FORTRAN by applying the versions of ELLPACK that were vectorized and installed by the authors on both the Syracuse University (NPAC) FX/80 Alliant supermini-computer and the Cornell Theory Center IBM 3090-600E supercomputer. The algorithm was designed such that the user specifies the accuracy required in the solution; hence, the fine grid selected was based on conservative requirements imposed by the authors (or users) in this investigation. The exact solution was simulated 'exactly,' *i.e.*, to within the precision of the

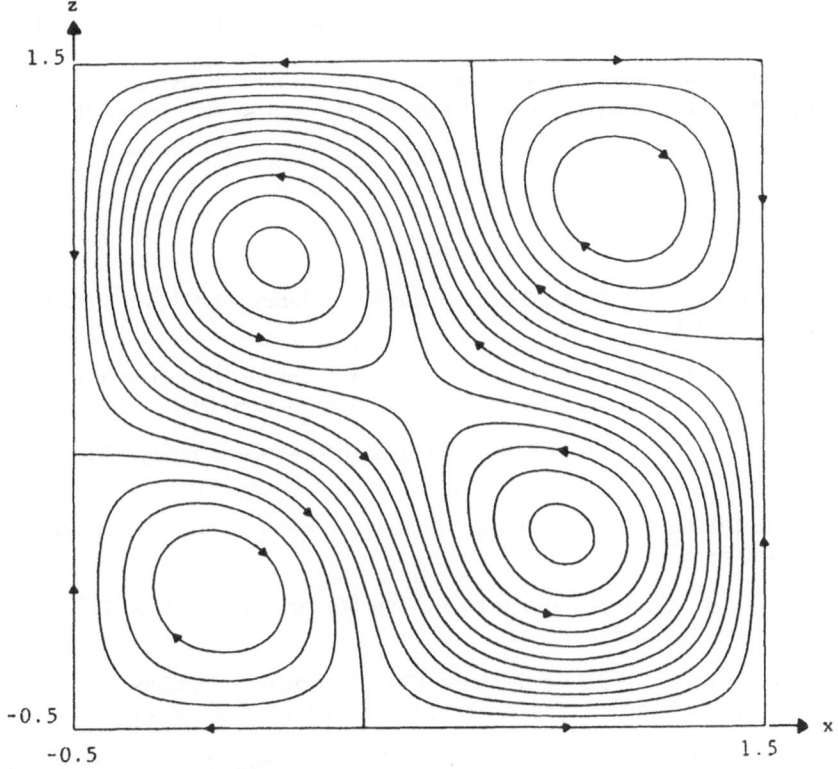

Figure 7. Illustration of the onset of merging of the vortices due to an appropriately selected small perturbation ($t = 0.484$, $-0.3 \times 10^{-4} \leq \psi \leq 0.11 \times 10^{-3}$, $\Delta\psi = 0.1 \times 10^{-4}$); the breakdown of the vortex array is expected based on the stability considerations reported by S. P. Lin and his colleagues (1986,1987,1988).

machine. Hence, the computational procedure was verified by successfully simulating Taylor's nonparallel flow exact solution.

Finally, the onset of the nonlinear breakdown of Taylor's vortex array when an appropriately selected small perturbation is imposed was simulated. The 'merging' of vortices of like sense was predicted. The merging process alters the original flow pattern without altering the time history of the energy decay. Hence, the time history of the decay of total energy is *not* an indicator of whether the vortex array persists or whether the original flow pattern breaks down. This result is consistent with recent theoretical results that have been reported S.P. Lin and his colleagues; see Lin and Tobak (1986,1987) and Lin and Chen (1988).

Acknowledgements

This work was supported in part by the National Science Foundation under grants CEE-84084239 and ECS-8515914. The first named grant provided partial support for the first author (D.T.V.) in his efforts to develop computational procedures to solve viscous flow problems in which buoyancy plays a role. The latter grant provided computer time on the Cornell National Supercomputer Facility (or the Cornell Theory Center), which is supported in part by the National Science Foundation, New York State, and IBM Corporation.

Additional support for this work was provided in part by The Northeast Parallel Architectures Center (NPAC) at Syracuse University, which is funded by DARPA, under contract to Rome Air Development Center (RADC), Griffis AFB, NY. NPAC provided computer time on their Alliant FX/80 superminicomputer.

Also additional support for this work was provided in part by the Military Technical College, Cairo, Egypt. Egypt provided the financial support for the second author to study for his Ph.D. at Clarkson University.

REFERENCES

Birkhoff, G., and Lynch, R. E., 1984: *Numerical Solution of Elliptic Problems*, SIAM, Philadelphia, PA.

Bradley, R. G. *et al*, 1986: "Current Capabilities and Future Directions in Computational Fluid Dynamics," National Research Council, Washington, DC.

Burden, R. L., and Faires, J. D., 1985: *Numerical Analysis*, Prindle, Weber & Schmidt, Boston, Mass.

Dahlquist, G., Björck, Å., and Anderson, N., 1974: *Numerical Methods*, Prentice-Hall, Englewood Cliffs, NJ.

Dongarra, J. J., Bunch, J. R., Moler, C. B., and Stewart, G. N., 1979: *LINPACK Users' Guide*, SIAM, Philadelphia, PA.

Hindmarsh, A. C., 1984: "ODE Solvers for Time-Dependent PDE Software," in *PDE Software: Modules, Interfaces and Systems,* North-Holland, Amsterdam.

Houstis, E. N., Lynch, R. E., Rice, J. R., and Papatheodorou, T. S., 1978: "Evaluation of Numerical Methods for Elliptic Partial Differential Equations", *J. Comp. Phys.* **27**, 323-350.

Isaacson, E., and Keller, H. B., 1966: *Analysis of Numerical Methods*, John Wiley & Sons, New York, NY.

Kim, J., Moin, P., and Moser, R., 1987: "Turbulence statistics in fully developed channel flow at low Reynolds number," *J. Fluid Mech.* **177**, 133-166.

Lin, S. P., and Chen, H. B., 1988: *personal communication*, in reference to the Ph.D. dissertation proposal to study the stability of nonparallel flows presented by Chen in July 1988.

Lin, S. P., and Tobak, M., 1986 "Spectral Stability of Taylor's Vortex Array," *Phys. Fluids* **29**, 3477-3478.

Lin, S. P., and Tobak, M., 1987: "Nonlinear Stability of Taylor's Vortex Array," *Phys. Fluids* **30**, 605-606.

Mikhlin, S. G., and Smolitskiy, K. L., 1967: *Approximate Methods for Solution of Differential and Integral Equations.*, American Elsevier Publishing, New York, NY.

Orszag, S. A., and M. Israeli, 1974: "Numerical Simulation of Viscous Incompressible Flows," *Annual Review of Fluid Mechanics*, Vol. 6, 281-318.

Papatheodorou, T. S., 1982: "Inverse for a class of banded matrices and applications to piecewise cubic approximation", *J. of Computational and Applied Mathematics* **8**.

Rice, J. R., and Boisvert, R. F., 1985: *Solving Elliptic Problems Using ELLPACK*, Springer-Verlag, New York, NY.

Taylor, G. I., 1923: "On the decay of vortices in a viscous fluid." *Phil. Mag.* **46**, 671-674.

Valentine, D. T., 1988: "Control-Volume Finite Difference Schemes to Solve Convection Diffusion Problems," *Proc. ASME Internl. Comp. Engrg. Conf.*, San Francisco, Calif.

Valentine, D. T., Mohamed, A. G., and Hessel, R. E., 1988: "Laminar Separation over an Annular Backstep: 4th-order FEM Solution," *Computers & Fluids*, (in press).

APPLICATION OF THE EBE (ELEMENT-BY-ELEMENT) PRECONDITIONED CONJUGATE
GRADIENT ALGORITHM TO THE FLUID (ACOUSTIC)/STRUCTURE PROBLEM

Robert T. Chapman and
Donald L. Cox

Engineering Mechanics Division
Naval Underwater Systems Center
New London, CT 06320

ABSTRACT

   This paper addresses the solution of the fluid (acoustic)/structure
interaction problem using the finite element method in conjunction with
an iterative solver, the Element-by-Element Preconditioned Conjugate
Gradient algorithm (EBE-PCG). The primary motivation behind this work
was to eliminate size constraints that exist when direct solution techniques, such as Gauss elimination, are applied to an assembled, global
stiffness matrix. After reviewing the element-by-element preconditioned
conjugate gradient algorithm and the finite element equations for the
fluid/structure interaction problem, a technique for using the EBE-PCG
method to solve the coupled problem is presented. It is demonstrated
that the EBE-PCG method is a viable approach to the solution of large
3-Dimensional problems in underwater acoustics.

1.0 INTRODUCTION

   The finite element method has long been recognized as both a general
and reliable approach for solving problems involving the acoustic radiation or scattering from a submerged structure. In this approach both
the fluid and the structure are represented as finite elements with the
coupled response achieved by double noding along the interface between
fluid and solid and requiring the normal component of velocity of each
to be equal. Its major shortcoming has been that for many problems of
interest the necessary discretization when assembled into a global stiffness matrix results in a system of linear equations that either tax or
overwhelm available computer resources. For example, degree of freedom
counts of 60,000 are not uncommon for problems where the assumption of
axisymmetry has been introduced. Extension to more accurate 3-D representations would require hundreds of thousands to upwards of one million
degrees of freedom essentially eliminating them from consideration.

   This paper addresses the use of the EBE-PCG (Element-by-Element
Preconditioned Conjugate Gradient) algorithm as an approach to circumvent
this problem. Originated by Hughes and his co-workers,[1] this method does
not require assembly of a global stiffness matrix and as a result allows

much larger problems to be addressed. Furthermore, for 3-D problems it appears to be more efficient than direct solvers which operate on an assembled matrix. The work follows that of Hughes, Ferencz, and Hallquist[2] in that a fully vectorized implementation of the algorithm was implemented into a new finite element program called NOMAD (No global Matrix Decomposition).

This paper begins with a brief presentation of the EBE-PCG algorithm including a description of its implementation into the NOMAD program. Although this work follows that of Ferencz and Hughes, it differs slightly in that the concept of using only unique elements in the EBE strategy is exploited.

A discussion of the solution of fluid/structure interaction problems using the finite element method is given in Section 3. Following a brief historical review of the finite element modeling of an acoustic media, the finite element equations for the coupled acoustic/structural are presented. The development of these equations is given in some detail as these details are germane to their solution using the EBE-PCG method.

Section 4.0 contains the discussion of the present approach of using the EBE-PCG method for solving the fluid/structure problem. The technique is outlined, and limitations of the present capability set forth. This section concludes with a set of both 2- and 3-dimensional example problems.

In the final section, conclusions based on the work to date are drawn and the direction of future work is given.

2.0 ELEMENT-BY-ELEMENT PRECONDITIONED CONJUGATE GRADIENT METHOD (EBE-PCG)

The concept of an element-by-element (EBE) algorithm originated in a paper by Hughes, Levit, and Winget[1] published in 1983. The intention of this work was "to reduce costs associated with the solution of large transient heat conduction problems." In this work, "the EBE concept was used to develop a non-iterative, second-order time accurate, unconditionally stable, transient algorithm for both linear and nonlinear problems." Because of accuracy problems, the original work was reformulated as an iterative linear equation solver[3] thereby leaving considerations of accuracy and stability to the technique chosen for time integration.

From this beginning, research into and exploitation of EBE methods has been vigorously pursued. A brief sampling includes research aimed at accelerating convergence for ill-conditioned problems,[4,5,6] work on a fully vectorized implementation of the algorithm,[2] and implementation of the algorithm on a coarse grained parallel computer.[7]

In what follows, use is made of the original Crout type EBE preconditioner coupled with the conjugate gradient driver. The implementation follows that of Ferencz in that a blocking algorithm is used that allows a fully vectorized implementation of the EBE preconditioner. Additionally, to further reduce storage requirements, a technique for using only unique elements was developed and incorporated into the code.

In the following paragraphs, the EBE-PCG algorithm is first presented. This is followed by a description of the blocking algorithm and a discussion of storage requirements. The section concludes with a description of the unique element storage concept and a comparison of its storage requirements with the usual technique.

2.1 GENERAL FORM OF THE EBE-PCG ALGORITHM

The linear system of equations to be solved by the EBE-PCG method can be generically defined as follows:

$$[A] \{x\} = \{b\} \tag{2.1}$$

To ensure convergence of the Conjugate Gradient algorithm, two characteristics of matrix [A] are required. Matrix $[A] \epsilon R^{n \times n}$ must be:

1. Positive Definite
2. Symmetric

The implications of these two properties, stated briefly and without proof are:[8]

1. All of the eigenvalues of [A] are real and there is an orthonormal basis of eigenvectors.
2. $\{x\}^T[A] \{x\} > 0$ for all nonzero $\{x\} \epsilon R^n$.
3. For [A] non-singular, there is not only an [L][D][M] decomposition of [A], where [L] and [M] are lower and upper triangular matrices with all entries of [D] positive, but an $[L][D][L]^T$ decomposition, where $[L] = [L]^T$.

(Note: The property of symmetry $[A] = [A]^T$ is only useful for $[A] \epsilon R^{n \times n}$.[9]

The preconditioned conjugate gradient algorithm can be represented as follows:

Initialize Variables.

(1) $\{p\}^T_i [A] \{p\}_i$ = a large number

(2) $\{x\}_i \qquad = \{0.0\}$

(3) $\{z\}_i \qquad = \{r\}_i$

for i = 1, max. number iterations or until convergence

(4) $\{z\}_i \qquad = \{z\}_i [W]^{-1/2}$

(5) :SOLVE $\{z'\}_i = [M]^{-1} \{z\}_i$

(6) $\{z'\}_i \qquad = \{z'\}_i [W]^{-1/2}$

(7) $\beta = \dfrac{\{z'\}^T_i [A] \{p\}_i}{\{p\}^T_i [A] \{p\}_i}$

(8) $\{p\}_{(i+1)} = \{z'\}_i - \beta \{p\}_i$

(9) $\alpha = \dfrac{\{z'\}^T_i \{r\}_i}{\{p\}^T_{(i+1)} [A] \{p\}_{(i+1)}}$

(10) $\{x\}_{(i+1)} = \{x\}_i + \alpha \{p\}_{(i+1)}$

(11) $\{r\}_{(i+1)} = \{r\}_i - \alpha [A] \{p\}_{(i+1)}$

(12) $\{z\}_{(i+1)} = \{r\}_{(i+1)}$

if $\|\{r\}_{(i+1)}\| / \|\{r\}_o\| < \Delta$

YES → STOP

NO → i = i+1

For the solution step (4), the preconditioning matrix [M] is of the Product Crout Factorization Element-by-Element form and is defined by:

$$[M] = [W]^{1/2} \times \prod_{j=i}^{n} [L(\bar{A})]_j \times \prod_{j=i}^{n} [D(\bar{A})]_j \times \prod_{n}^{j=1} [L(\bar{A})]_j \times [W]^{1/2} \quad (2.2)$$

and

$$[\bar{A}] = [I] + [W]^{-1/2} \times ([A^e] - [W^e]) \times [W]^{-1/2} \quad (2.3)$$

where:

[$\bar{A}$] is the Winget Regularized element matrix
[A^e] is the eth elements contribution to the global matrix
[W] is the assembled global diagonal
[$L(\bar{A})$] is the decomposed lower triangular matrix
[$D(\bar{A})$] is the decomposed element diagonal
[$U(\bar{A})$] is the decomposed upper triangular matrix
[W_e] is the element martrix diagonal

(Note: For the unpreconditioned conjugate gradient algorithm steps (4) - (6) are eliminated and the identity matrix [I] can be thought of as the preconditioner, or [M] = [I].)

The preconditioning steps of (4) - (6) can be rewritten as follows for the preconditioned case [10]:

(1) $[L]_1 [L]_2 \cdots [L]_n \ldots \ldots \{z'\} = [W]^{-1/2}\{r\}_o = \{z\}(1)$

$[L]_2 [L]_3 \cdots [L]_n \ldots \ldots \{z'\} = [L]_1^{-1}\{z\}(1) = \{z\}(2)$

(2) $[L]_{n-1} \cdots \{z'\} = [1]_{n-2}^{-1}\{z\}(2) = \{z\}(3)$

$[D]_1 \cdots [D]_n \cdots \{z'\} = [L]_n^{-1}\{z\}(3) = \{z\}(4)$

(3) $[D]_2[D]_3 \cdots [D]_n \ldots \ldots \{z'\} = [D]_1^{-1}\{z\}4 = \{z\}(5)$

$[U]_n \cdots [U]_1 \ldots \{z'\} = [D]_n^{-1}\{z\}(5) = \{z\}(6)$

$[U]_{n-1}[U]_{n-2} \cdots [U]_1[W]^{1/2}\{z'\} = [U]_n^{-1}\{z\}(7) = \{z\}(8)$

(4) $[U]_2 [W]^{1/2} \{z'\} = [U]_2^{-1} \{z\}(e) = \{z\}(9)$

$[W]^{1/2} \{z'\} = [U]_1^{-1} \{z\}(9)$

(5) $\{z'\} = [W]^{-1/2}\{z\}(9)$

Step (1) of the preconditioning step is a diagonal scaling of the system of equations by the square root of the assembled global stiffness matrix diagonal. Steps (2), (3), and (4) constitute the forward reduction, factored element diagonal scaling, and back substitution phases

respectively of the preconditioning. Step (4) of the preconditioning
phase is computed from the last element block to the first, maintaining
the symmetry of the preconditioner.[2] Step (5) is again a diagonal
scaling by the square root of the assembled global stiffness matrix
diagonal.

2.2 ELEMENT BLOCKING ALGORITHM

The most recent edition of the NOMAD program was written for use
primarily on a Cray-2 computer. To obtain a fully vectorized
implementation of the EBE-PCG algorithm, the preconditioning phase of
the algorithm requires special attention. A technique for achieving
vectorization of the EBE preconditioner was presented by Ferencz, Hughes
and Hallquist.[2] Presented herein is an abbreviated discussion of their
work. The reader is encouraged to consult reference [2] for a more
detailed explanation of this information.

Vectorization of the preconditioner phase of the algorithm can be
achieved by incorporating an element blocking algorithm that renumbers
the elements in the node/element connectivity array so that blocks of
internally disjoint elements can be formed. Each internally disjoint
element matrix block contains in it element matrices that share no
degrees of freedom. Element preconditioning then takes place an element block at a time, where the term 'preconditioning' implies performing
a Winget Regularization, [L][D][U] factorization, forward reduction, and
back substitution. Vectorization of the forward reduction and back
substitution phases of the preconditioning require that no two elements
in the same block share a common degree of freedom. This point is
emphasized because if two elements of the same block do share a nodal
degree of freedom, the resulting recursion would improperly update and
contaminate the recursive component of the intermediate solution vector.

2.3 STORAGE REQUIREMENTS; UNIQUE ELEMENT METHOD

The conjugate gradient algorithm requires the storage of five vectors
each dimensioned to the number of degrees of freedom in the system of
equations. Also required is the storage of a vector of boundary
conditions, two vectors of element and global diagonals, and a reordered
node/element connectivity array. Symbolically, using the notation of
Hughes and Winget,[11] where S() denotes storage:

    Solution vector: S(x) = Ndof words
    Residual vector: S(r) = Ndof words
    Temporary vector: S(z) = Ndof words
    [A] {p} vector: S(Ap) = Ndof words
    Conjugate search vector: S(p) = Ndof words
    Boundary condition vector: S(bc) = Ndof words
    Global diagonal vector: S(gdiag) = Ndof words
    Factored element diagonal vector: S(ediag) = Ndof words
    Reordered node/element conn. array: S(nec) = (Neln+1) X Nel words

where:

    Ndof is the global number of degrees of freedom in the model
    Neln is the number of nodes per element
    Nel is the global number of elements

The storage of these eight vectors and one array is constant for the implementation of both EBE-PCG and EBE-CG algorithms.

In an attempt to minimize storage demands, only unique element matrices are stored in the latest version of the NOMAD program. The criteria used for determining the uniqueness of an element is:

1. Geometric uniqueness
2. Material uniqueness
3. Formulation uniqueness (e.g., reduced integration, etc.)

With this technique the storage requirements of either EBE-PCG or EBE-CG algorithms becomes:*

    Element matrices:
        S(element) = Nuniqueel X (1/2) X (Neleq + 1) X Neleq words
    Factored matrices:
        S(f.element) = Maxblocksize X (1/2) X (Neleq + 1) X Neleq words
    Coordinate information:
        S(x,y,z) = Neln X Nuniqueel X Ndim words
    Localization matrix:
        S(lm) = Neleq X Maxblocksize words

*The EBE-CG algorithm does not require S(gdiag), S(ediag), S(f.element)

    Temporary vectors:
        S(temp) = Neleq X Maxlblocksize words

where:

    Nuniqueel is the number of unique elements stored
    Neleq is the number of element equations
    Maxblocksize is the maximum number of elements in a block
    Ndim is the number of spatial dimensions of model

To incorporate the unique element storage scheme into the preconditionng step of the algorithm, a subroutine which gathers element matrix entries for the current block of disjoint elements, defined by a reordered node/element connectivity array, must be used. The reordered node/element connectivity array has incorporated in it the particular unique element matrix identification information corresponding to the global element number. This gathering routine is implemented as follows:

```
kft = (lft)                establish the limits of the outermost loop
klt = (llt)                defining the global numbering of the elements
                           in the block initialize an indice counter for
j = 0                      the gathered element matrices
loop k = kft, klt
    j = j+1
    loop l = 1, number of entries in the element matrix
        [S]₁ⱼ = [S2₁([nec]ₖ)]
    endloop l
endloop k
nnpml = j                  assign variable nnpml to the final value of
                           j to be used as a loop counter in the
                           remaining preconditioner routines, and
                           define the number of elements in the current
                           block
```

where:

[S] is the gathered element stiffness matrices from the current element block
[S2] contains the unique element matrices
[nec] is the reordered node/element connectivity array
[kft] is the initial global element number of the current element block
(klt) is the last global element number of the current element block

(Note: Element matrices are symmetric in this case, and it is only required to store the upper triangle and diagonal entries. Matrix entries are numbered compactly in columns given by: $A(1,1) = S2(1,$ unique element #), $A(1,2) = S2(2,$ unique element #), $A(2,2) = S2(3,$ unique element #), etc.)

Because element matrices are gathered, and later processed an element block at a time, element matrix indexing is then not based on the global numbering of the element (indice k) but is based on the sequential ordering of the elements as they arise in the element block (indice j). The range of indices for the gathered element arrays are then from one up to the maximum number of elements in the block. The required degrees of freedom associated with the current block of elements being operated on must also be gathered using the reordered node/element connectivity array to define a permutation matrix, as discussed in [2]. The permutation matrix, which defines the activated dof in the current element block, need only be computated at the element block level and is initialized by the following:

```
j = 0                      initialize a counter
loop k = kft, klt
    j = j+1
    loop l = 1, number of dof per element matrix
        [lm]₁ⱼ = [nec]₁ₖ
    endloop l
endloop k
j = 0                      initialize boundary conditions
loop k = kft,klt
    j = j + 1
    loop l = 1, number of dof per element matrix
        if bc([nec]₁ₖ) = 1; [lm]₁ⱼ = nnd + 1
    endloop l
endloop k
```

(Note: The condition of $[lm]_{1j} = nnd+1$ defines a nodal constraint in the preconditioning routines that follow.[2]

The unique element storage implementation of either the EBE-PCG or the EBE-CG algorithms incorporates in it a significant reduction in storage requirements when compared to storing upper and diagonal entries of all element matrices, both factored and original. The storage commitment when using all element matrices would be:

Element matrices:
    S(element) = Nel X (1/2 X (Neleq + 1) X Neleq words
Factored matrices:
    S(f.element) = Nel X (1/2) X (Neleq + 1) X Neleq words
Coordinate information:
    S(x,y,z) = Nnn X Ndim words
Localization matrix:
    S(lm) = Neleq X Maxblocksize words
Temporary Vectors:
    S(temp) = Neleq X Nel words

As an example of the storage reduction gained using only unique elements, consider the storage requirements for a 56007 dof fluid/structure interaction problem (See Table 1):

STORAGE (WORDS) UNIQUE ELEMENTS		ALL ELEMENTS
S(element)	336	1848000
S(f.element)	19200	1848000
S(x,y,z)	48	162729
S(lm)	1536	1536
S(temp)	1536	393600
Total	22656	4253865

For this example the storage requirements of the unique element method are at a minimum, because for this example there are only two unique elements, one acoustic and one solid. For more complex geometries, the number of unique elements could increase appreciably. If for the previous example, forty percent of both the solid and fluid discretization incorporated geometrically unique elements, then the comparison becomes:

STORAGE (WORDS) UNIQUE ELEMENTS		ALL ELEMENTS
S(element)	739200	1848000
S(f.element)	19200	1848000
S(x,y,z)	464640	162729
S(lm)	1536	1536
S(temp)	1536	393600
Total	1226112	4253865

The factored element, localization matrix, and temporary storage do not change because entries of these arrays are internal to the algorithm and recomputed at the element block level when required.

The price for this savings in storage is that the unique element implementation of both the EBE-PCG and EBE-CG algorithms are computationally slower than the standard versions. This is because all elements of every element block must be gathered, scaled, and decomposed for the EBE-PCG method or gathered in the EBE-CG method at every iteration. The computational and memory fetch costs for extremely large problems using either algorithm are very high even with complete vectorization. Future work in the area of decreasing the computational time of either the EBE-PCG or the EBE-CG algorithms will be directed towards use of the Cray X-MP SSD and/or multi-tasking. By using the Cray X-MP SSD, blocks of factored element matrices could be written to and read from storage, eliminating the need to decompose element matrices at every iteration. Micro-tasking may be implemented in the preconditioning phase of the algorithm where one block of elements could be decomposed concurrently with the forward reduction and back substitution phase of another, already factored, block of element matrices. This would not decrease CPU time, but would significantly reduce "wall clock" time if CPU's were made available upon request.

A significant reduction in the computational work can be realized from the unique element storage method in that the time spent computing required element stiffness matrices is decreased, i.e., there are a reduced number of element stiffness matrices to be computed. The reduction in the number of element stiffness matrices to be computed is entirely mesh dependent, but experience has shown that most elements in a finite element discretization are the same.

3.0 FLUID/STRUCTURE INTERACTION

As noted in the introduction, the finite element method has been established as a useful tool for the acoustic analysis of fluid/structure systems. Because the finite element method had been customarily used for the solution of solid problems, the original challenges in applying the method to the coupled fluid/structural acoustic problem were in the representation of the fluid (acoustic) medium and the coupling of the two continua. A brief history of this work follows.

Initially, two approaches were taken in the finite element representation of the acoustic media. The first[12] developed the finite element equations directly from the wave equation for pressure using the Galerkin approach. This approach yields an acoustic element with the desired quantity, pressure, as the nodal unknown; unfortunately, when coupled with a solid or structural element, a nonsymmetric global stiffness matrix results. The second approach,[13] developed primarily to enable existing finite element programs to perform the analyses, used the material matrix to reformulate a continuum based solid element to make it appear to be an inviscid, compressible continua. This approach gives an acoustic element with either two or three degrees of freedom per node depending upon the dimension of the analysis. This simplifies the coupling of the acoustic media to the structural media, and at the same time gives a symmetric assembled stiffness matrix. The primary disadvantages of this approach are that storage requirements are much higher than the pressure approach due to the multiple degrees of freedom per node, and more importantly, for harmonic problems it reportedly[14] suffers a spurious resonance problem.

An alternate approach, popularly known as the pressure analogy, allows existing codes to model the acoustic media without need of a new pressure element, and eliminates the disadvantage of multiple degrees of freedom per node. This technique, developed by Everstine,[15] follows the same general path as the displacement approach in that the material matrix is altered to represent the acoustic media; however, it is done in a way that allows in the 3-D case removal of two degrees of freedom with the remaining displacement actually representing pressure. This technique is considered the "lumped equivalent" of the original pressure approach with the advantage of its implementation not requiring modification or addition of a new element to an existing program. It has been widely used and extended[16] to the axisymmetric case. Although a straight-forward implementation of the pressure analogy does result in a nonsymmetric stiffness matrix, it has been shown that for the steady state harmonic case,[17] by scaling the structural variables such that upon solution the structural unknowns are off by a constant factor, the assembled global matrix can be made symmetric. Furthermore, Everstine[18] has shown that by substituting for pressure a velocity potential as the principal unknown, a symmetric set of equations will result. In this approach, the coupling terms usually present in the mass and stiffness matrices are moved to the damping matrix. As a result, for problems that do not have any damping present, one must still solve a complex system of equations. This approach does however always yield a symmetric system of equations regardless of the analysis type, e.g. harmonic vs. transient.

3.1 FINITE ELEMENT EQUATIONS FOR THE ACOUSTIC MEDIA

In order to proceed with the finite element formulation of the fluid/ structure interaction problem, it was necessary to choose a technique from those described above for representing the acoustic media. Since the work described herein involves the development of a new finite element program, the advantage offered by the displacement approach and pressure analogy of not having to add a new element to an existing library disappears. As a result, the displacement approach which also suffers from a spurious resonance problem was discarded. Since the development of a pressure element from the wave equation and the pressure analogy essentially result in the same formulation, the pressure analogy was chosen for this work. This was partly due to a familiarity with that element formulation.

For ensuing discussions regarding the solution of the fluid/structure problem using the EBE-PCG algorithm, it is of value to review the finite element equations that result from the pressure analogy. As noted above, more complete discussions of this work are contained in references 15 and 16.

Modeling the fluid or acoustic media requires that pressure satisfy the wave equation,

$$\nabla^2 p = \frac{1}{c^2} \ddot{p} \,. \tag{3.1}$$

For an isotropic, linearly elastic material, the displacement form of the equations of motion are obtained by substituting the strain displacement ·relation.

$$
\begin{bmatrix} \varepsilon_{11} \\ \varepsilon_{22} \\ \varepsilon_{33} \\ \varepsilon_{12} \\ \varepsilon_{23} \\ \varepsilon_{31} \end{bmatrix} = \begin{bmatrix} \partial u_1/\partial x_1 \\ \partial u_2/\partial x_2 \\ \partial u_3/\partial x_3 \\ \partial u_1/\partial x_2 + \partial u_2/\partial x_1 \\ \partial u_2/\partial x_3 + \partial u_3/\partial x_2 \\ \partial u_3/\partial x_1 + \partial u_1/\partial x_3 \end{bmatrix} \tag{3.2}
$$

into the constitutive equation (Hooke's Law).

192

$$
\begin{bmatrix} T_{11} \\ T_{22} \\ T_{33} \\ T_{12} \\ T_{23} \\ T_{31} \end{bmatrix} = \begin{bmatrix} (\lambda+2\mu) & \lambda & \lambda & & & \\ \lambda & (\lambda+2\mu) & \lambda & & & \\ \lambda & \lambda & (\lambda+2\mu) & & \underline{0} & \\ & & & \mu & & \\ & \underline{0} & & & \mu & \\ & & & & & \mu \end{bmatrix} \begin{bmatrix} \varepsilon_{11} \\ \varepsilon_{22} \\ \varepsilon_{33} \\ \varepsilon_{12} \\ \varepsilon_{23} \\ \varepsilon_{31} \end{bmatrix} \qquad (3.3)
$$

and substituting that result into Cauchy's equation of motion (written using Cartesian tensor notation).

$$
T_{ij,j} + \rho b_i = \rho \ddot{u}_i . \qquad (3.4)
$$

The result is three second-order partial differential equations for the three displacements,

$$
(\lambda + \mu)\frac{\partial^2 u_k}{\partial x_i \partial x_k} + \mu \frac{\partial^2 u_i}{\partial x_k \partial x_k} + \rho b_i = \rho \ddot{u}_i . \qquad (3.5)
$$

The wave equation, (3.1), can then be obtained from these equations by setting

$$
u_1 = p
$$
$$
u_2 = u_3 = 0
$$
$$
b_i = 0 \qquad (3.6)
$$
$$
\lambda = -k
$$
$$
\mu = +k
$$
$$
\rho = \rho f .
$$

With these substitutions, the following matrix equations for the acoustic media result (using the notation of reference 16):

$$
\begin{bmatrix} m^{ii} & 0 & 0 \\ 0 & m^{ww} & 0 \\ 0 & 0 & m^{bb} \end{bmatrix} \begin{Bmatrix} \ddot{p}^i \\ \ddot{p}^w \\ \ddot{p}^b \end{Bmatrix} + \begin{bmatrix} c^{ii} & c^{iw} & c^{ib} \\ c^{wi} & c^{ww} & c^{wb} \\ c^{bi} & c^{bw} & c^{bb} \end{bmatrix} \begin{Bmatrix} \dot{p}^i \\ \dot{p}^w \\ \dot{p}^b \end{Bmatrix} +
$$

$$
+ \begin{bmatrix} k^{ii} & k^{iw} & k^{ib} \\ k^{wi} & k^{ww} & k^{wb} \\ k^{bi} & k^{bw} & k^{bb} \end{bmatrix} \begin{Bmatrix} p^i \\ p^w \\ p^b \end{Bmatrix} = \begin{Bmatrix} F_1^i \\ F_1^w \\ F_1^b \end{Bmatrix} \qquad (3.7)
$$

where the pressure and load vectors have been partitioned into values in the interior, at the structural interface, and on the outer boundary of the acoustic media.

To complete the analogy for the acoustic media, expressions for the surface loading, F_i, must be developed. In any continuous media, the surface traction T_i is related to the stress in the material by the expression

$$T_i = T_{ij} n_j \qquad (3.8)$$

where n is a positive outward unit normal. In the finite element formulation the load vector F_1, is obtained consistently from the following integral

$$F_1 = \int_S [N]^T \{T_1\} ds \qquad (3.9)$$

where N is the matrix of interpolation functions, or in lumped form as

$$F_1 = \Delta A T_1 \qquad (3.10)$$

where ΔA is an appropriate area surrounding the node where the force is to be applied. Expanding equation (3.8), and then in turn substituting into it the material relation (3.6), the strain displacement relation (3.2), and recalling that in the analogy u_1 and u_3 are set equal to zero gives

$$T_1 = \rho c^2 \frac{\partial p}{\partial n} . \qquad (3.11)$$

Using this equation, together with (3.8) or (3.9) expressions for the load vector in (3.7) can be derived.

With the exception of the force internal to the fluid, F_1^i which is set to zero in this work thereby ignoring body forces, the remaining forces, F_1^W and F_1^p, correspond to boundary conditions, at the acoustic/structure interface, and at the outer boundary of the acoustic media, respectively. At the interface between solid and fluid, it has been shown (refs. [15] and [16]) that

$$F_1^W = (\rho c)^2 \Delta A \ddot{u}_n \qquad (3.12)$$

where u_n is the normal component of structural acceleration. At the outer boundary of the fluid, the conditions of interest are:

$$p = 0 , \quad \text{a pressure release surface} \qquad (3.13)$$

$$\frac{\partial p}{\partial n} = 0 , \quad \text{a rigid surface} \qquad (3.14)$$

$$\frac{\partial p}{\partial n} = -\frac{1}{c}\frac{\partial p}{\partial t} , \quad \text{a 'non-reflecting boundary condition} \atop \text{(see Ref. 12).}} \qquad (3.15)$$

The pressure release surface is an essential boundary condition; therefore, F_1^p is not specified, but rather calculated as a reaction force. The rigid outer boundary, (3.14), is invoked by simply setting F_1^p equal to zero. By combining (3.15) with (3.11) and (3.10), an expression for the 'non-reflecting' boundary condition,

$$F_1^p = -\rho c \Delta A \dot{p} \qquad (3.16)$$

is obtained.

3.2 FINITE ELEMENT EQUATIONS FOR THE STRUCTURE

The matrix equations for the dynamic analysis of a structure can be written very generally as follows:

$$[M]\{\ddot{u}\} + [C]\{\dot{u}\} + [K]\{u\} = \{F\} \tag{3.17}$$

When the structure is surrounded by an acoustic media, these matrix equations can be written in an expanded and partitioned form as

$$
\begin{bmatrix} M^{II} & 0 \\ 0 & M^{WW} \end{bmatrix}
\begin{Bmatrix} \ddot{u}^I \\ \ddot{u}^W \end{Bmatrix} +
\begin{bmatrix} C^{II} & C^{IW} \\ C^{WI} & C^{WW} \end{bmatrix}
\begin{Bmatrix} \dot{u}^I \\ \dot{u}^W \end{Bmatrix} +
\begin{bmatrix} K^{II} & K^{IW} \\ K^{WI} & K^{WW} \end{bmatrix}
\begin{Bmatrix} u^I \\ u^W \end{Bmatrix} =
\begin{Bmatrix} F^I \\ F^W \end{Bmatrix} \tag{3.18}
$$

Again, as for equation (3.7), the notation of reference 16 is employed where 'I' implies degrees of freedom internal to the structure and 'W' implies degrees of freedom on the wetted interface surface.

The forces on the interface, F^W, includes the influence of the pressure on the structural motion, $p^W \Delta A^W$, as well as any explicit nodal point forces, F_e^W, that may be specified such that

$$F^W = p^W \Delta A^W + F_e^W \tag{3.19}$$

3.3 FINITE ELEMENT EQUATIONS FOR THE ACOUSTIC/STRUCTURE PROBLEM

Combining equations (3.7) and (3.18), and substituting the appropriate expressions for the external loads as given by (3.12), (3.16), and (3.19), gives the following matrix equations for the dynamic analysis of the acoustic/structure interaction problem.

$$
\begin{bmatrix}
M^{II} & 0 & 0 & 0 & 0 \\
0 & M^{WW} & 0 & 0 & 0 \\
0 & 0 & m^{ii} & 0 & 0 \\
0 & -(\rho c)^2 \Delta A & 0 & m^{WW} & 0 \\
0 & 0 & 0 & 0 & m^{bb}
\end{bmatrix}
\begin{Bmatrix} \ddot{u}^I \\ \ddot{u}^W \\ \ddot{p}^i \\ \ddot{p}^W \\ \ddot{p}^b \end{Bmatrix} +
\begin{bmatrix}
C^{II} & C^{IW} & 0 & 0 & 0 \\
C^{WI} & C^{WW} & 0 & 0 & 0 \\
0 & 0 & c^{ii} & c^{iw} & c^{ib} \\
0 & 0 & c^{wi} & c^{iw} & c^{wb} \\
0 & 0 & c^{bi} & c^{bw} & c^{bb}+\rho c \Delta A
\end{bmatrix}
\begin{Bmatrix} \dot{u}^I \\ \dot{u}^W \\ \dot{p}^i \\ \dot{p}^W \\ \dot{p}^b \end{Bmatrix}
$$

$$
+
\begin{bmatrix}
K^{II} & 0 & 0 & 0 \\
K^{WI} & 0 & \Delta A & 0 \\
0 & k^{ii} & k^{iw} & k^{ib} \\
0 & k^{wi} & k^{ww} & k^{wb} \\
0 & k^{bi} & k^{bw} & k^{bb}
\end{bmatrix}
\begin{Bmatrix} u^I \\ u^W \\ p^I \\ p^W \\ p^b \end{Bmatrix}
=
\begin{Bmatrix} F^I \\ F^W \\ 0 \\ 0 \\ F_e^W \end{Bmatrix} \tag{3.20}
$$

If the dynamic analysis is restricted to the steady state case where

$$p = p_o e^{i\omega t}$$

$$u = u_o e^{i\omega t} \tag{3.21}$$

these equations become

$$
\begin{bmatrix}
(-\omega^2 M^{II}+i\omega C^{II}) & (i\omega C^{Iw}+k^{Iw}) & 0 & 0 & 0 \\
\quad +k^{II} & & & & \\
(i\omega C^{wI}+k^{wI}) & -\omega^2 M^{ww}+i\omega C^{ww}) & 0 & \Delta A & 0 \\
 & \quad +k^{ww} & & & \\
0 & 0 & \left(\begin{array}{l}-\omega^2 m^{ii}+i\omega c^{ii}\\ +k^{ii}\end{array}\right) & (i\omega c^{iw}+k^{iw}) & (i\omega c^{ib}+k^{ib}) \\
0 & (\rho c)^2\omega^2\Delta A & (i\omega c^{wi}+k^{wi}) & \left(\begin{array}{l}-\omega^2 m^{ww}+i\omega c^{ww}\\ +k^{ww}\end{array}\right) & (i\omega c^{wb}+k^{wb}) \\
0 & 0 & (i\omega c^{bi}+k^{bi}) & (i\omega c^{bw}+k^{bw}) & \left(\begin{array}{l}-\omega^2 m^{bb}+i\omega(c^{bb}+\rho c\Delta A)\\ +k^{bb}\end{array}\right)
\end{bmatrix}
$$

$$
\begin{bmatrix} u^I \\ u^w \\ p^i \\ p^w \\ p^b \end{bmatrix} =
\begin{bmatrix} F^I \\ F^w \\ 0 \\ 0 \\ F^w_e \end{bmatrix}
\tag{3.22}
$$

From Eq. (3.22), it can be noted that when assembled, the finite element equations for the general, steady-state case are neither symmetric nor positive definite, the desired properties of the Conjugate Gradient algorithm (recall Section 2.1). To overcome these difficulties, a globally iterative technique was employed where the solid and acoustic equations are solved separately. This is discussed in the following section.

4.0 THE SOLUTION OF THE FLUID (ACOUSTIC)/STRUCTURE INTERACTION PROBLEM USING THE EBE-PCG ALGORITHM

The solution of solid displacements and fluid pressures involved in the fluid (acoustic)/structure interaction problem is accomplished in two phases. As shown in Appendix A, the algorithm developed consists of an outer 'global iteration' and two 'subiterate' levels for the solution of either solid displacements or acoustic fluid pressures. The algorithm shown is for use in the solution of radiation problems. The iterative solution of the fluid (acoustic)/structure interaction problem is defined by the two coupled systems of equations,[19] (4.1) and (4.2).

$$
([K_s] + i\omega(C_s] - \omega^2[M_s]){\delta} = {F} - ([M_s]{a_g} - [S_{fs}]{p}
\tag{4.1}
$$

$$
([K_f] + i (D_f] - \omega^2[M_f]){p} = \rho[S_{fs}]({a_g} - \omega^2{g})
\tag{4.2}
$$

where

$[K_s]$ is the stiffness matrix for the solid
$[K_f]$ is the stiffness matrix for the fluid
$[C_s]$ is the damping matrix for the solid
$[D_f]$ is the damping matrix for the fluid
$[M_s]$ is the mass matrix for the solid
$[M_f]$ is the mass matrix for the fluid
$\{\delta\}$ is the vector of displacements for the solid
$\{p\}$ is the vector of scalar pressures for the fluid
$[S_{sf}]$ is the coupling matrix for solid to fluid
$[S_{fs}]$ is the coupling matrix for fluid to solid
$\{a_g\}$ is the vector of structural accelerations
ζ is a damping factor of ω/c
$\{F\}$ is the initial forces on the structure

Allowing:

$$[K_s] = [K_s] + i\omega[C_s] - \omega^2[M_s]$$

$$[K_f] = (K_f] + i\zeta[D_f] - \omega^2[M]$$

(4.3)

and substituting into equations (4.1) and (4.2) yield the equations that are iteratively solved for both solid displacements and fluid pressures (Equations (4.4) and (4.5)). At the present time, the damping terms presented in the equations of (4.3) are not considered. Then:

$$[\widetilde{K}_s] = [K_s] - \omega^2[M_s]$$

$$[\widetilde{K}_f] = [K_f] - \omega^2[M_f]$$

(4.3)

$$[\widetilde{K}_s]\{\delta\} = \{F\} - [M_s]\{a_g\} + [S_{fs}]\{p\}$$

(4.4)

$$[\widetilde{K}_f]\{p\} = -\rho[S_{fs}]^T(\{a_g\} - \omega^2\{\delta\})$$

(4.5)

Both equations (4.4) and (4.5) are solved separately at the subiterate level, using the EBE-PCG method. At the subiterate level, the right-hand sides of either equations (4.4) or (4.5) are updated at the end of each subiteration using the previously computed solid displacements, $\{\delta\}$, or fluid pressures, $\{p\}$, computed at the wetted interface. For example, for a given radiation problem where only the solid is loaded and initial acoustic pressures are zero, for the first global iteration, equation (4.6) is solved to obtain the solid displacements $\{\delta\}$.

$$[\widetilde{K}_s]\{\delta\} = \{F\}$$

or

(4.6)

$$\{\delta\} = [K_s]^{-1}\{F\}$$

For the first global iteration then, equation (4.6) is solved in an un-coupled condition and thus at least two global iterations must be per-formed to solve the coupled fluid (acoustic)/structure radiation problem. The displacements of the solid at the wetted perimeter, multiplied by a predetermined coupling factor, $[S_{fs}]^T$, are applied as nodal forces to the right-hand side of equation (4.5), assuming that no additional structural acceleration excitations exist (i.e. $\{a_g\} = \{0.0\}$) at the wetted perimeter.

$$[\widetilde{K}_f]\,p = \rho\omega^2[S_{fs}]^T\{\delta\}$$

or

(4.7)

$$\{p\} = \rho\omega^2[K_f]^{-1}[S_{fs}]^T\{\delta\}$$

Equation (4.7) is then solved using the EBE-PCG method to obtain acoustic fluid pressures. From the solution of (4.7), with acoustic fluid pressures at the wetted perimeter now available, nodal forces due to acoustic fluid pressures are applied back to the structure, beginning the second global iteration. Global iteration number two begins by solving the coupled equation (4.8) for $\{\delta\}$ with the right-hand side consisting of the initial loading of the structure plus the load due to the acoustic fluid pressures $\{p\}$.

or
$$[\widetilde{K_s}]\{\delta\} = \{F\} + [S_{fs}]\{p\}$$
$$\{\delta\} = [\widetilde{K_s}]^{-1}(\{F\} + [S_{fs}]\{p\}) \tag{4.8}$$

With the solution of (4.8), using the new vector of solid displacements, nodal forces are again computed and applied to the fluid and the coupled equation (4.9) is solved for $\{p\}$.

or
$$[\widetilde{K_f}]\{p\} = \rho\omega^2[S_{fs}]^T\{\delta\}$$
$$\{p\} = \rho\omega^2[\widetilde{K_f}]^{-1}[S_{fs}]^T\{\delta\} \tag{4.9}$$

With the solution of equation (4.9) complete, the convergence criterion for the global iteration is checked. The global convergence criteria was taken to be the ratio of the norm of the (i+1)th global solution vector, $\{x\}_{(i+1)}$, to the norm of the ith global solution vector, $\{x\}_i$, (equation (4.10)). Another check for convergence could be to check the ratio of the norm of the (i+1)th load vector, $\{r\}_{(i+1)}$, with that of the ith load vector, $\{r\}_i$, knowing that if the global iteration (i+1)th and ith loading conditions are the same, the (i+1)th and ith global solution vector will also remain unchanged (equation (4.11)).

$$||\{x_g\}_{(i+1)}|| \; / \; ||\{x_g\}_i|| < \Delta_g \tag{4.10}$$

$$||\{r\}_{(i+1)}|| / \; ||\{r\}_i|| < \Delta_g \tag{4.11}$$

As was briefly mentioned above, the solution for either solid displacements, $\{\delta\}$, or acoustic fluid pressured, $\{p\}$, is obtained separately. This allows solid and fluid finite element discretization to be separately 'blocked' and all fluid or solid element by element computations (i.e. forward reduction, back substitution, the $[A]\{p\}$ product, etc.) performed separately. This method of computation follows naturally from the previous derivation.

5.0 RESULTS AND CONCLUSIONS

Both 2-dimensional and 3-dimensional acoustic fluid/structure interaction problems have been examined. Figures 1 and 2 represent the model where the exterior dimensions, and analysis variables are as follows:

    Fluid - 6" long in the x-direction
            1" width in the y-direction
            1" depth in the z-direction (3-D case only)

    Solid - variable length in the x-direction
            1" width in the y-direction
            1" depth in the z-direction (3-D case only)

Frequency of Analysis - 1000 Hz
Youngs Modulus of Steel - .300E+08
Poissons Ration of Steel - .300E+00
Mass Density of Steel - .783-03
Bulk Modulus of Acoustic Media - .34560E+06
Mass Density of Acoustic Media - .9600E-05

Uniform loading of the solid x = 0.0" of 10. units

Reflective boundary condition of the fluid at x = 6"

Constrained y translation of all nodes in solid
Constrained x translation along y = 0.0" and y = 1.0" of solid.

THe 3-D test runs encompassed degree of freedom counts which numbered
from 400 dof (96 solid displacements; 304 acoustic pressure) to 56007 dof
(2646 solid displacements; 53361 acoustic pressures). The 2-D test runs
encompassed degree of freedom counts which numbered from 287 dof (28
solid displacements; 259 acoustic pressures) to 12650 dof (184 solid
displacements; 12466 acoustic pressures). Subiterate and global
convergence criteria used for the examples were Δ_g = .10E-02 and Δ_g
= 0.10E-07. All example runs were performed on the NUSC Cray X-MP728.
The example problems do not, at the time of this writing, include damping
at the fluid boundaries, but use a reflective boundary condition for the
termination of the fluid. In both 2-D and 3-D cases the solid is
modeled using continua type elements as opposed to generalized stress
type elements.

Tables 1 and 2 present the total degree of freedom counts, solid dof,
fluid dof, and the number of elements in each coordinate axis for the
solid and fluid discretization. Tables 3 - 6 present the total CPU time
(sec) for the example problems for the two convergence criteria, and the
associated iteration counts for each subiteration. Most of this CPU
(sec) data was obtained using the 'flowtrace' option of the CF177
compiler. Because this option slows computation somewhat, some limited
timings were obtained without it in effect. The data in parentheses
represents these timings and is considered more representative of the
necessary CPU time.

Comparing the computational efficiency of the fluid (acoustic)/
structure interaction algorithm of the EBE-PCG method, given in Tables
3 - 6, shows that the EBE-CG algorithm is computationally faster than
the EBE-PCG method even with its reduced iteration count. An examination
of the case where all element and factored element matrices are kept in
storage reveals why this is so. In general, if all elements of the mesh
are stored, the requirement for the EBE-PCG method to be faster than the
EBE-CG method is that the time spent in the forward reduction, back
substitution, and the [A]{p} product in the EBE-PCG method must be less
than the time spent in the [A]{p} product of the EBE-CG method. For the
unique element storage technique, however, the decomposition phase is
internal to the EBE-PCG algorithm. Table 7 presents a comparison of the
time spent in the [A]{p} product phase of the EBE-CG algorithm versus the
time spent in the forward reduction and back substitution phases of the
EBE-PCG method using the unique element storage technique. The data
verifies that the decreased computational efficiency of the preconditioned method lies in the decomposition step of the algorithm, a deficiency which may be corrected. A continuing effort of this project will
be to eliminate this deficiency of the EBE-PCG algorithm by making use
of the Cray X-MP SSD and micro-tasking, as stated previously.

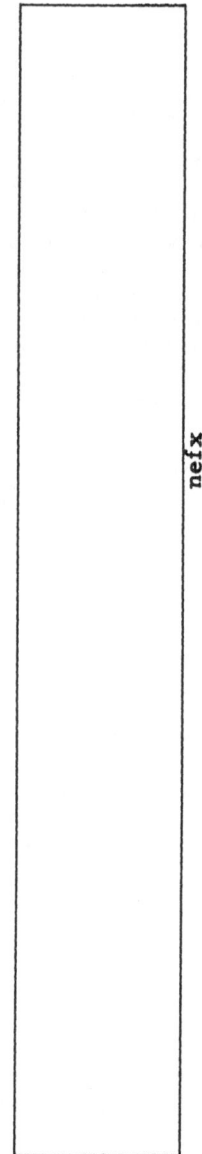

nefx

nefy

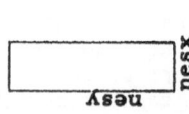

nesx

nesy

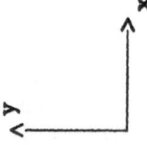

x

y

Figure 1. 2-D Fluid (Acoustic) / Structue Interaction Model.

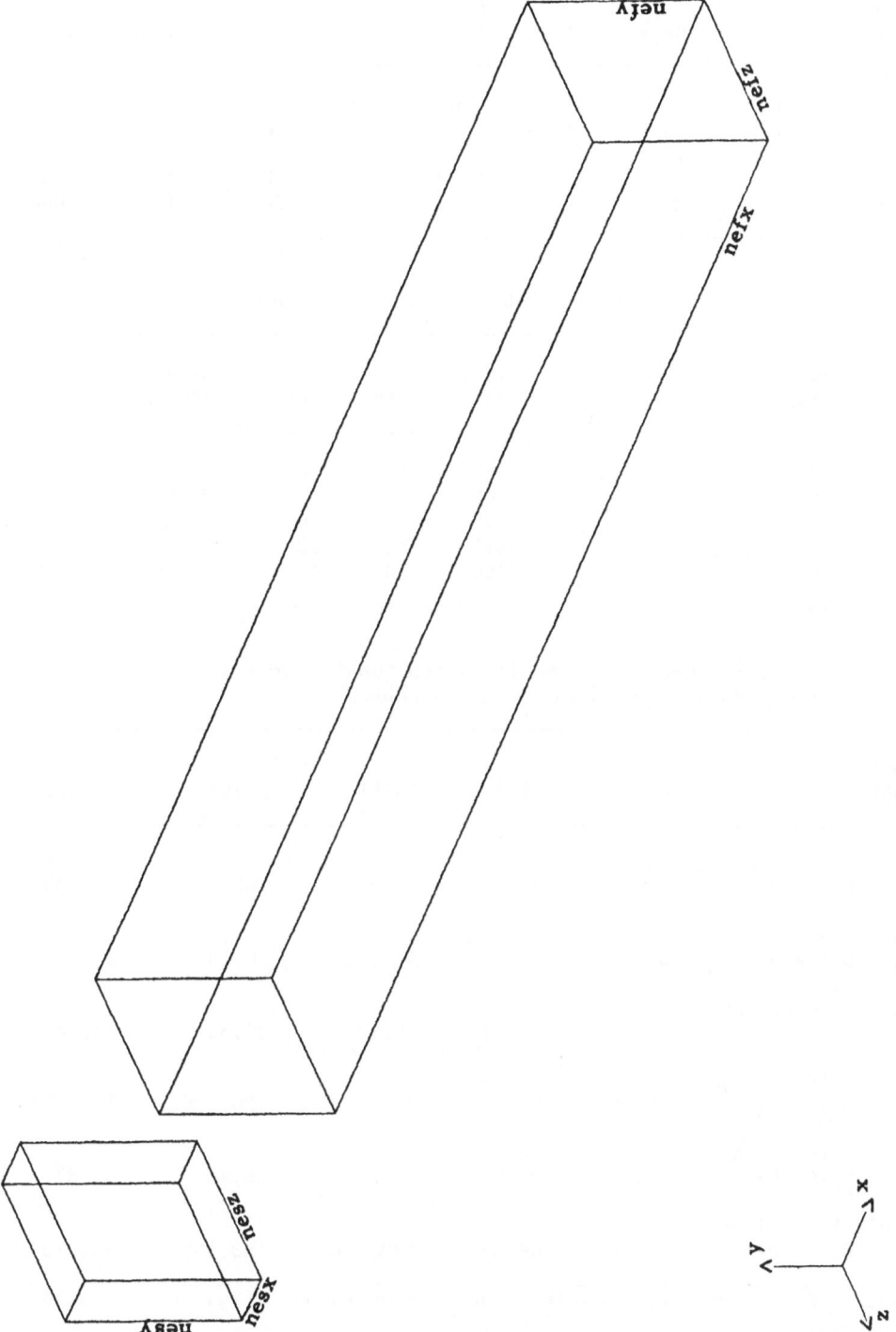

Figure 2.   3-D Fluid (Acoustic) / Structure Interaction Model.

Table 1. 3-Dimensional Case Runs

Solid Disp.	nesx	nesy	nesz	Fluid Press.	nefx	nefy	nefz	Total DOF
96	1	3	3	304	18	3	3	400
294	1	6	6	1813	36	6	6	2107
1014	1	12	12	12337	72	12	12	13351
1734	1	16	16	28033	96	16	16	29767
2646	1	20	20	53361	120	20	20	56007

Table 2. 2-Dimensional Case Runs

Solid Disp.	nesx	nesy	Fluid Press.	nefx	nefy	Total DOF
28	1	6	259	36	6	287
52	1	12	949	72	12	1001
68	1	16	1649	96	16	1717
124	1	30	5611	180	30	5735
184	1	45	12466	270	45	12650

Table 3. CPU (sec) and Iteration Counts for Test Problems
Using EBE-CG Algorithm at Two Residual Criteria (3-D Cases)

TOTAL DOF	400	2107	13351	29767	56007
Total CPU (sec) $\Delta_g = 0.10e-02$	0.46	3.15	31.97	88.71	(157.25) 193.08
Total CPU (sec) $\Delta_g = 0.10e-07$	0.56	4.09	44.16	101.49	230.11
Solid Iterations $\Delta_s = 0.10e-02$	5,6	12,13	26,27	34,35	42,42
Fluid Iterations $\Delta_f = 0.10e-02$	30,30	52,52	93,93	120,120	148,149
Solid Iterations $\Delta_s = 0.10e-07$	5,12	23,31	55,64	56,72	68,87
Fluid Iterations $\Delta_f = 0.10e-07$	39,39	68,68	138,138	153,153	188,188

*Note: All cases shown required the minimum number of global iterations.

Table 4. CPU (sec) and Iteration Counts for Test Problems
Using EBE-PCG Algorithm at Two Residual Criteria (3-D Cases)

TOTAL DOF	400	2107	13351	29767	56007
Total CPU (sec) $\Delta_g = 0.10e-02$	0.80	5.10	49.28	131.48	(218.23) 288.52
Total CPU (sec) $\Delta_g = 0.10e-07$	1.12	7.54	68.53	180.26	402.24
Solid Iterations $\Delta_s = 0.10e-02$	4,4	5,5	9,9	12,12	18,18
Fluid Iterations $\Delta_f = 0.10e-02$	11,11	21,21	38,38	49,49	60,60
Solid Iterations $\Delta_s = 0.10e-07$	8,10	12,14	22,25	28,34	40,45
Fluid Iterations $\Delta_f = 0.10e-07$	16,16	31,31	56,56	74,74	92,92

*Note: All cases shown required the minimum number of global iterations.

Table 5. CPU (sec) and Iteration Counts for Test Problems
Using EBE-PCG Algorithm at Two Residual Criteria (2-D Cases)

TOTAL DOF	287	1001	1713	5735	12646
Total CPU (sec) $\Delta_g = 0.10e-02$	0.37	1.80	3.36	17.96	54.07
Total CPU (sec) $\Delta_g = 0.10e-07$	0.50	2.48	4.77	25.51	77.50
Solid Iterations $\Delta_s = 0.10e-02$	4,4	6,6	7,7	12,12	16,16
Fluid Iterations $\Delta_f = 0.10e-02$	22,22	44,44	55,55	109,109	160,164
Solid Iterations $\Delta_s = 0.10e-07$	9,10	13,16	15,20	25,33	34,47
Fluid Iterations $\Delta_f = 0.10e-07$	31,31	63,63	84,84	160,160	240,240

*Note: All cases shown required the minimum number of global iterations.

Table 6. CPU (sec) and Iteration Counts for Test Problems
Using EBE-CG Algorithm at Two Residual Criteria (2-D Cases)

TOTAL DOF	287	1001	1713	5735	12646
Total CPU (sec) $\Delta_g = 0.10e-02$	0.24	1.06	2.00	9.86	29.03
Total CPU (sec) $\Delta_g = 0.10e-07$	0.27	1.25	2.40	11.91	35.73
Solid Iterations $\Delta_s = 0.10e-02$	6,7	12,13	15,16	27,29	39,41
Fluid Iterations $\Delta_f = 0.10e-02$	46,46	87,87	115,115	214,215	320,320
Solid Iterations $\Delta_s = 0.10e-07$	6,13	12,23	16,31	31,55	45,81
Fluid Iterations $\Delta_f = 0.10e-07$	57,57	111,111	146,146	270,270	402,402

*Note: All cases shown required the minimum number of global iterations.

APPENDIX A

THE ELEMENT BY ELEMENT PRECONDITIONED CONJUGATE GRADIENT ALGORITHM
FOR THE ACOUSTIC FLUID/STRUCTURE INTERACTION PROBLEM

for j = 1, max. number of global iterations of convergence (1)

    ! begin solution of structural displacements
    ! initialize variables.

$\{p_s\}_i^T [A_s] \{p_s\}_i$ = a large number (2)

$\{x_s\}_i = \{0.0\}$ (3)

$\{z_s\}_i = \{r_s\}_0$ (4)

for 1 = 1, max. number iterations or until convergence

$\{z_s\}_i = \{z_s\}_i [W_s]^{-1/2}$ (5)

:SOLVE $\{z'_s\}_i = [M_s]^{-1} \{z_s\}$ (6)

$\{z'_s\}_i = \{z'_s\}_i [W_s]^{-1/2}$ (7)

$$\beta_s = \frac{\{z'_s\}_i^T [A_s] \{p_s\}_i}{\{p_s\}_i^T [A_s] \{p_s\}_i}$$ (8)

$\{p_s\}_{(i+1)} = \{z'_s\}_i - \beta_s \{p_s\}_i$ (9)

$$\alpha_s = \frac{\{z'_s\}_i^T[A_s]\{r_s\}_i}{\{p_s\}_{(i+1)}^T[A_s]\{p_s\}_{(i+1)}} \tag{10}$$

$$\{x_s\}_{(i+1)} = \{x_s\}_i + \alpha_s\{p_s\}_{(i+1)} \tag{11}$$

$$\{r_s\}_{(i+1)} = \{r_s\}_i - \alpha_s[A_s]\{p_s\}_{(i+1)} \tag{12}$$

$$\{z_s\}_{(i+1)} = \{r_s\}_{(i+1)} \tag{13}$$

$$\text{if } ||\{r_s\}_{(i+1)}|| \ / \ ||\{r_s\}_i|| < \Delta_s \tag{14}$$

$$\text{NO} \rightarrow i = i=1! \text{return to step (4)}$$

$$\text{YES} \rightarrow !\text{begin solution of acoustic pressures}$$

$$\{z_f\}_i = 0.0 \quad ! \text{ initialize acoustic load} \tag{15}$$

$$\{z_f\}_i = \rho\omega^2[S_{fs}]^T\{x_s\}_{(i+1)} \ !\text{compute interaction} \tag{16}$$
$$\text{load @ wetted interface}$$

$$\{r_f\}_0 = \{z_f\}_i \ !\text{define initial acoustic residual and } || \{r_f\}_0 || \tag{17}$$

$$!\text{initialize variables for solution of acoustic pressures}$$

$$\{p_f\}_i^T[A_f]\{p_f\}_i = \text{a large number} \tag{18}$$

$$\{x_f\}_i \qquad\qquad = 0.0 \tag{19}$$

$$\text{for } l = 1, \text{ max. number iterations or until convergence}$$

$$\{z_f\}_i = \{z_f\}_i[W_f]^{-1/2} \tag{20}$$

$$:\text{SOLVE } \{z'_f\}_i = [M_f]^{-1}\{z_f\}_i \tag{21}$$

$$\{z'_f\}_i = \{z'_f\}_i[W_f]^{-1/2} \tag{22}$$

$$\beta_f = \frac{\{z'_f\}_i^T[A_f]\{p_f\}_i}{\{p_f\}_i^T[A_f]\{p_f\}_i} \tag{23}$$

$$\{p_f\}_{(i+1)} = \{z'_f\}_i - \beta_f\{p_f\}_i \tag{24}$$

$$\alpha_f = \frac{\{z'_f\}_i^T[A_f]\{r_f\}_i}{\{p_f\}_{(i+1)}^T[A_f]\{p_f\}_{(i+1)}} \tag{25}$$

$$\{x_f\}_{(i+1)} = \{x_f\}_i + \alpha_f\{p_f\}_{(i+1)} \tag{26}$$

$$\{r_f\}_{(i+1)} = \{r_f\}_i - \alpha_f[A_f]\{p_f\}_{(i+1)} \tag{27}$$

$$\{z_f\}_{(i+1)} = \{r_f\}_{(i+1)} \tag{28}$$

$$\text{if } ||\{r_f\}_{(i+1)}|| \ / \ ||\{r_f\}_0|| < \Delta_f \tag{29}$$

$$\text{NO} \rightarrow i = i=1! \text{return to step (20)}$$

$$\text{YES} \rightarrow \{z_s\}_i = [S_{fs}]\{x_f\}_{(i+1)} \ !\text{compute interaction load @ wetted}$$
$$\text{interface}$$

$$\{z_s\}_i = \{z_s\}_i + \{r_s\}_o \tag{32}$$

$$\text{if } j > 0 \tag{33}$$

$$\text{NO} \rightarrow \{r_s\}_o = \{z_s\}_i \tag{34}$$

$$!\text{define the new } ||\{r_s\}|| \tag{35}$$

$$\{x_g\}_i = \{x_s\}(i+1) + \{x_f\}(i+1) \tag{36}$$

$$!\text{repeat step } (1), (2), (3) \tag{37}$$

$$j = j+1 \ !\text{return to step } (4) \tag{38}$$

$$\text{YES} \rightarrow \{x_g\}(j+1) = \{x_s\}(i+1) + \{x_f\}(i+1) \tag{39}$$

$$\text{if } ||\{x_g\}(j+1)|| \ / \ ||\{x_{g1}\}_j \ ||<\{\Delta_g\} \tag{40}$$

$$\text{YES} \rightarrow \text{STOP} \tag{41}$$

$$\text{NO} \rightarrow \tag{42}$$

$$!\text{repeat steps } (1), (2), (3) \tag{43}$$

$$j = j+1 \ !\text{return to step } (4) \tag{44}$$

REFERENCES

1. T. J. R. Hughes, I. Levit, and J. Winget, Implicit Unconditionally Stable, Element by Element Algorithm for Heat Conduction Analysis, J. Engng. Mech. Div., ASCE 36, 576-585, 1983.

2. T. J. R. Hughes, R. M. Ferencz, and J. O. Hallquist, Large-Scale Vectorized Implicit Calculations in Solid Mechanics on a CRAY X-MP/48 Utilizing EBE Preconditioned Conjugate Gradients, Computer Methods in Applied Mechanics and Engineering, 61, 215-248, 1987.

3. T. J. R. Hughes, I. Levit, and J. Winget, An Element by Element Algorithm for Problems of Structural and Solid Mechanics, Computer Methods in Applied Mechanics and Engineering, 36, 241-254, 1983.

4. A. Muller, Element-by-Element Iterative Procedures in Structural Finite Element Analysis, Ph.D. Dissertation, Stanford University, 1985.

5. B. Nour-Omid, B. N. Parlett, A. Raefsky, "Comparison of Lanczos With Conjugate Gradient Using Element Preconditioning," Center for Pure and Applied Mathematics, University of California, Berkeley, 1987.

6. I. Levit, Element-by-Element Solvers of Order N, Computers & Structures, Vol. 27, No. 3, 357-360, 1987.

7. R. B. King and V. Sonnad, Implementation of an Element-by-Element Solution Algorithm for the Finite Element Method on a Coarse-Grained Parallel Computer, Computer Methods in Applied Mechanics and Engineering, 65, 47-59, 1987.

8. G. H. Golub and C. F. Van Loan, Matrix Computations, The Johns Hopkins University Press, 1983.

9. D. M. Young, <u>Iterative Solution of Large Linear Systems</u>, Academic Press, 1971.

10. M. A. Crisfield, <u>Finite Elements and Solution Procedures for Structural Analysis - Vol. 1: Linear Analysis</u>, Pineridge Press, 1986.

11. J. M. Winget and T. J. R. Hughes, Solutions Algorithms for Nonlinear Transient Heat Conduction Analysis Employing Element-by-Element Iterative Strategies, <u>Computer Methods in Applied Mechanics and Engineering</u>, 52, 711-815, 1985.

12. O. C. Zienkiewicz and R. E. Newton, Coupled Vibrations of a Structure Submerged in a Compressible Fluid, Proc. Int. Symp. on Finite Element Techniques, Stuttgart, 1969.

13. A. J. Kalinowski, Fluid-Structure Interaction Problems Using Finite Elements, Proceedings of the Fifth Navy-NASTRAN Colloquim, CMD-32-74, 71-86, 1974.

14. M. A. Hamdi and Y. Ousset, A Displacement Method for the Analysis of Vibrations of Coupled Fluid-Structure Systems, <u>International Journal for Numerical Methods in Engineering</u>, 13, 139-150, 1978.

15. G. C. Everstine, E. A. Schroeder, and M. S. Marcus, The Dynamic Analysis of Submerged Structures, NASTRAN User's Experiences, NASA TM X-3278, National Aeronautics and Space Administration, Washington, D.C., 419-429, 1975.

16. A. J. Kalinowski, Solution of Axisymmetric Fluid/Structure Interaction Problems With NASTRAN, 10th NASTRAN Colloquim, 1982.

17. A. J. Kalinowski, Steady State 3-D Symmetric Fluid/Structure Interaction Formulation for Finite Element Implementation, NUSC Technical Report 7151, 1984.

18. G. C. Everstine, A. Symmetric Potential Formulation for Fluid-Structure Interaction, <u>Journal of Sound and Vibration</u>, 79, 157-160, 1981.

19. S. K. Sharan, A Non-Reflecting Boundary in Fluid-Structure Interaction, <u>Computers & Structures</u>, Vol. 26, No. 5, 842-846, 1987.

NAVIER STOKES COMPUTATIONS OF A TURBULENT HORSESHOE VORTEX

AROUND AN ENDWALL MOUNTED CYLINDER

Craig Wagner

Engineering Mechanics Division
Naval Underwater Systems Center
New London, CT

INTRODUCTION

The Reynolds averaged Navier Stokes equations have been numerically solved for the case of a turbulent horseshoe vortex flow around an endwall mounted cylinder. Two different models were used, one grid containing 265,320 mesh points, and then a larger grid with 442,200 points. The two solutions obtained are compared to each other and against the experimental setup of Eckerle and Langston[1] for a turbulent endwall cylinder at Reynolds number 550,000 based on cylinder diameter. Comparisons between total pressure loss, endwall static pressures, and velocity vectors show good agreement for the smaller mesh over most of the compared flow field. For the larger mesh, the position and size of the horseshoe vortex indicate the solution has not fully converged. The reasons for the difficulty in getting the larger mesh solution to converge are discussed.

GOVERNING EQUATIONS AND SOLUTION TECHNIQUE

The computer code NAV3D was used to solve the Reynolds averaged Navier Stokes equations with the pseudo-compressibility approximation used in the continuity equation. NAV3D was developed by the author from the original INS3D[2,3] program and the details describing code corrections and enhancements are discussed in detail in Wagner.[4]

Pseudo-compressibility was first proposed by Chorin[5] and consists of adding a pressure-time derivative term to the incompressible continuity equation, thereby introducing an explicit expression for pressure in the governing equations as shown in equation (1).

$$\frac{\partial p}{\partial t} = - \beta \frac{\partial u_i}{\partial x_i}$$

$$\frac{\partial u_i}{\partial t} = - \frac{\partial u_i u_j}{\partial x_j} - \frac{1}{\rho} \frac{\partial p}{\partial x_i} + \frac{\partial}{\partial x_j}(v + v_t)\frac{\partial u_i}{\partial x_j} \ .$$

(1)

With explicit time expressions for all the primitive variables p, u, v, and w, the equations can now be advanced using a direct method. As long as the right-hand side of the continuity equation in (1) approaches zero as the solution is advanced, the true incompressible steady solution is obtained. It is important to note at this point that the time advancement of the solved equations represents only a numerical time evolution to achieve a steady solution and the flow field computed during the convergence process may not have any physical relevance to a physically time dependent flow field.

Alternating Direction Implicit or ADI methods are used to advance these equations in time. The ADI methods are especially useful for the solutions of extremely large matrices, because the resulting matrix problem consists only of solving systems of block 4 x 4 tridiagonal matrices. The tridiagonal matrices are solved in full vectorized fashion using straight Gaussian elimination. However, the ADI method introduces an approximation error which can be significant if too large a time step is used to advance the equations. Proper choice of the time step at each local grid cell can help to minimize the factorization error. The local time stepping scheme used here is discussed in Wagner.

EXPERIMENTAL CONFIGURATION

The endwall mounted cylinder experiment is described in detail in Eckerle and Langston and will be only briefly discussed here. The

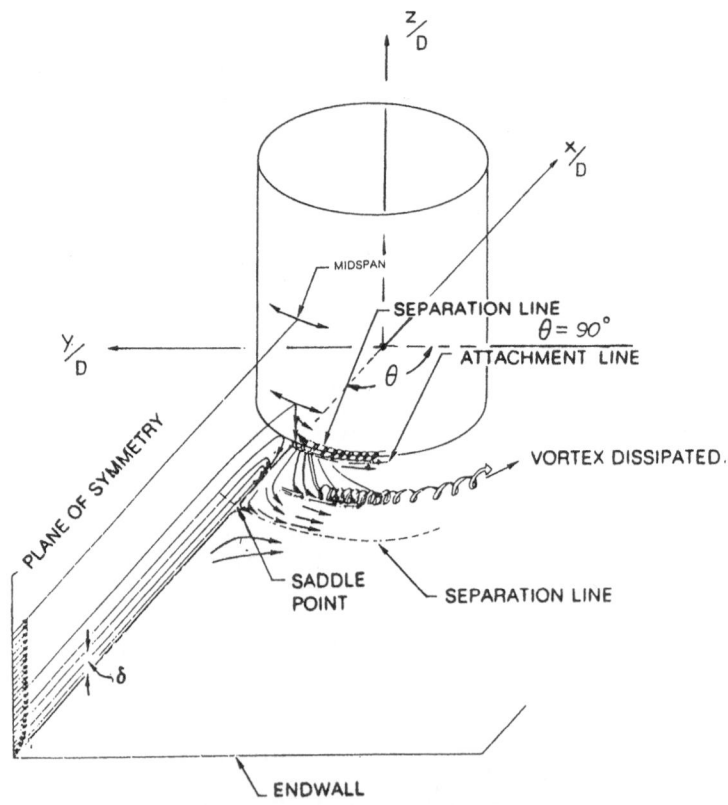

FIGURE 1.   EXPERIMENTAL CYLINDER GEOMETRY

cylinder is one foot in diameter being one foot high and is mounted in a wind tunnel between an upper and lower wall. The distance between side walls is approximately six feet with the cylinder located midway between them. At full steady speed the flow conditions correspond to a Reynolds number of 550,000 and is incompressible.

Velocity and pressure measurements were obtained using a five hole pressure probe at many stations in the flow field. In addition, endwall pressure measurements and cylinder surface pressure measurements were taken. Figure 1 shows the cylinder configuration and the coordinate system used in discussions here. The angle, theta, is taken as 0 degrees in the plane of the oncoming flow, as shown, and is swept through to 90 degrees at which point this plane is perpendicular to the inlet flow field as shown in the figure.

Eckerle and Langston measured pressures and velocities at theta planes of -5, 0, 5, 25, 45, and 90 degrees. In this paper comparisons are made for endwall pressures and velocities in the 0, 25, 45, and 90 degree planes. An additional quantity, total pressure loss, was also measured and will be compared to the computed values. Total pressure loss is not generally discussed in comparisons between numerical and experimental data. Yet any disagreement between the computed and measured flow field will be most noticeable in the total pressure loss comparisons. In particular, if the flow field is contaminated by a large amount of artificial dissipation (used in computer codes to control spurious high frequency oscillations in the solution process), the losses in the flow field will be significantly different. Also, if the turbulence model used does not correctly predict the viscous losses (either too much or too little) this will also show up in the total pressure loss comparisons.

NUMERICAL MODEL

The geometry used for the computations was constructed to depict the physical experiment exactly. Eckerle and Langston took reference velocity and pressure measurements at a station 2.73 diameters upstream of the cylinder center, therefore, our model starts at this location. Downstream of the 90 degree plane, measurements were not taken, so the computational mesh was terminated at 3 diameters past the cylinder center. This location was chosen to minimize the stretch of the mesh as one sweeps around the cylinder in theta planes. Our numerical side wall was located at three cylinder diameters from reference location, also to simulate the experimental setup.

Since the experimental measurements indicated a certain degree of symmetry across the 0 degree plane, our numerical model is constructed with this symmetry plane in mind, as shown in figure 2. In addition, symmetry is applied in the vertical direction since the horseshoe vortex structure is symmetric from top to bottom wall. This considerably reduces the number of mesh points required to accurately model the flow field.

Both meshes used for computations contained the same horizontal mesh construction of 88 x 67 grid points. The largest mesh used 75 of these planes stacked vertically with considerable stretch used from the endwall region to the vertical symmetry plane. the smaller mesh consisted of 45 of these stacked planes, again with considerable stretch used to get from endwall to vertical symmetry plane. Details of the actual mesh construction are given in Wagner.

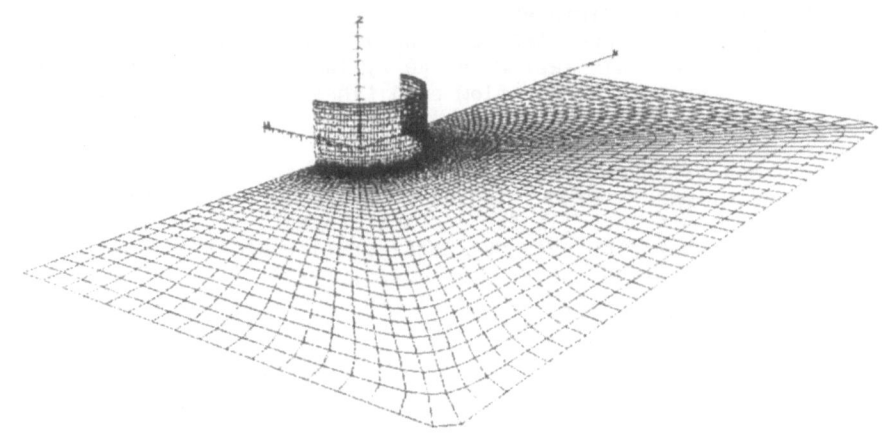

FIGURE 2.   NUMERICAL CYLINDER GEOMETRY

BOUNDARY AND INITIAL CONDITIONS

The inlet conditions were applied at the reference location measured by Eckerle and Langston. In this solution process, the total pressure is held fixed in the inlet core flow region, adjusting the velocity and pressure to maintain constant total pressure. The boundary layer region of the inlet flow is maintained at the shape measured by Eckerle and Langston smoothed by a Musker[6] profile fit to the measured data. A zero pressure gradient condition is enforced through the inlet boundary layer.

For the cylinder surface and endwall, zero velocity and zero normal gradient of pressure were enforced. At the sidewall and the vertical symmetry plane the normal component of velocity across these planes was constrained to zero and the in-plane velocities as well as pressure were held at zero normal derivative across these planes. This sidewall condition does not quite match the experimental setup as the experiment contained a developing boundary layer and corner flows at the sidewall surface. Considerably more mesh points would have been required if this condition were to be modeled exactly. It is important to note that this numerical approximation does change certain characteristics of the flow field. The primary difference is the effective blockage area of the experimental setup will be slightly greater than that computed. This additional blockage will correspond to the momentum displacement due to the boundary layer development on the sidewall. The influence on the computed solution due to not including the corner vortices and boundary layer on this wall are not known, but the effects are believed to be minimal since the boundary layer there is rather thin.

Downstream boundary conditions for subsonic computations are the most difficult to specify. Originally the downstream static pressure was held fixed across the entire outflow plane and second order extrapolation used to compute the velocities at this boundary. Considerable difficulty in converging the solution occurred when the entire downstream plane was held at the single value of static pressure which Eckerle and Langston measured on the sidewall. Investigation of the trouble area in the solution showed severe pressure crinkling at the 180 degree symmetry plane near the outlet boundary. This was attributed to the constraint on the cylinder wake caused by fixing one value of outlet static pressure across the entire plane. It is not reasonable to believe that the static pressure through the cylinder wake is constant at a given plane when one is only three diameters away, especially since it is known that the cylinder wake may persist for over 150 diameters downstream. Rather than build a mesh with a downstream plane further away to allow wake development, the boundary condition was altered to extrapolate to first order the outlet static pressure through some of the wake region. Over the portion of the outlet plane not in the wake region, the pressure was maintained at the measured value. This method proved to be considerably more stable than the constant outlet pressure across the entire plane. However, as evidenced by the difficulty in converging the larger mesh solution, this wake region is still highly unstable and is most likely unsteady in the physical world. Since our numerical approach is not developed for unsteady flows, absolute convergence of the solution is never obtained.

To start the three-dimensional solution, a converged solution to a two-dimensional infinite extent cylinder at right angle to uniform flow at the same Reynolds number was used. The velocities computed in the 2-D solution were scaled in the vertical direction by the measured shape of the incoming velocity profile. In other words, at the endwall, the velocities were set to zero and brought up to the two-dimensional solution value at the height corresponding to the measured height of the incoming boundary layer. This method provided a good starting point for the three-dimensional solution process.

TURBULENCE MODEL AND ARTIFICIAL DISSIPATION MODEL

A simple algebraic eddy viscosity model was used for both the large and smaller mesh computations. This model is patterned after that of McDonald and Fish.[7] It's ease in implementation and stability characteristics for this particular flow problem made it a reasonable choice. Two adjustable parameters are required in this model, the local skin friction coefficient and boundary layer height. Since the boundary layer height and boundary layer itself are undefined in areas of separation, a constant value for boundary layer height equal to that at the inlet plane was used. Any scheme to actually try to compute a local boundary layer thickness was considered to be no more accurate than using this approximation.

To compute the local skin friction value, the distance to each of the walls had to be considered. Skin friction values based on the normal distance from the mesh point in question to both the endwall and cylinder surface were computed. Then the distance to each wall was compared and the skin friction corresponding to the closest wall was used.

Artificial dissipation is required to stabilize the solution in areas where the mesh insufficiently resolves the gradients of the computed primitive variables. The artificial dissipation model used is that of Briley, Buggeln, and McDonald.[8] One of the favorable attributes of this model is the fact that artificial dissipation is added only when the local grid cell Reynolds number exceeds 2. For both meshes the artificial dissipation coefficient was used with a value of 1. No attempt was made to investigate the effects of lowering the value of the coefficient. However, several test runs by Wagner show that this particular artificial dissipation model, with the coefficient set at 1, provided the greatest convergence speed and stability.

CONVERGENCE CHARACTERISTICS

Figure 3 shows the convergence of the NAV3D program to a simple three-dimensional channel with periodic spanwise boundary conditions. This channel corresponds to the physical channel that would exist if the cylinder were removed from the wind tunnel. The Reynolds number and inlet conditions correspond to that in Eckerle and Langston. The most striking characteristic of this figure is the log linear convergence of the continuity and momentum equations. The parameter delta Q corresponds to the maximum value of the change in pressure or velocities at a given time step. All of the residuals plotted are the absolute maximum in the entire flow field at the time step indicated. Therefore, from time step to time step the mesh location of the plotted maximums may not be the same.

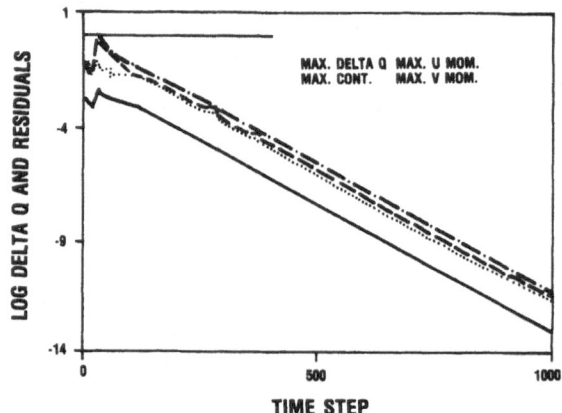

FIGURE 3.   CHANNEL FLOW CONVERGENCE

Figures 4 and 5 show the convergence characteristics of the smaller and then larger mesh. Notice the larger mesh residuals eventually started to increase after a certain point. The long ramp in this residual plot between time step 1200 and 1500 is due to the loss of convergence history for this time period. The behavior was oscillatory, slowly working its way up to the level shown in the region between 1500 and 1650. The solution for the larger mesh after time step 1650 was unstable to any attempt to go any further. The problem was, as previously discussed, the inability of the wake region to stabilize. That the larger mesh solution should prove more unstable than the smaller mesh solution may be due to the improved resolution of the wake region present in the larger mesh. Since the physical wake is unsteady, the better resolution provided by the larger mesh brings more of the actual physics into play causing the code to attempt to better predict the physical unsteadiness. The smaller mesh may be diffusing more of the unsteady processes by increased levels of artificial dissipation which tend to stabilize the solution. In any case the solution for the large mesh is used at the 1650 time step level for comparison purposes. For the compared region between 0 and 90 degrees, the wake instability did not appear to contaminate the solution that was obtained at this point.

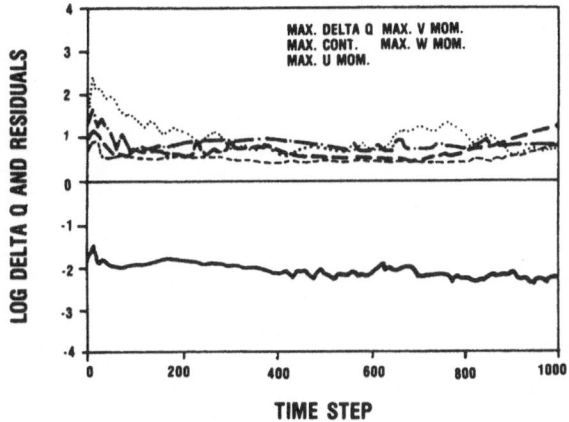

FIGURE 4.   265,320 MESH POINT CONVERGENCE

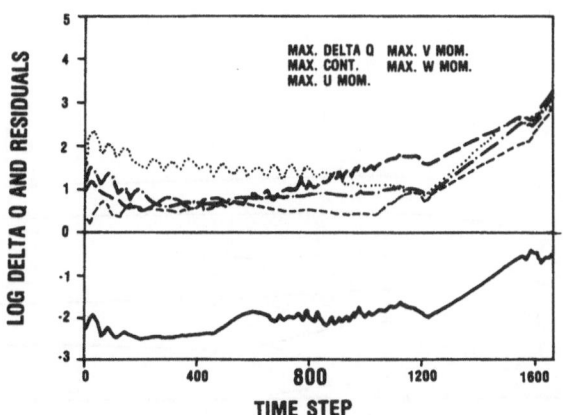

FIGURE 5.   442,200 MESH POINT CONVERGENCE

The smaller mesh solution was taken at time step 1000, but this solution was stable for a time step size five times that used for the larger mesh. In addition, the smaller mesh was run out an additional 2000 time steps after the first 1000 steps to check for consistency and stability of solution. The position of the horseshoe vortex and the location of the forward separation point, as well as the compared velocities and pressure, did not change from that computed at time step 1000.

TOTAL PRESSURE LOSS COMPARISONS

The total pressure loss comparisons for the 0, 25, 45, and 90 degree planes are shown in figures 6, 7, 8, and 9. These figures actually show the total pressure loss coefficient C_{P_T} which is defined as

$$(P_{T_0} - P_T)/\rho U_\infty^2 .$$

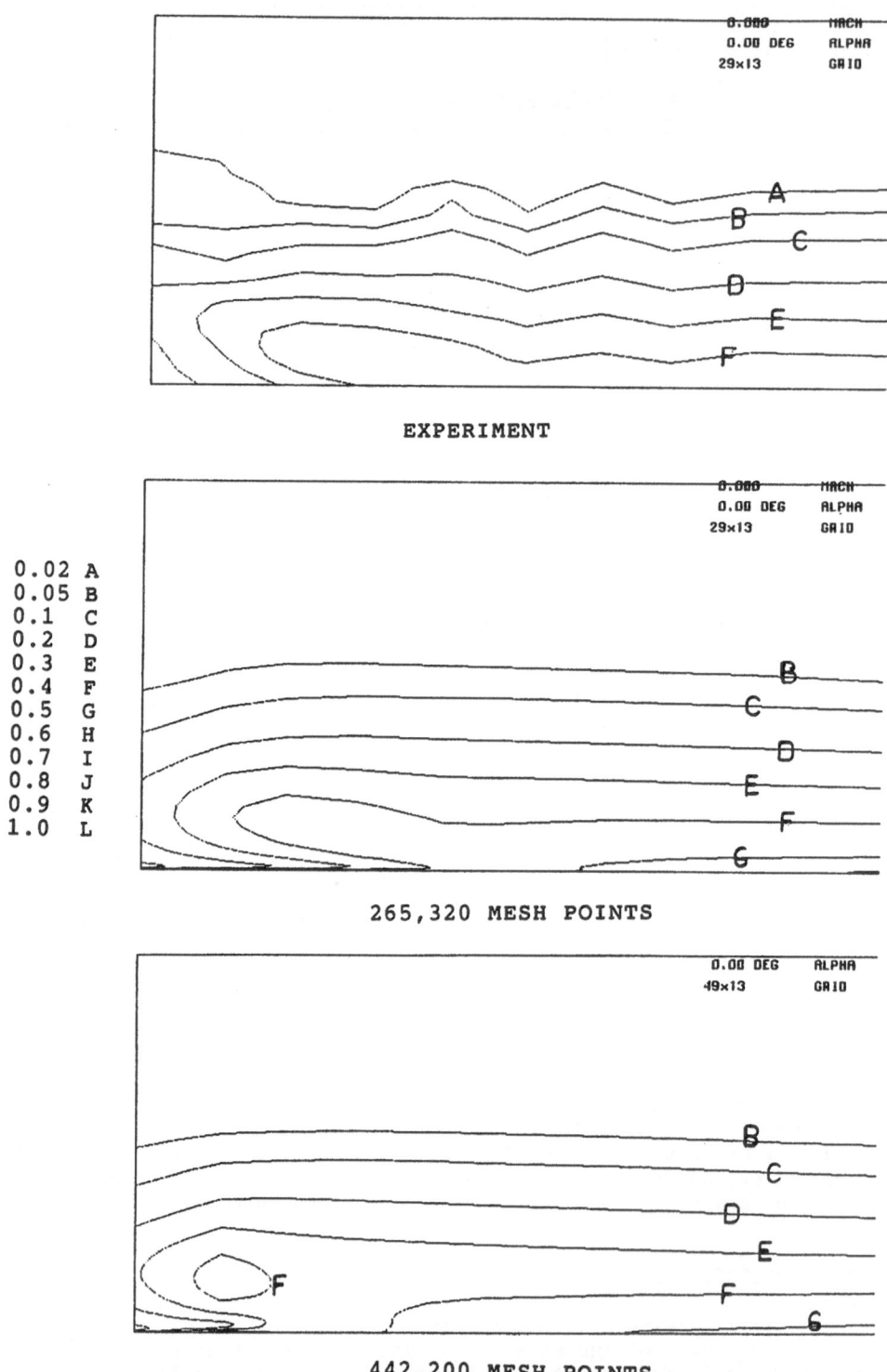

EXPERIMENT

0.02 A
0.05 B
0.1   C
0.2   D
0.3   E
0.4   F
0.5   G
0.6   H
0.7   I
0.8   J
0.9   K
1.0   L

265,320 MESH POINTS

442,200 MESH POINTS

FIGURE 6.   TOTAL PRESSURE LOSS 0 DEGREE PLANE

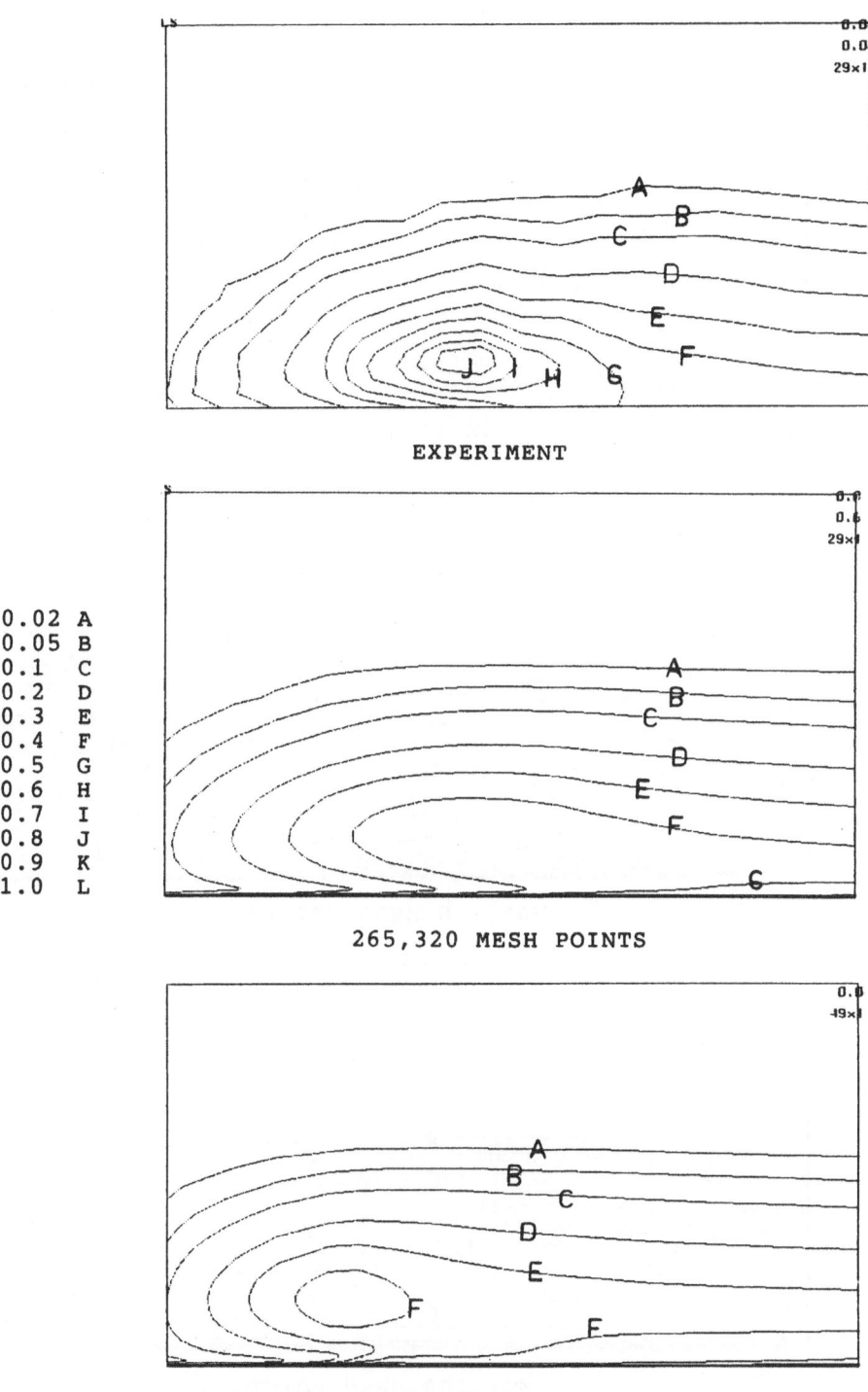

EXPERIMENT

0.02	A
0.05	B
0.1	C
0.2	D
0.3	E
0.4	F
0.5	G
0.6	H
0.7	I
0.8	J
0.9	K
1.0	L

265,320 MESH POINTS

442,200 MESH POINTS

FIGURE 7.   TOTAL PRESSURE LOSS 25 DEGREE PLANE

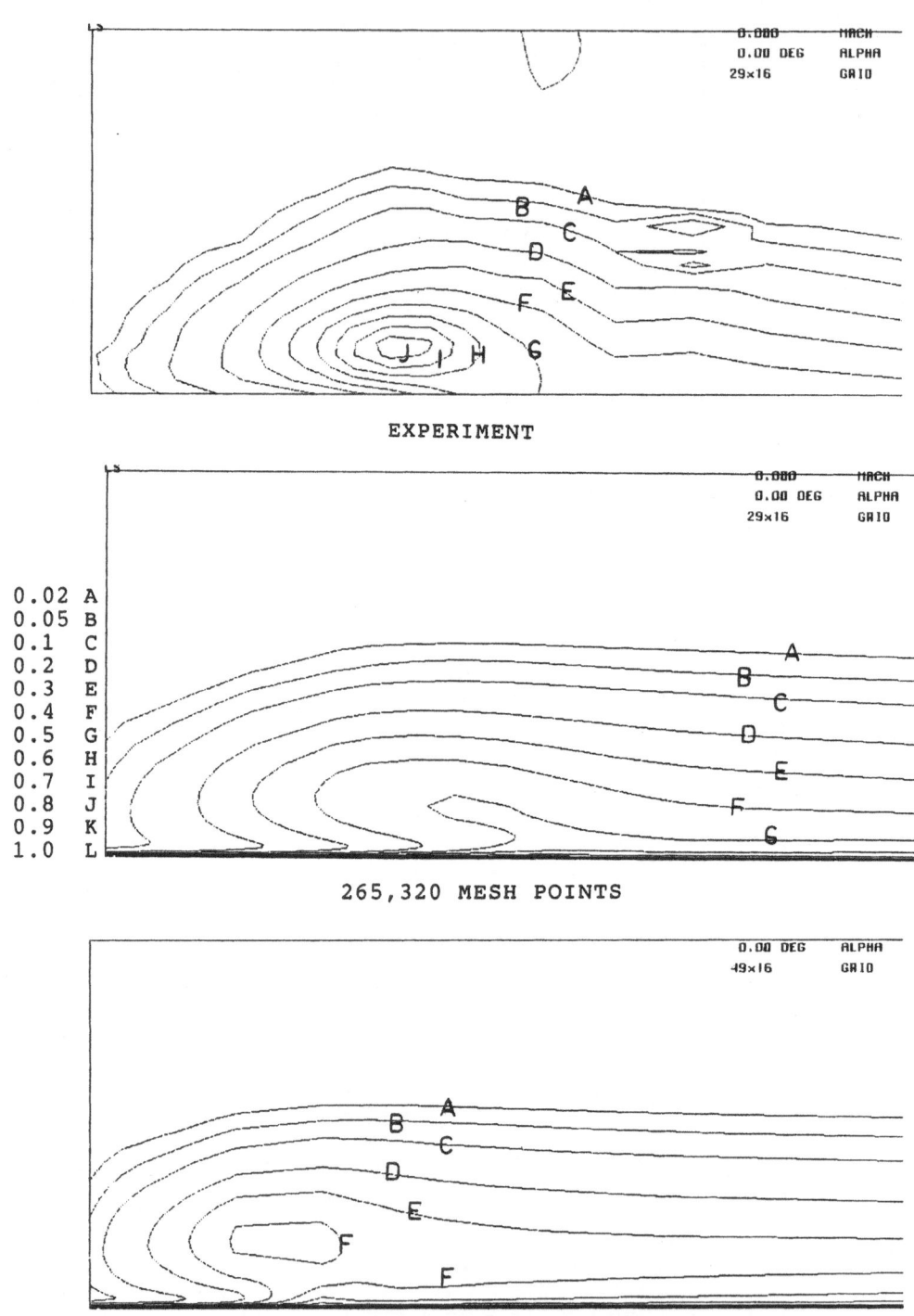

EXPERIMENT

0.02	A
0.05	B
0.1	C
0.2	D
0.3	E
0.4	F
0.5	G
0.6	H
0.7	I
0.8	J
0.9	K
1.0	L

265,320 MESH POINTS

442,200 MESH POINTS

FIGURE 8.   TOTAL PRESSURE LOSS 45 DEGREE PLANE

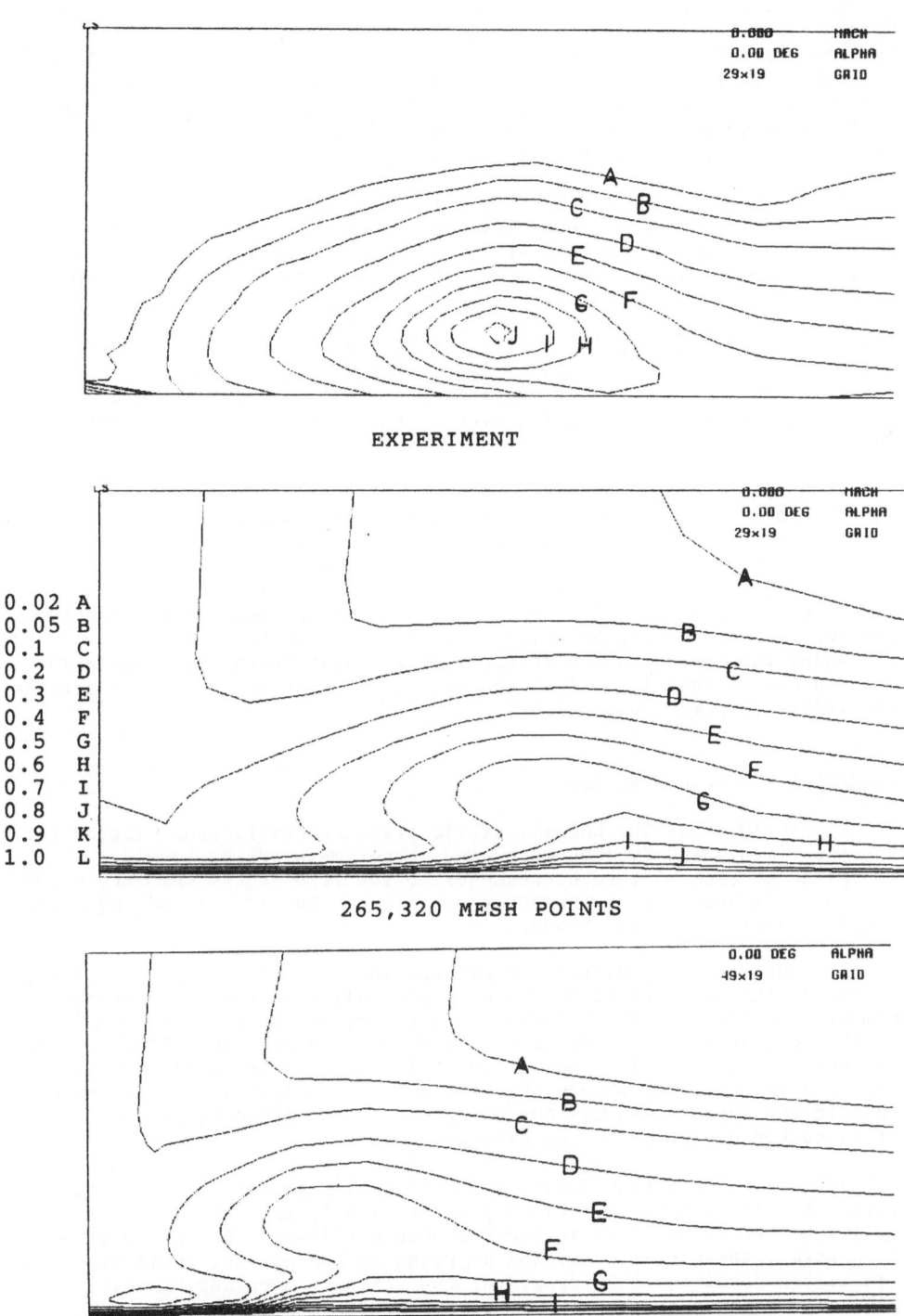

EXPERIMENT

265,320 MESH POINTS

442,200 MESH POINTS

FIGURE 9.   TOTAL PRESSURE LOSS 90 DEGREE PLANE

It is clear from all four figures that the computed vortex field does not contain the local high total pressure loss as seen in the experimental data. Even the more developed smaller mesh flow field does not match exactly the total pressure loss in the vortex core region. This is not as clear in the 0 degree plane, but as one moves through the different angular planes the comparison between the computed total pressure loss and measured loss becomes less favorable.

At the 90 degree plane the computed total pressure loss does not display the region of enclosed total pressure loss as seen in the experiment. Rather, the loss area is closer to the wall, away from the vortex core, and is more diffused than the experimental losses. In general, the numerical solution seems to indicate that the total pressure losses, overall, are not as high locally as the experiment indicates. This is especially surprising since it is expected that the artificial dissipation models used in these types of codes would tend to create more viscous losses than are physically present. However, these additional losses may be acting over a greater area tending to diffuse the vortex field.

The fact that the loss field is underpredicted may also be attributable to the turbulence model used. The algebraic turbulence model is really formulated for non-separating two-dimensional boundary layer type flows were local equilibrium and the universal velocity profile apply. This is certainly not the case for this three-dimensional, separated flow field. Both the turbulence model effects and the presence of the artificial dissipation may be smearing or diffusing the concentration of the vortex itself which would cause the total pressure loss to be less concentrated. This can be better seen in the velocity vector comparison.

ENDWALL PRESSURE COMPARISON

Figure 10 shows the endwall static pressure coefficient plotted for the experiment, the smaller mesh, and larger mesh, respectively. The top plane of each picture corresponds to the zero degree symmetry plane and the left boundary is the 90 degree plane: The flow is moving from right to left across the figure.

The most striking difference between the measured and predicted flow fields is the shallowness of the pressure valley in the computations as compared to the measured valley. This valley is indicated by the 'kink' in the contour lines of pressure as one moves toward the cylinder. For the smaller mesh the location of the valley correlates well with the measured values, even though the degree of the 'kink' is not the same. Because the larger mesh was not fully converged, the pressure valley is still tucked closer in to the cylinder.

The vortex formation during the computation process is such that the vortex starts as a small recirculation region tucked very close to the cylinder surface and deep inside the boundary layer. As the solution progresses, the vortex grows and migrates up and outward until it reaches its stable location. The growth of the vortex region during this migration process can be seen by comparing the greater extent of the pressure valley in the converged, smaller mesh plot, over that shown for the non-converged larger mesh.

Of interest, also, is the difference in the size of the pressure valley between experiment and computation. While the computed pressure valley does not have the depth of the measured one, the valley spans a much greater area. From this plot it is clear that the computed location of the horseshoe vortex is not precisely where indicated by the measurements, especially for the larger mesh. It appears, again, that the numerical predictions are showing a more diffused vortex region than was measured.

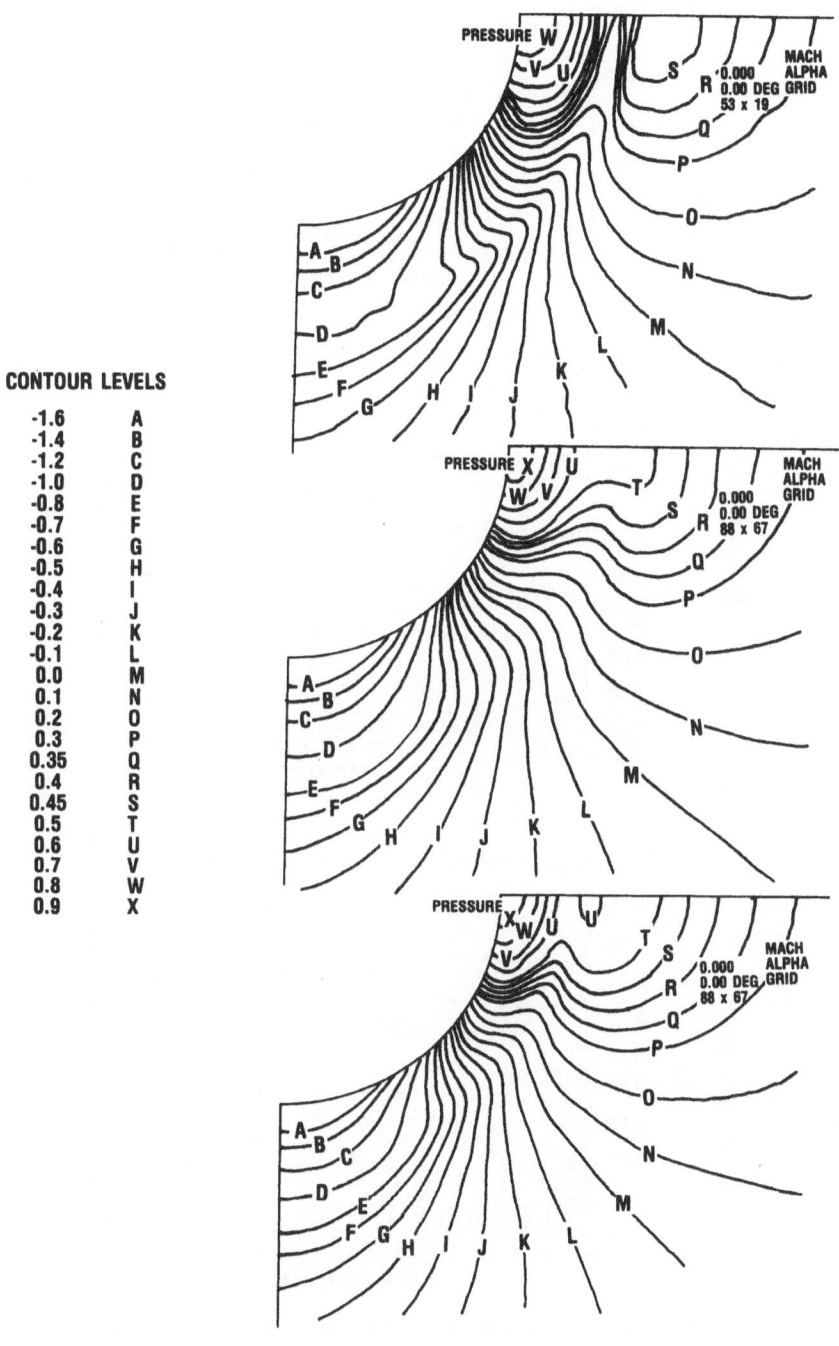

CONTOUR LEVELS

-1.6	A
-1.4	B
-1.2	C
-1.0	D
-0.8	E
-0.7	F
-0.6	G
-0.5	H
-0.4	I
-0.3	J
-0.2	K
-0.1	L
0.0	M
0.1	N
0.2	O
0.3	P
0.35	Q
0.4	R
0.45	S
0.5	T
0.6	U
0.7	V
0.8	W
0.9	X

FIGURE 10. ENDWALL STATIC PRESSURE CONTOURS

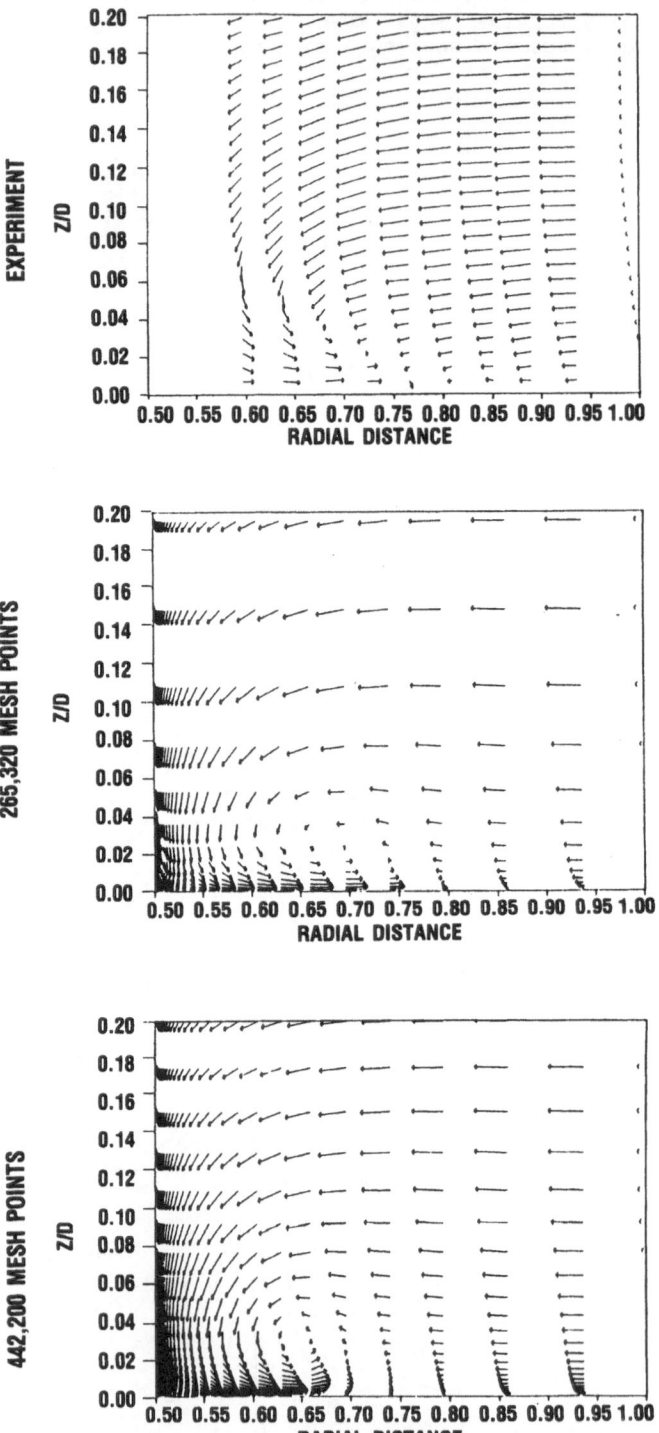

FIGURE 11.   VELOCITY VECTORS 0 DEGREE PLANE

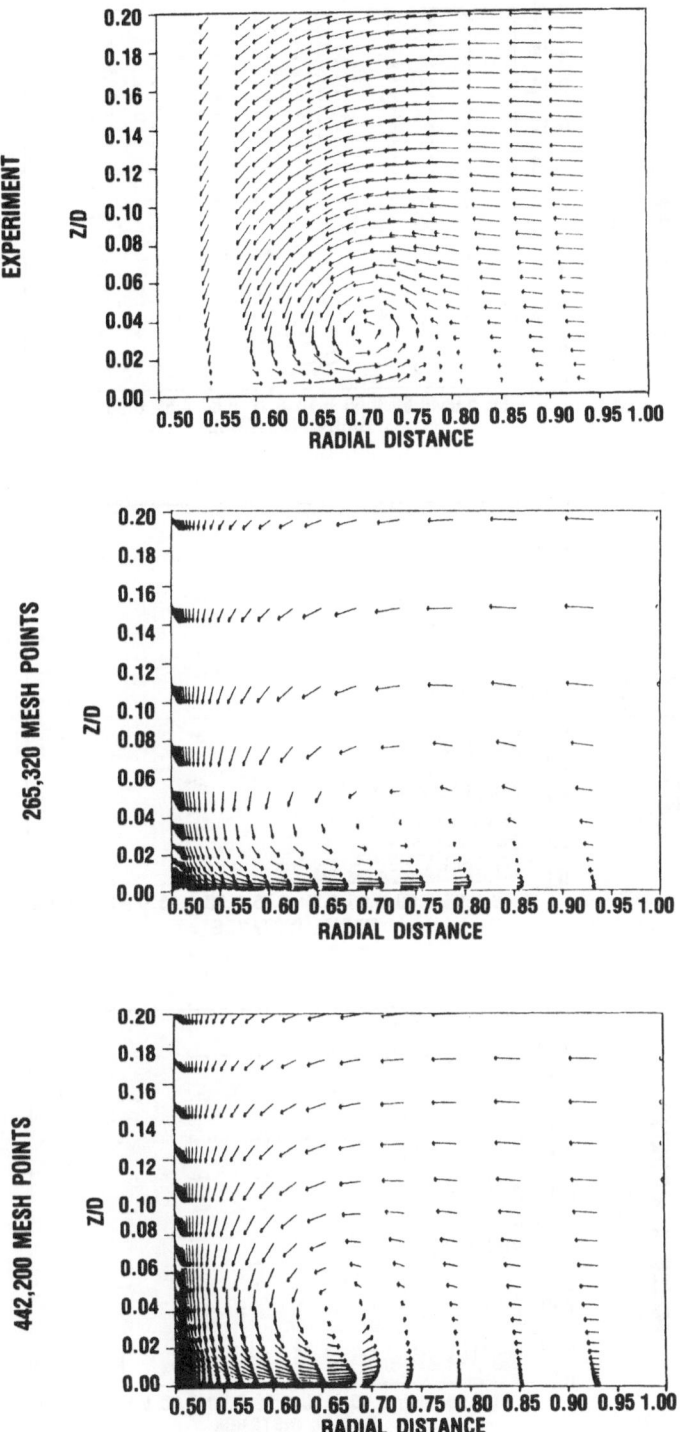

FIGURE 12.   VELOCITY VECTORS 25 DEGREE PLANE

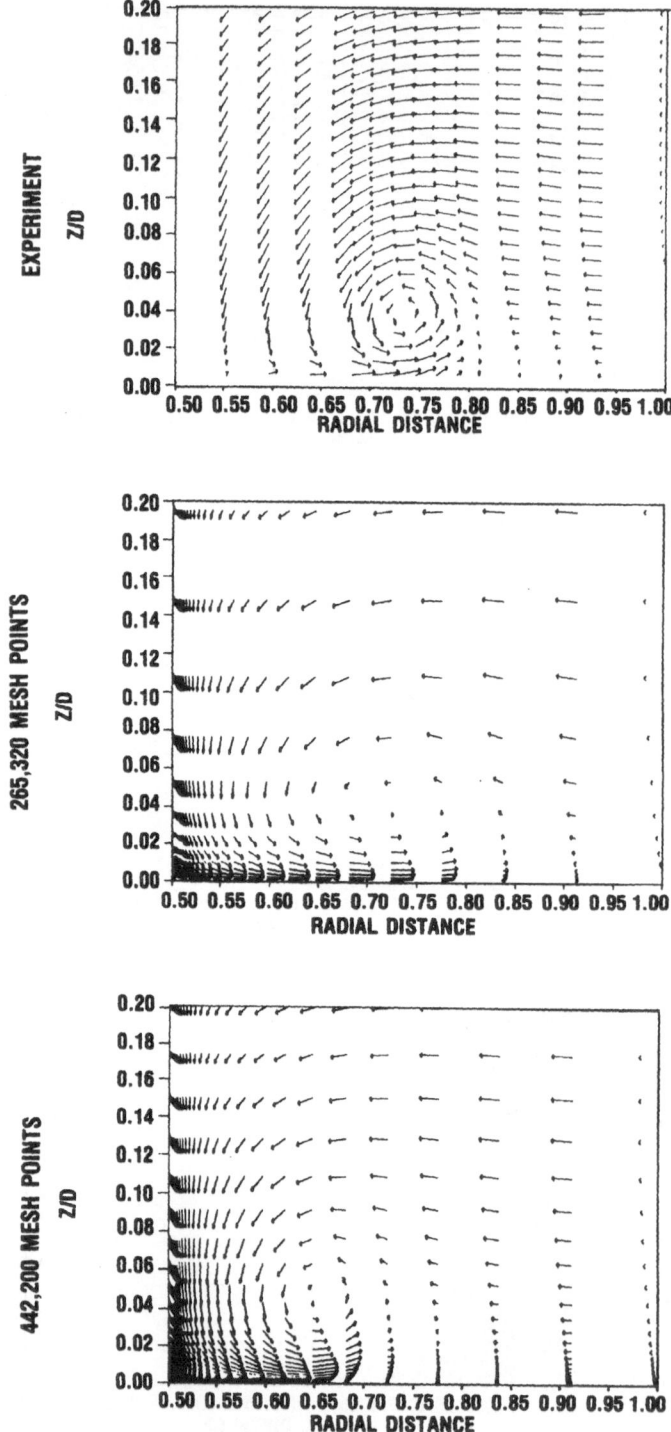

FIGURE 13. VELOCITY VECTORS 45 DEGREE PLANE

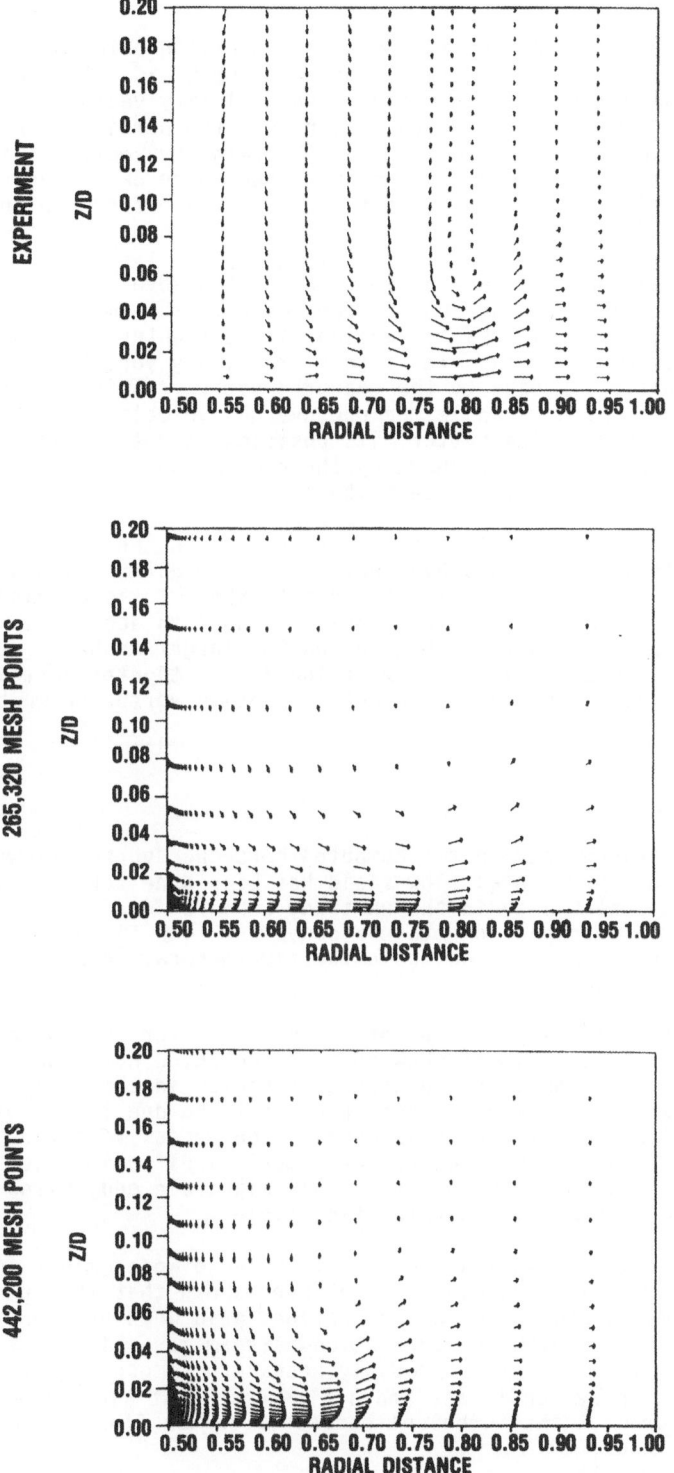

FIGURE 14.   VELOCITY VECTORS 90 DEGREE PLANE

VELOCITY COMPARISONS

Figures 11, 12, 13, and 14 show the comparison between the two computed and measured velocity vectors for the 0, 25, 45, and 90 degree planes, respectively. In each figure the velocity vector shown in a projection of the total velocity vector onto the particular plane indicated. The relative magnitude of the velocity vector is indicated by the length of the arrow. For each of the three cases shown this arrow length can be compared directly to see the difference in absolute velocity magnitude.

One of the more interesting features is the development of the horse-shoe vortex from the larger unconverged mesh to the smaller converged mesh. As the vortex starts to develop, it is tighter and smaller than the final computed vortex. It is this transitory vortex which actually correlates better in size to the measured vortex. This is clearly seen in the 0 degree plane. While the unconverged velocity field matches the experimental vortex size better, its position is not nearly as far from the cylinder as the vortex shown by the converged solution. This is consistent between all the planes shown.

Note also the same diffusion of the vortex in the final computed solution. Again it is possible that the overly diffuse appearance of the computed vortex may be attributable to the artificial dissipation present and/or the lack of the turbulence model to accurately predict this flow. Since we were unable to run the larger mesh out any further in time, it is not possible to determine if the tighter vortex shown does eventually diffuse into the more expansive vortex shown by the plots for the smaller mesh.

CONCLUSIONS

The comparison between two computed horseshoe vortex flows and the corresponding experimental flow field has been done for two different mesh sizes. Instability in the unsteady wake region for the larger mesh used made it impossible to obtain a converged flow field. However, the results can be used to demonstrate the time-generation process of the computed horseshoe vortex.

The overall total pressure loss comparisons for the converged solution and the experiment show that the computed horseshoe vortex does not have the same concentrated high loss region in the vortex core. This is shown in the velocity vector plots to be due to the overly diffuse vortex predicted by this model. The excessive diffusion can be attributed to either the presence of numerical diffusion used for stability or the inability of the simple algebraic eddy viscosity model to predict this complex separated flow field.

A more refined mesh and a better turbulence model might improve the computed flow field. However, it has been shown that this numerical method becomes unstable for a more refined grid when used on this particular problem with the current downstream mesh distribution and boundary conditions. It is probably necessary to use full time dependent boundary conditions and a time dependent algorithm to more accurately predict the turbulent horseshoe vortex flow around an endwall mounted cylinder.

REFERENCES

1. W. A. Eckerle and L. S. Langston, "Measurements of a Turbulent Horseshoe Vortex Formed Around a Cylinder," NASA Contractor Report 3986, June 1986.

2. D. Kwak, J. L. C. Chang, S. P. Shanks, and S. R. Chakravarthy, "An Incompressible Navier-Stokes Flow Solver in Three-Dimensional Curvilinear Coordinate System Using Primitive Variables," AIAA Paper No. 84-0253, AIAA 22nd Aerospace Sciences Meeting, Reno, Nevada, January 9-12, 1984.

3. J. L. C. Chang and D. Kwak, "On the Method of Pseudo-Compressibilty for Numerically Solving Incompressible Flows," AIAA Paper No. 84-0252, AIAA 22nd Aerospace Sciences Meeting, Reno, Nevada, January 9-12, 1984.

4. C. A. Wagner, "Computations of a Horseshoe Vortex Around an Endwall Mounted Cylinder," Ph.D Thesis, University of Connecticut, Storrs, Connecticut (in preparation).

5. A. J. Chorin, "A Numerical Method for Solving Incompressible Viscous Flow Problems," J. Comp. Phys., vol. 2, pp. 12-16, 1967.

6. A. J. Musker, "Explicit Expression for the Smooth Wall Velocity Distribution in a Turbulent Boundary Layer," AIAA J., vol. 17, no. 6, June 1979.

7. H. McDonald and R. W. Fish, "Practical Calculations of Transitional Boundary Layer," Int. J. Heat Mass Transfer, vol. 16, pp. 1729-1744, 1973.

8. W. R. Briley, R. C. Buggeln, and H. McDonald, "Solution of the Three-Dimensional Navier-Stokes Equations for a Steady Laminar Horseshoe Vortex Flow," AIAA Paper No. 85-1029-CP, AIAA 7th Computational Fluid Dynamics Conference, Cincinnati, Ohio, July 15-17, 1985.

NUMERICAL INTEGRATION AND SPARSE BLOCKED EQUATION SOLUTION TECHNIQUES FOR LARGE SCALE BOUNDARY ELEMENT ANALYSIS

James H. Kane and Sunil Saigal

Computer Aided Engineering Center
Worcester Polytechnic Institute, Worcester MA 01609

INTRODUCTION

The Boundary Element Method (BEM) has steadily developed a niche for itself as a viable approach to the solution of problems in computational mechanics. Starting with the successful modeling of bulky objects in structural and thermal analysis, accomplishments of boundary element researchers in the simulation of physical behavior in linear acoustics, elastodynamics and electromagnetics, and fluid dynamics are well documented. Recently, nonlinear response predictions in the areas of metal plasticity and creep, large strains and deflections, contact and friction, and solidification have been demonstrated. Coupling of this method with the more mature Finite Element Method (FEM) has also been shown to be an important option for problems with a certain duality in character. The advantage that the BEM in obviating the requirement for the discretization of the entire domain of certain models and also for the subsequent calculation of fundamental solution variables at all discretization points in these domains is compelling, as is the fact that the BEM can naturally handle boundary conditions at infinite boundaries in an exact fashion. The utilization of the BEM on extremely large problems has been inhibited, however, by the lack of effective implementations of the formulations. Major strides in the performance of the computational tasks that consume the most computer resources in BEM can be made, and are required before the positive attributes of the BEM can be exploited in the analysis of very large problems. Considered together, numerical integration and equation solving can amount to 90 % of the computer resources expended in a Boundary Element Analysis (BEA).

This paper contains a discussion of some techniques in equation solving and numerical integration that can significantly impact the computer resources required to perform large scale boundary element analysis. The concept of multi-zone boundary element analysis is first introduced including an example of the reduction in computer storage and processing time that can result due to the corresponding imposition of a sparse blocked character on the resulting overall boundary element system equations. Improvements in the time required for the numerical integration step in multi-zone analysis is also demonstrated, along with the improvement in the accuracy of multi-zone model overall solutions due to better conditioning of the matrices involved. Techniques to condense (substructure) unchanging portions of nonlinear boundary element models are also discussed. This technique is shown to naturally allow for the parallel processing of the individual boundary element zone condensations, and for the solution of the overall boundary element system equations in drastically less high speed memory. For problems with localized nonlinearity, this methodology is shown to provide a strategy that can make extremely large problems tractable on today's computers. Some techniques to produce an order of magnitude improvement in the efficiency of the numerical integration of the boundary integral equations are presented. These techniques involve the strategic selection and subsequent reuse of sample point geometric information in the numerical integration process. These techniques can be thought of as enabling techniques that begin to bring the BEM to a more competitive position in relation to the FEM in regard to computational efficiency, and allow for effective utilization of the BEM on extremely large problems in computational mechanics.

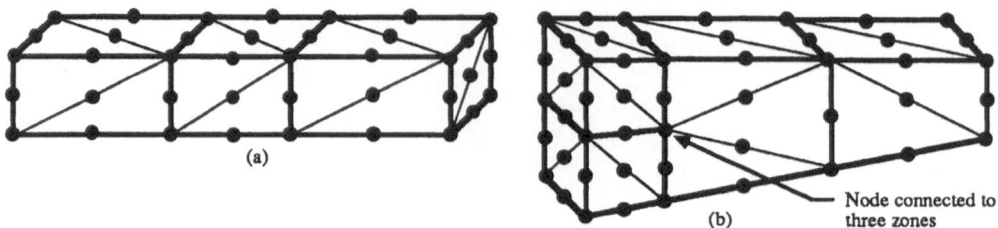

Figure 1 . Multi-zone boundary element models.

MULTI-ZONE BOUNDARY ELEMENT ANALYSIS

Literature survey

Boundary element analysis techniques that give rise to overall system matrices that have a blocked and sparse character due to the employment of multi-zone models are introduced in textbooks by Banerjee and Butterfield[1], Brebbia and Walker[2], and Brebbia, Telles, and Wrobel[3]. Articles concerned with the assembly and solution of these types of unsymmetric, sparse blocked matrix equations include Lachat and Watson[4,5] and Lachat[6], Crotty[7] and Das[8], in which the multi-zone strategy is shown to provide the added benefit of significantly extending the range of model shapes and characteristics that can be successfully treated by the BEA method, while simultaneously producing substantive computational economy in the numerical integration, equation solving, and response recovery phases of the overall analysis procedure. Bialecki and Nahlik[9] and Bialecki[10] discuss what they term an unsymmetric sparse blocked frontal equation solving algorithm in the context of solving BEA nonlinear heat transfer problems, while presenting a treatment that extends the applicability of the Kirchhoff transformation concept to multiple zones with different material properties. Tomlin[11] and Butterfield and Tomlin[12] present very similar approaches that are motivated by the solution of nonhomogeneous continuum problems modeled by multiple piecewize homogeneous boundary element zones, and Chang[13] employed multiple piecewize homogeneous zones and coupled these zones to finite elements to solve seepage problems.

Beer[14], Mustoe[15], and Davies[16] describe procedures coupling boundary elements and finite elements in the same analysis and condense the boundary element matrices to just the degrees of freedom on the boundary element - finite element interface. Condensation of boundary element degrees of freedom with known boundary conditions is also presented by Jin, Runesson, and Sammulson[17] and Margenov, Georgiev, Hadjikov, and Novakova[18] in conjunction with solving elastic contact problems, in an approach similar to those taken in some finite element contact analysis strategies exemplified by the publications of Nour-Omid and Wriggers[19]. Kane, Saigal, et al.[20-26] have demonstrated how multi-zone boundary element analysis and sensitivity analysis significantly impacts the ability to exploit the additional matrix sparsity present in the design sensitivity analysis step occurring during shape optimization of objects with partial geometric sensitivity. They have also shown how the multi-zone capability can facilitate the effective utilization of reanalysis techniques in the shape optimization context. Kline[27] and his co-workers developed a multi-zone analysis approach that always condenses the degrees of freedom exclusively in a single zone, and this condensation approach was demonstrated[28] to be quite amenable to parallel computation. The results given by Bettess[29] on mixed direct - iterative methods for equation solving in boundary element analysis is also significant, although this treatment has not been extended to multi-zone analysis. In another publication Kane and Saigal[30] describe a formulation that allows for the arbitrary selection of boundary element zones to be either condensed into substructures, or left at their original size, and also for their subsequent optional expansion, as desired, is presented. This approach allows for both condensed and uncondensed boundary element zones to consistently coexist in the same multi-zone analysis. For shape optimization problems, they[31] have also demonstrated the ability to perform sensitivity analysis of models that are composed totally or in part of substructures.

The Formulation

Multi-zone boundary element analysis[1-8] is accomplished by first breaking up an entire boundary element model into zones as illustrated, for example, by the three zone model shown in Figure 1a. The surface of each boundary element zone totally encloses a sub-volume of the entire model and may generally contain boundary elements and nodes that belong exclusively in this single zone, and also boundary elements and nodes that are on an interface with other zones. The governing boundary integral relationship can then be written for each zone. In elastostatics, for example, Somigliana's identity[1-3] is the appropriate relationship. Substituting for the actual surface response, an approximate surface response interpolated from the node point values of traction and displacement using boundary element interpolation functions, one obtains the discretized boundary integral equation[1-3]. By evaluating this expression at a set of locations of the load point of the fundamental solutions occurring in the boundary integral equation corresponding to the node point locations for the zone in question, one can generate a matrix system of equations for each zone.

$$[F^i] \{u^i\} = [G^i] \{t^i\} \tag{1}$$

In this equation, $[F^i]$ and $[G^i]$ are square and rectangular coefficient matrices, respectively, and $\{u^i\}$ and $\{t^i\}$ are column vectors of node point displacement and traction components, respectively, of conformable size. The rectangular nature of the $[G^i]$ matrix is due to the fact that jumps in specified tractions can exist when continuous boundary elements are employed, and these additional traction components are collected and put at the end of the vector $\{t^i\}$. In all symbols shown in Equation (1), and below, the "i" superscript denotes that the superscripted quantities are those associated with the i'th zone in a multi-zone BEA model.

The matrix relations written for each of the individual zones can be put together for use in an overall analysis[1-8] by considering the conditions of displacement compatibility and equilibrium of the traction components at all zone interfaces. The continuity condition requires that the nodal displacements calculated for zone-i at an interface between zone-i and zone-j, must equal the nodal displacement components calculated in zone-j at that same interface. A similar relationship exists due to equilibrium considerations for the components of the traction vector at interface nodes between two zones, except that a negative sign must be present to account for the opposite directions of the outward surface normals in the two zones in question.

$$\{u_{ij}^i\} = \{u_{ij}^j\} \tag{2}$$

$$\{t_{ij}^i\} = -\{t_{ij}^j\} \tag{3}$$

In these compatibility and equilibrium relations, the double subscript notation is used to convey that the vector in question is a column vector of components entirely on the interface between zone-i and zone-j. Expanding the size of the boundary element zone matrix equations to the size of the overall problem, bringing the unknown tractions at zone interfaces to the left hand side of the equation, and using the compatibility and equilibrium relations, one can form the boundary element system equations for the overall multi-zone BEA problem. For example, the equations for the three zone problem shown in Figure 1a are given below. It should be noted that this model has no interface between zone-1 and zone-3. In this instance, the final multi-zone BEA system of equations can be produced by simply removing the blocks associated with this 1-3 interface shown in Equation (4).

$$
\begin{bmatrix}
[F_{11}^1] & [F_{12}^1] & [0] & -[G_{12}^1] & [0] & [0] & [0] & [0] & [0] \\
[0] & [F_{12}^2] & [0] & [G_{12}^2] & [F_{22}^2] & [F_{23}^2] & [0] & -[G_{23}^2] & [0] \\
[0] & [0] & [0] & [0] & [0] & [F_{23}^3] & [0] & [G_{23}^3] & [F_{33}^3]
\end{bmatrix}
\begin{Bmatrix}
\{u_{11}^1\} \\
\{u_{12}^1\} \\
\{u_{13}^1\} \\
\{u_{21}^1\} \\
\{u_{22}^2\} \\
\{u_{23}^2\} \\
\{t_{31}^1\} \\
\{t_{32}^2\} \\
\{u_{33}^3\}
\end{Bmatrix}
$$

$$
=
\begin{bmatrix}
[G_{11}^1] & [0] & [0] & [0] & [0] & [0] & [0] & [0] & [0] \\
[0] & [0] & [0] & [0] & [G_{22}^2] & [0] & [0] & [0] & [0] \\
[0] & [0] & [0] & [0] & [0] & [0] & [0] & [0] & [G_{33}^3]
\end{bmatrix}
\begin{Bmatrix}
\{t_{11}^1\} \\
\{0\} \\
\{0\} \\
\{0\} \\
\{t_{22}^2\} \\
\{0\} \\
\{0\} \\
\{0\} \\
\{t_{33}^3\}
\end{Bmatrix}
\tag{4}
$$

The previous example points out the basic features of multi-zone boundary element matrix equations. The matrix equation shown above is actually a hypermatrix with matrices as its entries, and generally these matrix entries are called blocks or partitions. Likewise, the overall vectors shown have vectors for their entries and these entries are also referred to as blocks or partitions. The blocked sparsity characteristic of the matrices that

```
START.        Start block triangular factorization;
              I = 0
DIAG_LOOP.       I = I + 1
                 IF ( I > N-1 ) GO TO LAST_BLOCK
                 Factor  $A_{II} = L_{II} U_{II}$  by Gauss elimination with partial pivoting
                 J = 0
COL_LOOP.           J = J + 1
                    IF ( J > N ) GO TO DIAG_LOOP
                    IF ( $A_{JI}$ = 0 ) GO TO COL_LOOP
                    Solve  $A_{II} D = A_{JI}$  by forward reduction, back substitution of the columns of D
                    K = I
ROW_LOOP.              K = K + 1
                       IF ( K > N ) GO TO COL_LOOP
                       IF ( $A_{KI}$ = 0 ) GO TO ROW_LOOP
                       Form  $A_{KJ} = A_{KJ} - A_{KI} D$
                       GO TO ROW_LOOP
LAST_BLOCK.   Factor $A_{II} = L_{II} U_{II}$
END_PROC.     Return
```

Figure 2 . Sparse blocked matrix factorization algorithm.

result from the multi-zone BEA approach is clearly evident from the zero blocks present in Equation (4). The order of the occurrence of the partitions shown in Equation (4) effects the amount of computer storage and the number of computations required to solve the overall matrix system of equations by direct methods. This is due to a phenomenon associated with the block triangular factorization step where zero blocks above the main diagonal block, but located below a populated block in a block column experience block fill in[4-8]. A similar block fill in occurs in block rows to the left of the main diagonal. The particular order of the partitions presented in the example shown above has been purposely selected to cluster the non-zero blocks as near to the main diagonal of the matrix as possible. For arbitrarily connected zones in multi-zone, continuous, boundary element analysis, nodes may very often be present in the model that are connected to more than two zones. Figure 1b, for example, illustrates a situation where a multi-zone boundary element model has some nodes that are simultaneously connected to elements in three different zones. Each distinct combination of nodes sharing particular zones present in a multi-zone analysis determines a partition in the overall blocked matrix equations. Thus, one must generally consider the development of an algorithm whereby it is possible to have nodes present in a model connected to elements in as many zones as are present in the overall model, and yet still be able to automatically determine an optimal partition ordering scheme.

Sparse Blocked Triangular Factorization Procedures

To exploit the significant sparsity imposed by the multi-zone analysis procedure presented above, an effective blocked matrix factorization strategy must be employed. The blocked matrix triangular factorization procedure used in this study is shown in Figure 2. The procedure starts with the triangular factorization of the first diagonal block. This is performed using a Gauss elimination algorithm with partial pivoting. The triangular factors of this diagonal block are stored in the same location that the original diagonal block was located. This matrix factorization is then used to alter the second block column. This is accomplished by forward reduction and backward substitution of the columns of the matrix block A_{12} to form the matrix D_{12}, (in the figure this matrix is symbolized by D). This matrix is then used to alter all the blocks below A_{12} in block column two by the matrix multiplication and subtraction step shown in Figure 2. All blocks below block A_{13} in column three is then processed in a similar fashion and then the fourth column and so forth until the entire matrix has been altered. This entire process is then repeated using the submatrix consisting all blocks except those in block row and column one. The second major phase in the algorithm causes the alteration of the submatrix consisting of all blocks except those in block rows and columns one and two. At every stage of this process, checks are made concerning the sparsity of the matrix. Any block operation that can be avoided due to the block sparsity present in the matrix, is avoided. Except for the need to store the largest D_{ij} block that ever occurs in the overall procedure, the entire operation can be done 'in place', with the updated matrix blocks replacing the original blocks in computer memory. The fundamental characteristic of this block triangular factorization algorithm described above and in Figure 2 is that it is sequential in nature. For example, the triangular factorization of the second and third diagonal blocks can not be performed in parallel because the second and third diagonal blocks must first be altered by the algorithm and are therefore not in the appropriate form at the beginning of the procedure.

Performance of Multi-zone Procedures

In order to illustrate the advantages associated with multi-zone analysis relative to single-zone boundary element analysis, an example geometry that presents definite pathological problems for single zone analysis has been selected for study and modeled as both a single zone and as a multi-zone model. A twenty to one aspect ratio (ratio of length to lateral dimension) beam problem was run as a single zone BEA model and the resulting timings and storage information displayed in Figure 3. The same problem was modeled using a four zone boundary element mesh also shown in Figure 3. Note that the four zone model has alternating zones shown shaded. Corresponding computer storage space and CPU time required for the multi-zone analysis is also shown in this figure. In this figure, the CPU times for certain major operations are given inside the rectangular boxes shown. Comparison of the resources consumed during the equation solving step for both the single zone and the multi-zone analyses shows the dramatic improvements that can be obtained by employing the multi-zone approach in these class of problems. The single zone analysis used about five times the CPU time and three times as much main memory as the multi-zone analysis. The third case shown in this figure depicts the results of a multi-zone analysis where the individual zone boundary element matrices are condensed as described later on in this presentation, and elsewhere[30]. This third example ran in about the same time as the second, but the analysis was performed using considerably less computer memory due to the condensation strategy adopted.

The time spent in performing the matrix factorization step is shown to undergo the most significant improvement. In addition, numerical integration times tend to also be less for multi-zone models because only load points corresponding to nodes in the particular zone being integrated are used in the sequence of integrations used to form zone coefficient matrices. The accuracy of the boundary element method when applied to model such slender objects is can also be dramatically improved in some instances when multi-zone techniques are utilized. In the beam problem described above, the accuracy of the predicted end deflection produced by the single-zone model was comparable to the corresponding multi-zone mesh. In a subsequent study, the same BEA models were 'stretched' to a forty to one aspect ratio shape. In these analyses, the multi-zone model remained stable, while the single zone model yielded erroneous results. When one contrasts the resources corresponding to the first example (single zone model) with the CPU time and memory resources utilized in the third example (multi-zone model utilizing condensation), the substantive impact of the techniques begins to become apparent. In a subsequent part of this presentation, the impact of the substructuring approach applied to a nonlinear problem will be shown to be even more compelling.

BOUNDARY ELEMENT SUBSTRUCTURES

The concept of condensation of degrees of freedom in the BEA context can be introduced by considering the boundary element matrix equations for a single zone. In the equations that follow, the superscript notation indicating the number of the particular zone being considered will be implied but explicitly omitted. Reordering the degrees of freedom and partitioning Equation (1) into blocks that correspond to master degrees of freedom and also into blocks that correspond to degrees of freedom that could be condensed, one can arrive at the matrix equation shown below. In this equation, the additional right hand side vector is included to consistently account for any body force type of loading that might be present in the analysis, such as gravity, centrifugal, of thermal loading.

$$
\begin{bmatrix} [F_{MM}] & [F_{MC}] \\ [F_{CM}] & [F_{CC}] \end{bmatrix} \begin{Bmatrix} \{u_M\} \\ \{u_C\} \end{Bmatrix} = \begin{bmatrix} [G_{MM}] & [G_{MC}] \\ [G_{CM}] & [G_{CC}] \end{bmatrix} \begin{Bmatrix} \{t_M\} \\ \{t_C\} \end{Bmatrix} + \begin{Bmatrix} \{f_M\} \\ \{f_C\} \end{Bmatrix}
$$

$$\text{(5a)}$$
$$\text{(5b)}$$

Solving the matrix Equation (5b) for $\{u_C\}$ gives

$$
\{u_C\} = [F_{CC}]^{-1} [G_{CM}] \{t_M\} + [F_{CC}]^{-1} [G_{CC}] \{t_C\} - [F_{CC}]^{-1} [F_{CM}] \{u_M\} + [F_{CC}]^{-1} \{f_C\} \qquad (6)
$$

Substituting Equation (6) into the matrix Equation (5a) and collecting terms yields

$$
[M_1] \{u_M\} = [M_2] \{t_M\} + [M_3] \{t_C\} + [M_4] \{f_C\} + \{f_M\} \qquad (7)
$$

Where

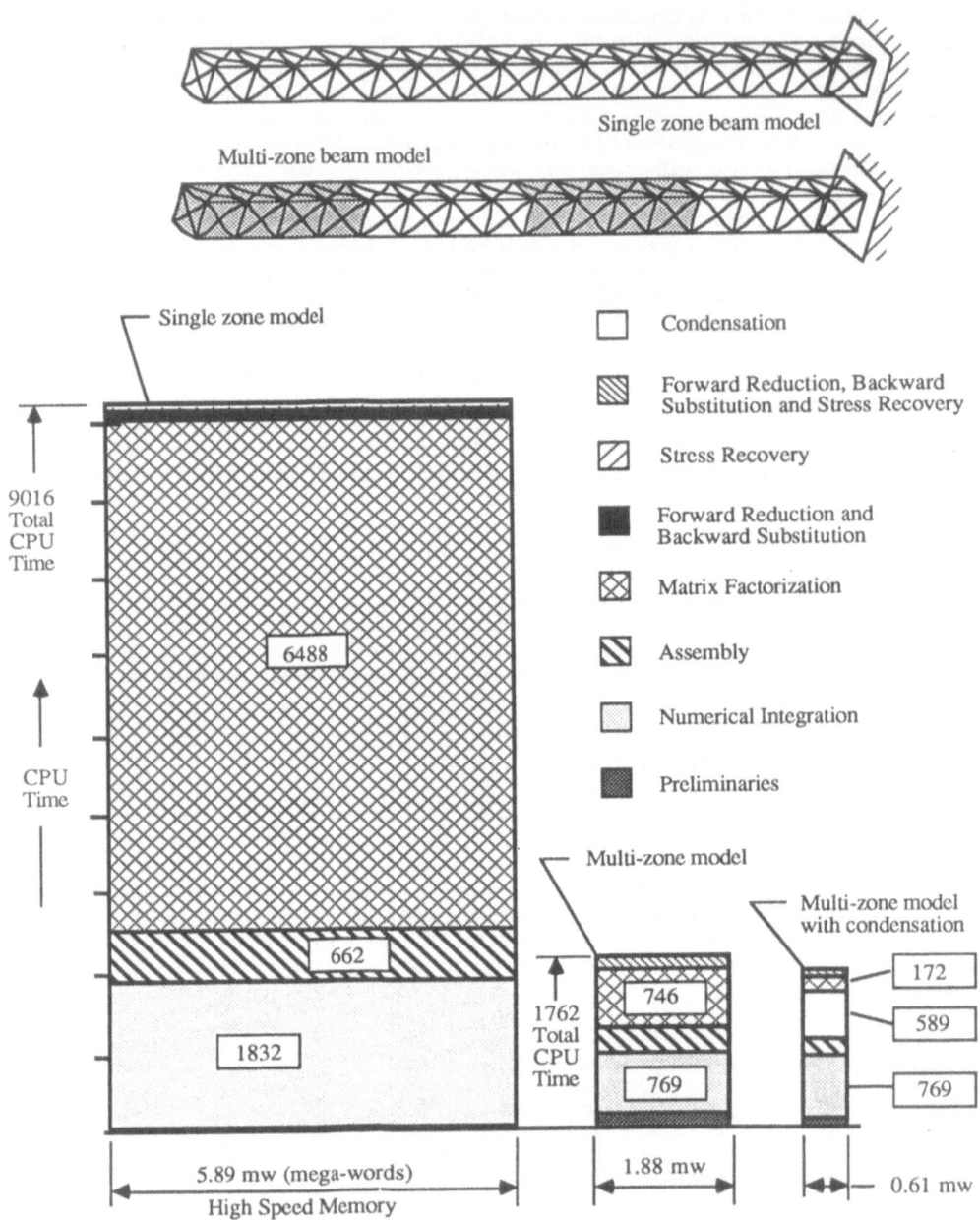

Figure 3 . Performance Study of Single and Multi-zone BEA Models

$$[M_1] = [F_{MM}] - [F_{MC}] [F_{CC}]^{-1} [F_{CM}] \tag{8}$$

$$[M_2] = [G_{MM}] - [F_{MC}] [F_{CC}]^{-1} [G_{CM}] \tag{9}$$

$$[M_3] = [G_{MC}] - [F_{MC}] [F_{CC}]^{-1} [G_{CC}] \tag{10}$$

$$[M_4] = - [F_{MC}] [F_{CC}]^{-1} \tag{11}$$

Equation (7) can be appropriately called a condensed boundary element zone matrix equation, while Equation (6) referred to as a boundary element zone matrix expansion equation. Note that this procedure is general, in that no restriction has been placed on which degrees of freedom are chosen as master or condensed degrees of freedom. The zone matrix assembly procedure must be able to determine the destinations of all element contributions in the partitions shown above. It should also be noted that the condensation procedure embodied in the equations presented above is an exact formulation, in that, no terms have been neglected, nor has any approximation been made.

Figure 4 illustrates the relative sizes of the matrices and vectors present in a typical boundary element zone substructuring process. This figure, along with the substructuring equations, can be used to help describe and understand the individual boundary element zone, first level assembly algorithm used in this overall analysis procedure, and also the subsequent, optional condensation step. The first level assembly algorithm requires that accounting be performed to indicate the appropriate partition destinations for all boundary element contributions to the individual zone system matrices. These contributions can then be assembled with due

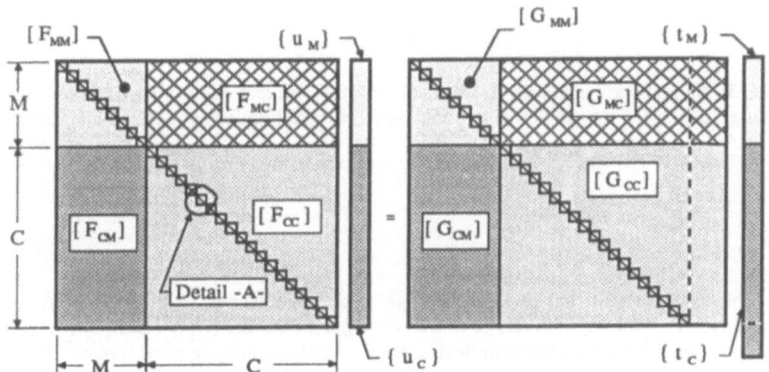

Assembled zone matrix showing master and condensed partitions

Detail -A- ; The 3 x 3 block of the zone
matrices corresponding to singular integration

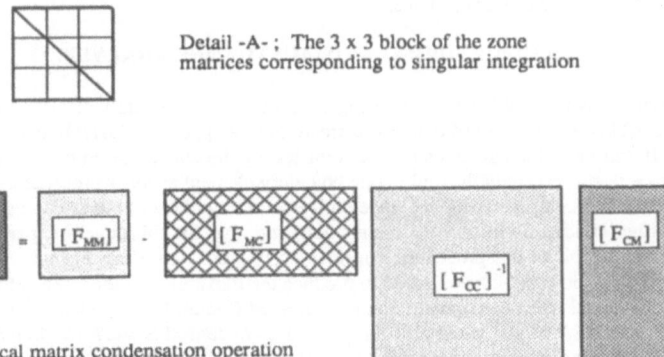

Typical matrix condensation operation
with corresponding matrix sizes

Figure 4 . Details associated with the assembly, reordering, partitioning, and condensation
of a boundary element zone matrix

regard for the boundary conditions present in the model, and by requiring that the individual zone's row sum be used to determine the diagonal block entries in the [F] matrix shown in Figure 4. These diagonal entries correspond to the integration of the singular traction kernel, shape function, and Jacobian products that occur when the load point of the fundamental solutions used in the boundary element analysis coincides with a node point of the element currently being integrated[1-3]. Note further that, due to the possible presents of specified displacement boundary conditions, corresponding column exchanges could necessitate the assembly of this row sum information into the [G] matrix, instead of the [F] matrix.

At first inspection, the substructuring process seems to require the inversion of the matrix block $[F_{CC}]$. A closer examination of the formulation, however, reveals that this is not the case. Whenever the matrix $[F_{CC}]^{-1}$ appears in these equations, it always premultiplies either a column vector or a rectangular matrix. As shown below, the use of the matrix inversion notation is purely symbolic, and in the computer implementation of this substructuring approach, no matrix inversion is ever actually performed. Instead, the triangular factorization of the matrix block $[F_{CC}]$ is performed once, and subsequently these factors are used to solve the matrix equations shown below by forward reduction and backward substitution of the right hand side vector or group of vectors shown below. This is possible because the result of the product of the matrix inverse $[F_{CC}]^{-1}$ and a vector or matrix can be premultiplied by $[F_{CC}]$ as shown below to yield a system of linear equations.

$$\{d\} = [F_{CC}]^{-1} \{v\}$$

$$[F_{CC}] \{d\} = \{v\} \tag{12}$$

$$\{D\} = [F_{CC}]^{-1} \{V\}$$

$$[F_{CC}] \{D\} = \{V\} \tag{13}$$

In Equation (13) the rectangular matrix [D] is a collection of column vectors $\{d_i\}$ that are formed from forward reduction and backward substitution of the column vectors $\{v_i\}$ collected in the rectangular matrix [V], as shown below.

$$[D] = [\{d_1\}, \{d_2\}, \ldots, \{d_N\}]$$

$$[V] = [\{v_1\}, \{v_2\}, \ldots, \{v_N\}] \tag{14}$$

It is also possible to save significant computer storage space[30] and data movement in the implementation of the above formulation by avoiding the storage of the rectangular matrices $[G_{MC}]$ and $[G_{CC}]$. Examination of this condensation formulation quite clearly indicates that the fundamentally important operation involved in the condensing process is the triangular factorization of the matrix block $[F_{CC}]$. This is followed by a series of forward and backward substitution operations using this factorization of the columns of the matrix blocks shown to form an intermediate matrix [D]. For each [D], the matrix multiply and subtract operations indicated in the figure must also be performed. In these equations, there are also a number of operations on an individual vector instead of the matrices discussed above.

COMBINATION OF SUBSTRUCTURING AND MULTI-ZONE ANALYSIS

A very natural way to combine substructuring with multi-zone boundary element analysis capability is to allow for the possible condensation of degrees of freedom that appear exclusively in any particular boundary element zone. In this case, the partitions to be eliminated by the condensation process coincide exactly with certain partitions already present in the multi-zone boundary element analysis procedure. This approach is also very natural from a modeling perspective, since entire boundary element zones can be easily and arbitrarily identified for either condensation or no condensation, and also for subsequent expansion if they are to be condensed. As an example of this procedure, consider the four zone boundary element model shown in Figure 1b. An automated graphic capability was used to produce the illustration of the left hand side matrix associated with this four zone model when no condensation is performed (Figure 5a). The left hand side matrix associated with this exact same BEA model when all four zones are condensed is shown in Figure 5b. These graphic illustrations of the two matrix populations were generated in the identical scale. In the overall matrix with no condensation performed, (Figure 5a), the block entries shown correspond to blocks present in the individual zone [F] and [G] matrices. In the overall matrix formed from the four condensed zone contributions, (Figure 5a), the block entries shown correspond to the zone [M] matrices shown in Equation (7). It should also be

236

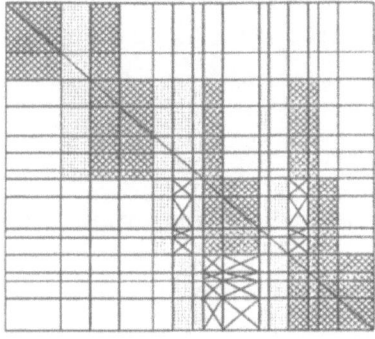

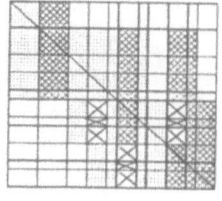

b) MATRIX POPULATION
   WHEN ALL ZONES ARE
   CONDENSED

a) MATRIX POPULATION WHEN NO
   CONDENSATION IS PERFORMED

= BLOCK CONTAINING INFORMATION FROM [F] OR [M_1] MATRIX

= BLOCK CONTAINING INFORMATION FROM [G] OR [M_2] MATRIX

= BLOCK THAT IS INITIALLY EMPTY BUT EXPERIENCES FILL IN
  DURING THE BLOCK TRIANGULAR FACTORIZATION STEP

Figure 5. Left hand side boundary element system matrix populations for
the four zone mesh shown in Figure 1.

noted that many other combinations of zone condensations are possible. For example, zone-3 and zone-4 could be left at there original size, while zone-1 and zone-2 are condensed.

When zones can be arbitrarily selected for condensation, a second level assembly procedure for the formation of the overall sparse blocked system of equations must be able to assemble condensed or uncondensed zone contributions to the overall matrix. The algorithm for this second assembly step is very similar to the assembly procedures described in References [4-8], except for the additional complication of dealing with both condensed and uncondensed zone contributions. This procedure takes each column of a zone matrix partitions shown in Figure 4, determines its block destination and column destination within the block of the overall system matrix, and proceeds to assemble this column. The entire overall system matrix is stored in a one dimensional array with appropriate accounting arrays (twenty nine of them) used to indicate the locations and sizes of the individual blocks, along with indicators regarding whether each block is full, empty, or to-be-filled-in in the subsequent block triangular factorization step. Other information in these arrays includes the block numbers associated with each zone, the blocks associated with each node, the order of each degree of freedom in a particular block, and weather a block is a [F], [G], or [M] type of block.

It is important to note that the boundary element zone condensation step, described above, can be performed in parallel. The major computational tasks associated with this process include the triangular factorization of the blocks [F_{CC}], the forward reductions and backward substitutions indicated in Equation (13), and the matrix multiplication and subtraction operations indicated in Equations (8) through (11). These processing steps for the individual zones are totally independent of the other zones and can therefore be done simultaneously on separate processors[29]. Thus, the incorporation of the condensation of boundary element zones in multi-zone BEA, shifts the proportions of the computational resources expended in the overall equation solving process towards operations that lend themselves to parallel execution. The condensation approach also produces a favorable data structure for most parallel computing environments. This is due to the fact that the overall assembled system matrix is smaller in size and has less blocks. The fact that less block fill in generally occurs when condensation is performed, also leads to the supposition that with this resulting reduction in the amount of computation should also be accompanied by less roundoff error.

One additional consideration regarding the utilization of condensation in multi-zone analysis involves the extent to which partial pivoting can be employed. In the algorithm presented in this paper, partial pivoting is done exclusively within the diagonal blocks present in the overall system left hand side matrix, during block triangular factorization. There are rare cases, however, where pivoting beyond the boundaries of the diagonal blocks is warranted. For these cases, at least two alternative approaches have been used successfully by the

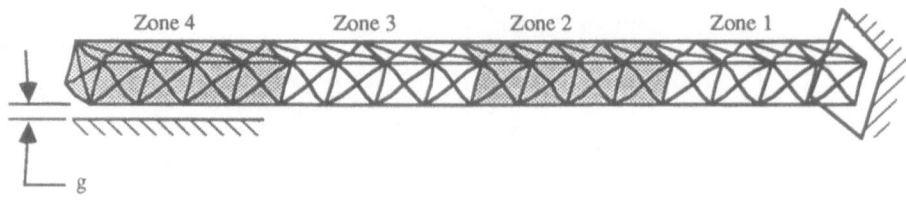

Figure 6. Contact Problem Studied With and Without Friction

authors. Because the condensation approach described herein can result in such relatively small overall system matrices, these system matrices can be factored by a full gauss elimination procedure with partial pivoting. For multi-zone models that have every zone selected for condensation, the penalty paid for treating the overall matrix as if it were fully populated is not very severe. For mixed problems, however, containing both full and condensed zones, this is not a viable strategy. For these cases, block pivoting is also possible, where an entire diagonal block is exchanged with another block of the same size in the block row. In the use of this approach, an effective criterion for the selection of the pivot blocks is required. To date the authors have used the approximate condition number of the modified diagonal block at the time of its triangular factorization as the criterion for its exchange with another block. While this requires a very marginal increase in storage and computation, it does facilitate the block pivoting strategy providing the necessary robustness in the overall procedure. Another benefit of this block pivoting strategy, is that the condition numbers of the diagonal blocks are available and can be used by the analyst to asses the goodness of the overall solution and the accuracy of the equation solving step.

A Nonlinear Example with Substructuring

As an example of the employment of the arbitrary condensing, noncondensing multi-zone equation solver in a nonlinear problem, the twenty to one aspect ratio, four zone beam model was used to simulate the contact problem shown in Figure 6. This problem is a geometrically nonlinear problem where the nonlinearity is localized. The beam is built in at its right end and subject to a uniformly distributed vertical traction along its entire top surface. Under its left end there is a flexible support located a small gap distance, g, below the undeflected beam. After the beam deflects due to its applied loading, a portion of the left end of the beam will come into contact with the flexible support, but it is not known a-priori which nodes will be in contact and which nodes will not. Therefore, an iterative process must be used to solve this problem. Many approaches could be used to solve this problem. For example, the application of a traction component proportional to the deflection of the flexible support could be used in each iteration until the problem converges. For this physical problem and contact algorithm described above, the computer resources required to perform ten iterations have been compiled in Table 1. Note that the solution to this problem has been approached in the following three ways: 1.) without any condensation, 2.) with condensation of the first three zones with subsequent expansion of the response to the full number of degrees of freedom in each iteration, and 3.) with condensation of the first three zones but without subsequent expansion of the response to the full number of degrees of freedom in each iteration.

Because the condensation approach discussed herein is exact, all approaches considered will produced identical converged results in the same number of iterations. The major benefit to be derived from this condensation can be identified with the help of Figure 3. In this figure, the third example (with condensation) is shown to require a much reduced amount of time in the matrix factorization step due to is smaller size. A similar type of improvement occures in the solution of these nonlinear problems using iterative processes. The CPU times required in the three runs of 8642, 3358 and 3798 however, are significantly different. Thus the condensation of unchanging portions in the nonlinear problem described above is shown to produce a two to one improvement in the computational efficiency of the overall iterative solution strategy. The third column in the table is included to show that the penalty paid for expanding to the full degrees of freedom at each stage of the iterative solution process is marginal. It is to be noted that this improvement is over and above the improvement described in the multi-zone studies depicted in Figure 3. The authors have presented other examples of the computational efficiency associated with the condensation of unchanging portions of nonlinear problems including the employment of this technique in shape optimization problems[30,31].

Table 1. Computer times for various condensation strategies in the contact problem

step description	no condensation	condensation	condensation and expansion
Accounting	16.1	13.5	13.5
Integration	769.4	769.4	769.4
First assembly	211.3	211.3	211.3
Condensation	----	589.5	589.5
Second assembly	4.3	1.2	1.2
Matrix factorization	746.0	172.9	172.9
Solution	14.2	3.3	3.3
Expansion	---	---	44.0
Total (first iteration)	1761.3	1761.1	1805.1
Second assembly	38.7	10.8	10.8
Matrix factorization	6714.0	1556.1	1556.1
Solution	127.8	29.7	29.7
Expansion	---	---	396.0
Total (9 more iterations)	6880.5	1596.6	1992.6
Total (10 iterations)	8641.8	3357.7	3797.7

NUMERICAL INTEGRATION OF BOUNDARY INTEGRAL EQUATIONS

The numerical integration step in three dimensional boundary integral structural or thermal analysis using curved boundary elements is currently one of the most computationally burdensome steps in the overall solution procedure. Existing algorithms for numerical integration of such elements have strategically clustered integration sample points based on the relative proximity of the load points to the boundary element being integrated using element subdivision or element coordinate transformation techniques. The emphasis in these techniques has been on minimizing the number of sample points required to obtain a given level of accuracy. The algorithm discussed herein, while closely following the spirit of these earlier approaches, takes a different approach, one that employs a discrete number of sets of predetermined, customized, near-optimum, sample point quantities associated with the intrinsic boundary element. The ability created by this approach to reuse sample point geometric information of the actual element allows for the realization of substantive computational economy. This algorithm provides accurate and efficient numerical results both when load points are far from, and when they are on the boundary element being integrated. Numerical results are provided to demonstrate the substantial economy achieved through the use of such an algorithm.

Another Literature Survey

Numerical integration accuracy can be maintained with some degree of computational economy by methods that test on the relative proximity of load points and boundary elements being integrated, and strategically assign the number and location of integration points to be employed in the integrations. To date, algorithms that exploit this concept have relied on either an element subdivision[4,32,33] or a coordinate transformation technique[34]. Early treatment of effective numerical integration techniques was given by Cruse[35], where flat elements were treated analytically. Aliabadi, Hall, and Phemister[36] have extended the work of Lean and Wexler[37] to analytically integrate the singular terms associated with curved boundary elements. Li, Han, and Mang[38] and Schwab[39] have used polar triangular coordinates to remove singularities from surface integrals and have also presented procedures for extending these techniques to the volume integrations associated with nonlinear analysis. The element subdivision approach requires that an element be divided into subelements and that each subelement be integrated separately. The contributions from each subelement are then combined to form the required overall element matrices. The criterion for the selection of the number of subelements required, their sizes, and the respective integration orders to be utilized within subelements is based on an approximate estimate of the accuracy produced by such a scheme. Recently, coordinate transformation techniques[34] have been used to efficiently cluster integration sample point locations without the need for element subdivision. This coordinate transformation procedure also, however, repetitively generates customized sets of integration rules, in real time, for each boundary element - load point pair involved in the analysis.

The Reusable Intrinsic Sample Point (RISP) Concept

The RISP algorithm discussed in this paper adheres closely to the spirit of the approaches mentioned above for concentrating integration points in areas of pronounced variation of the integrands under consideration. Unlike previous approaches, elements are not repetitively subdivided in the boundary element computer program. Instead, a discrete number of customized sets of intrinsic sample point locations and their associated weighting coefficients are predetermined by a combination of numerical experiment and standard accuracy criteria. In this discussion, the term intrinsic element refers to the parent element in the parametric space. The actual or real element is formed by a mapping (in this case isoparametric mapping) of the parent element to the real geometry of the model using the element's node point geometry and the boundary element interpolation functions shown elsewhere[44]. A sample point in the intrinsic element is then called an intrinsic sample point. These sets of integration rules are then appropriately selected and employed, based upon the relative proximity of boundary element - load point pairs. The manifold advantages of this approach include the items listed below: 1.) The inner loop of the numerical integration procedure does not have the burdensome subroutine call to a subroutine that takes the geometry of a subelement and computes its contribution to the overall matrices of the element. In addition, this characteristic facilitates the straightforward exploitation of vectorization and parallel computing facilities increasingly available on contemporary computers. 2.) The a-priori determination of the sets of sample points and corresponding weights is performed only once and is therefore not repeated for each boundary element - load point pair, thus avoiding significant redundant computations. 3.) These predetermined sets of sample points and weights are computed once, outside the program in quadruple precision, along with their associated isoparametric shape functions and shape function derivatives. This avoids further significant redundant computations and also postpones the onset of roundoff errors. 4.) The algorithm can handle both outside integration (i.e. load point not on the element being integrated), and inside integration (i.e. load point on the element being integrated). For inside integration, removable singularities are removed using a well known transformation to a polar coordinate system[1,3]. However, as with all of the above mentioned techniques, the transformation is performed once to the intrinsic (isoparametric) element and its effect is reflected in a customized set of sample point quantities. This approach makes the inside and outside integration procedures identical in form. 5.) The resulting computer procedure, while somewhat sophisticated in its concept, is extremely simple in its final manifestation. 6.) The computational economy associated with the approach is substantive as demonstrated by the test cases presented. A discussion on how the the integrations may be configured to maintain any desired level of accuracy is also presented.

Element Formulation

In this presentation attention is focused on a six-noded isoparametric triangular element, although the approach is general and could be applied to other types of elements. The discretization of the boundary integral equations for elastostatic analysis[1,3] produces the linear system of simultaneous equations, as shown for an individual zone in the previous discussion in Equation (1). The contribution to the system matrices [H] and [G] of an element, E, due to a load point at, P, is determined by the evaluation of the integrals[1,3] that follow.

$$[H]^{(P,E)} = \int_0^{+1} \int_0^{1-l_2} [T]^T [h] |g| \, dl_1 \, dl_2 \;\; ; \;\; [G]^{(P,E)} = \int_0^{+1} \int_0^{1-l_2} [U]^T [h] |g| \, dl_1 \, dl_2 \qquad (16), (17)$$

[T] and [U] are the matrices of the fundamental traction and displacement solutions, respectively, of the elasticity problem; [h] is a matrix of isoparametric interpolation functions; |g| is the determinant of the Jacobian matrix for the transformation from the intrinsic (isoparametric) coordinates l_i to the real element coordinates x_i. It is noted that |g| involves combinations of the derivatives of the isoparametric shape functions with respect to the isoparametric coordinates. The expressions for these quantities are given elsewhere[44]. The evaluation of the integrals in equations (16) and (17) is performed numerically by replacing the integrals with summations of weighted series as shown below.

$$[H]^{(P,E)} = \sum_{K=1}^{NR} [T^{(K)}]^T [h^{(K)}] |g^{(K)}| \, w^{(K)} \;\; ; \;\; [G]^{(P,E)} = \sum_{K=1}^{NR} [U^{(K)}]^T [h^{(K)}] |g^{(K)}| \, w^{(K)} \qquad (18), (19)$$

The superscript (K) denotes the K'th sample point, NR is the number of sample points in the integration rule, and w is the weighting coefficient associated with a particular integration sample point.

240

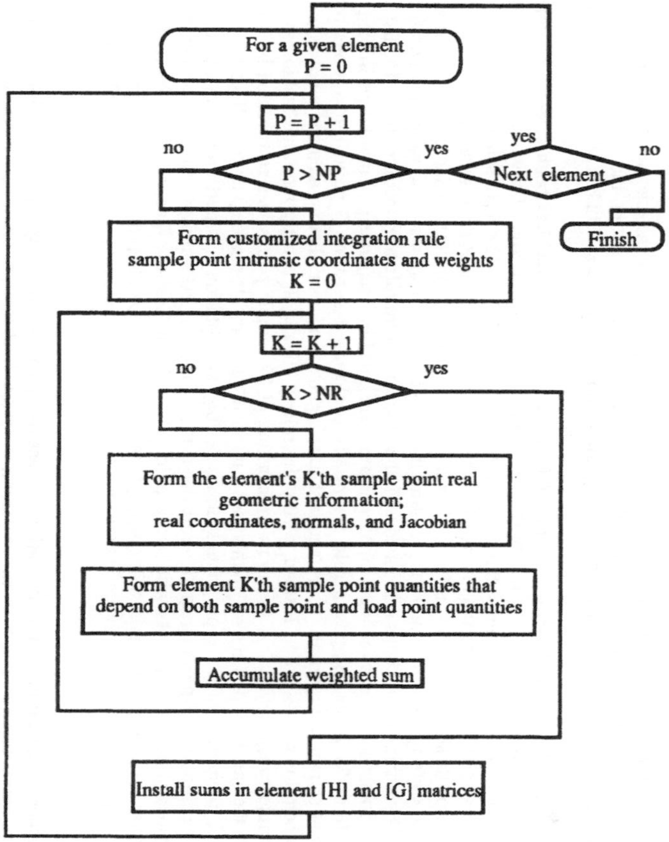

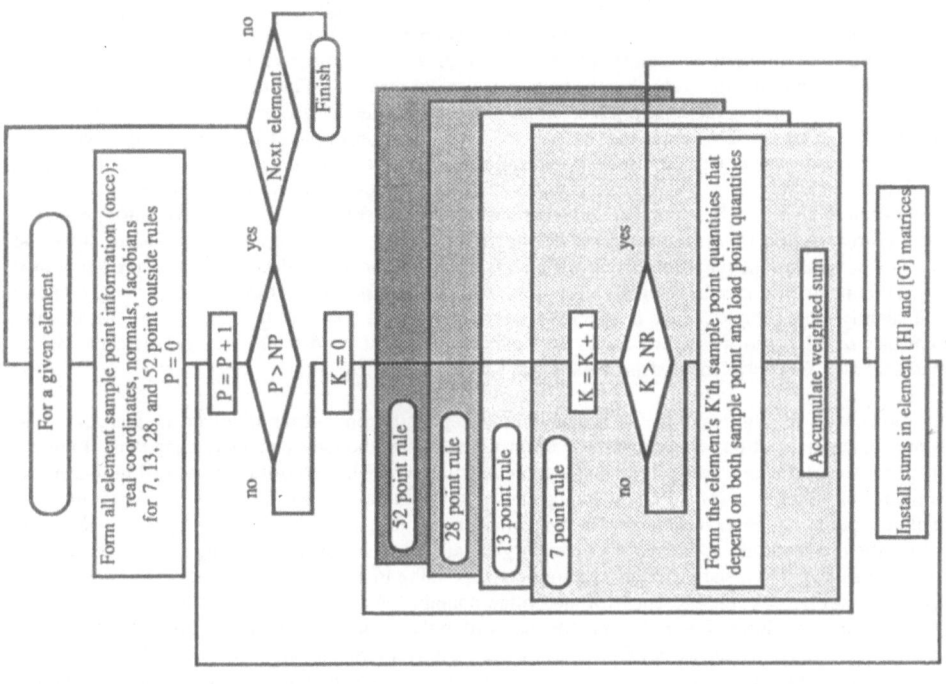

Figure 8. Schematic diagram of the operations performed in this new numerical integration algorithm for outside integration.

Figure 7. Schematic diagram of the operations performed in a traditional numerical integration algorithm.

241

Efficient Algorithm Considerations

The objective for all boundary element numerical integration algorithms is to accurately perform the summations shown in equations (18) and (19) in the most economical fashion. Both the element subdivision and the coordinate transformation algorithms seek to minimize the number of integration sample points required in an integration rule for evaluating the boundary element matrices. This objective is motivated by the premise that a lesser number of sample points in an integration scheme will lead to better computational efficiency for this scheme. The factor that has not received attention in the efficient determination of the boundary element matrices is that the use of optimized rules[32] for each boundary element - load point pair requires the regeneration of all geometric data (sample point coordinates, normals, and Jacobians) of the actual element, for each of a very large number of combinations. In the present study, the geometric information for an element corresponding to a preselected set of integration rules is first computed. Depending upon the proximity of a particular load point for this element, an integration scheme is selected from this predetermined set that will accurately evaluate the integrals of interest. Although the new RISP method will generally require the use of slightly more sample points, this increase in computation is dramatically offset by the fact that the entire sequence of calculations associated with the formation of geometric data of the actual element need be performed only once for each element using a particular scheme.

The flowcharts in Figures 7 and 8 schematically depict the operations involved in the numerical integration of the contribution of a boundary element to the overall boundary element equations. Figure 7 shows the conventional approach while Figure 8 represents the new RISP algorithm presented in this study. In these figures the outer loop is over all the load points involved in the analysis and the inner loop is over the number of integration sample points chosen for the particular boundary element - load point pair. As shown in Figure 7 for the conventional approach, the element geometric data is constantly being recomputed for the changing groups of optimum sample point locations corresponding to each load point location. On the other hand, as shown in Figure 8, the present RISP algorithm contains absolutely no geometric computations for the element inside either loop. The only computations within these loops are those that involve both sample point and load point data, which are required since they are unique to each boundary element - load point pair. This is possible due to the fact that the integration rules are being reused and therefore the same sample point information associated with these rules can be computed once outside these loops and reused as required.

Integration Schemes

Integration procedures have been developed for both the cases when the load point is coincident with one of the nodes of the element being integrated, and when the load point is not on the element being integrated. The former case will be referred to as inside integration while the latter case will be called outside integration.

Outside integration schemes. A set of four integration schemes has been selected or constructed for the cases when the load point is not on the element being integrated. The first two rules are the standard 7 and 13 point Gauss rules for integration over triangular regions[40,41]. The third scheme is a composite rule that is generated by considering the intrinsic element as being composed of the 4 subelements shown in Figure 9. In each of these subelements, a 7 point Gauss rule is adapted from the standard 7 point Gauss rule by scaling and shifting, and in some cases rotating the sample point locations. The weighting coefficients are generated by scaling the weights associated with the original 7 point rule by a factor of 0.25. The list of values of sample points and corresponding weights has been computed in a separate program in quadruple precision. The resulting 28 point composite rule essentially performs 4 separate 7-point integrations over each subelement shown in Figure 9. It is important to recognize, however, that no element subdivision actually occurs when this rule is used. Instead, a 28-point rule is utilized in exactly the same fashion as the 7-and 13-point rules, respectively. A fourth rule consisting of 52 sample points was constructed by a procedure similar to the one described above. In this case 4 intrinsic subelements make up an overall element, each with a 13-point integration rule. Appropriate scaling, shifting, and rotation was performed on these quantities to produce a composite 52-point integration rule that acts as if it were performing 4 separate 13-point integrations over portions of the elements. As with the 28-point rule, a list of 52 shape function values for each node, and their derivatives was predetermined in quadruple precision in a separate program for use by the algorithm as required. Again, no actual subdivision of the elements occurs when the rule is applied in the boundary element program.

Further examination of the integration process revealed that the sample point locations in the intrinsic element are only used for is to form the isoparametric shape functions and their derivatives with respect to the sample point intrinsic coordinates. The fact that only a discrete number of sets of sample points are used in the RISP approach allows the intrinsic coordinates of each sample point to be determined in advance. It follows

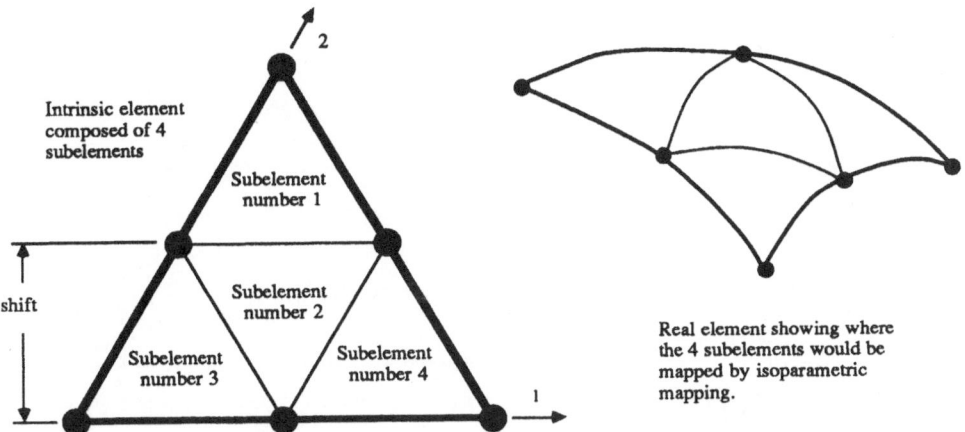

Figure 9. Intrinsic (isoparametric) element with the 4 subelements used to generate the composite outside integration rules.

that, the corresponding shape functions and their derivatives can also be evaluated in advance. These quantities were pre-computed in quadruple precision in a separate program, and need not be repetitively recomputed in the numerical integration step.

The selection of a particular outside integration rule from this predetermined set is made for each load point based upon the relative proximity of the load point to the element being integrated. Thus for each boundary element, the geometric information need only be computed once for each of the 7, 13, 28, and 52 point rules, respectively. All this information can then be subsequently reused by all outside integrations that use any of these 4 rules. The flowchart in Figure 8 illustrates this fact by depicting the 7, 13, 28, and 52 point rules as the shaded 'layers' shown. For any given load point, P , the program branches to one of these layers and the corresponding set of predetermined sample point lists and pre-computed element geometric information are employed to perform the numerical integrations.

Inside integration schemes. When the load point for the fundamental solutions is directly at a node for a particular element being integrated, an alternate procedure must be followed to obtain suitable integration rules. For these cases, locations in the element where the integrands will experience pronounced variations in their values are known a-priori. This knowledge can be exploited to strategically cluster the sample points in such a fashion as to economically and accurately account for this behavior. A twofold approach was taken in the development of this part of the algorithm. First, an intrinsic subelement technique similar to the one discussed above for outside integration was employed. This can be done for all subelements in the element except the one that contains the node on which the point load for the fundamental solution is applied. The second step in the process involved the treatment of the remaining subelement in which integrands become infinite.

Figure 10 depicts the intrinsic (isoparametric) element and the subelements used to position sample points and scale weighting coefficients in portions of the element where the integrands remain well behaved. Quadrilateral subelements with a 3 X 3 Gauss rule have been used. This can be considered as a specific manifestation of the general use of intrinsic subelements in an intrinsic element. For example, in earlier studies by the authors, 3 triangular subelements were used to fill each of the shaded regions shown in Figure 10. In each triangle, a 7-point triangular Gauss rule was utilized, thus requiring 21 sample points to perform the same task accomplished by the 9 sample points associated with the quadrilateral rule. The quadrilateral subelements were ultimately chosen because of the essentially quadrilateral shape of the shaded regions shown in figure 4, and because less sample points were then required to adequately treat the intrinsic element.

The sizing of the subelements shown in Figure 10, the Gauss rules employed within the subelement (both in the radial and the transverse direction), and the number of subelements, are the parameters that can be adjusted to produce an entire family of integration rules. After considerable numerical experiments, a sizing factor of 0.5, the 3 X 3 rule , and 3 intrinsic subelements in the nonsingular part of the element were found to provide accurate results in a computationally efficient manner, and this configuration is referred to as a near optimal configuration in this paper. The sizing factor refers to the fraction of the element length remaining to be divided that is taken up by the next subelement in the sequence of subdivisions shown in Figures 4 and 5.

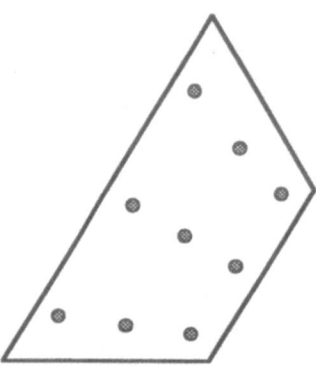

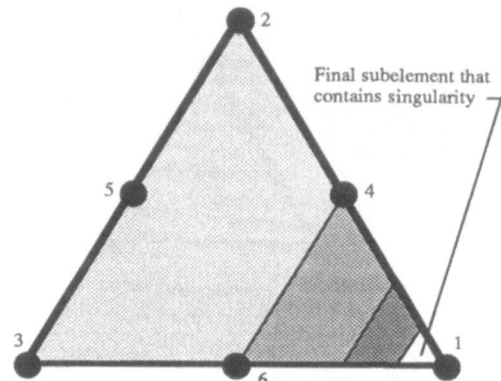

Intrinsic subelement showing Gauss integration points for a 3 X 3 rule.

Intrinsic (isoparametric) triangle showing the quadrilateral subelements used to form composite integration rule.

Figure 10. Illustration of Method Used to Cluster the Locations of the sampel points in the Intrinsic (Isoparametric) Element for Elements That Have Load Points at Corner Nodes

This particular combination of parameters results in 27 intrinsic sample points in the non-singular part of the element. The sample points for the non-singular part of the element and their associated weights were again predetermined in quadruple precision in a separate program along with the corresponding isoparametric shape functions and their derivatives. A similar strategy has been developed for the vase when the load point is no a midside node[44].

In both cases; i.e.; when the load point is at a vertex node, and when the load point is at a midside node, an intrinsic triangular sub-region in the element with the load point at one of its vertices remains to be treated. It is well known[1,3] that the strength of the singularity present in a 3-D boundary element analysis can be reduced by a factor of ρ via a transformation to a polar coordinate system. This reduction in the order of the singularity allows for the possibility of successful numerical integration of all terms in the element [G] matrix and all but 9 terms in the element [H] matrix. These 9 terms are associated with the node in the element with the load point applied at that node. This presents no difficulty because these 9 terms can be obtained by the negative row sum property of the overall boundary element matrices commonly referred to as the rigid body motion property[1,3].

The equations that follow contain the essential details for reducing the singularity in the traction kernel [H]. The procedure for the displacement kernel [G] is identical. For the remaining triangular sub-region, the contribution to the boundary element matrix [H] for element, E, and load point, P, can be written after a transformation to polar coordinates as shown below.

$$[\text{H}]^{(P,E)} = \int_0^{\frac{\pi}{4}} \int_0^{\frac{d}{\cos\theta}} [\text{T}]^T \, [\text{h}] \, |\text{g}| \, \rho \, d\rho \, d\theta \qquad (20)$$

In order to normalize the limits of integration in the above expression, let

$$\rho = \frac{d}{2\cos\theta} s + \frac{d}{2\cos\theta} \; ; \; \theta = \frac{\pi}{8} t + \frac{\pi}{8}$$

Then $[\text{H}]^{(P,E)} = \displaystyle\int_{-1}^{+1} \int_{-1}^{+1} [\text{T}]^T \, [\text{h}] \, |\text{g}| \, \rho \, |\text{J}| \, ds \, dt \qquad (21)$

Where $|\text{J}| = \dfrac{\pi \, d}{16 \cos\theta}$

It is now a very straightforward matter to generate a numerical integration rule for this case. The key feature of this approach is to recognize that this second transformation (the one characterized by the Jacobian |J| and the distance ρ) is a transformation from one intrinsic element portion to the same intrinsic element portion. As such, this second transformation does not involve any information concerning the geometry of the real element and can therefore be performed just once and its effect totally and exactly reflected in the weighting coefficient. The details of this operation are given below.

$$[H]^{(P,E)} = \sum_{K=1}^{NR} [T^{(K)}]^T [h^{(K)}] \, |g^{(K)}| \, \{ \rho^{(K)} |J^{(K)}| \, w^{(K)} \} \qquad (22)$$

$$[H]^{(P,E)} = \sum_{K=1}^{NR} [T^{(K)}]^T [h^{(K)}] \, |g^{(K)}| \, w_1^{(K)} \qquad (23)$$

where $w_1^{(K)} = \rho^{(K)} |J^{(K)}| \, w^{(K)}$ \qquad (24)

The new weighting coefficient w_1 accounts for w, ρ, and $|J|$. The sample points for this singular part of the element and their associated weights were, as with all other cases, predetermined in quadruple precision in a separate program along with the isoparametric shape functions and their derivatives. This then makes the inside and outside integration schemes appear identical in all respects except the number of points used in the rule. The singular part of the rule for the midside node singularity case is constructed in an analogous fashion.

The construction of these inside rules did requires additional numerical experimentation to obtain a rationale for the selection of the number of sample points to be employed in the radial and the transverse directions, respectively. The details of the actual study are presented elsewhere[44]. Integrals with known values were numerically computed, along with numerical tests on the actual boundary element integrands revealed that 3 points in the radial direction, and 5 in the transverse direction were required in the triangular sub-region containing the singularity. This leads to the requirement for a total of 15 sample points in this region. Instead of developing 6 inside integration rules (one for each node in the 6 noded triangular element), only 2 rules have been constructed. The rules for the load point of the fundamental solutions occurring at corner node 1, and at midside node 4, respectively, were the only rules introduced in the algorithm. For the cases when a load point occurs at a node other than element node 1 or 4, the element node numbers are subject to a cyclic permutation of their order to position the element as if the singularity was occurring at nodes 1 or 4. After the integration of this permuted element has been accomplished, the entries in the permuted element matrices are shifted to their appropriate (unpermuted) locations corresponding to the original node numbering of the element.

It should be mentioned that the RISP algorithm performs the inside integration in such a fashion that the intrinsic sample point coordinates, shape functions, and shape function derivatives are all reused for each element. Therefore, these quantities can be pre-computed in a manner completely analogous to the other pre-computed quantities described above. However, the actual geometric sample point information must be regenerated in the boundary element program for each case when a load point is on the element itself. This is due to the fact that, for a load point on a corner node or on a midside node, the element is integrated with the permuted node-point numbering described above, and not the actual node-point numbering. Since this occurs only 6 times for each element, compared to the possibly hundreds of times that an element reuses a particular outside integration rule, it represents a necessary, but rather insignificant, computational burden.

Throughout the previous discussions, frequent reference has been made to the use of subelements in determining the locations of sample points and values of the corresponding weighting coefficients. All this treatment, however, is a preliminary step. In the actual boundary element program there are absolutely no subelements. In the boundary element program there are only entire elements such as the intrinsic element shown in Figure 11. In this figure, the intrinsic sample point locations for inside integration of the element are shown for the case when the load point of the fundamental solutions is at a corner node. The RISP algorithm treats every element in its entirety and handles all special cases by processing whole elements using custom sample point data selected from a finite number of predetermined sets.

Computational Performance of the RISP Algorithm

The RISP algorithm has been implemented in a computer program for the three dimensional analysis of solid objects using boundary elements. The response of a hollow circular cylinder loaded with an internal hydrostatic pressure was chosen to be the subject of this study. The model used is shown in Figure 12. A quarter symmetry model of a flat circular plate with a circular hole, loaded with an internal constant pressure was created, using 176 continuous 6-noded triangular boundary elements in this model and 354 nodes. This model was chosen because an exact solution is available[42], it contains some elements that are curved, and because it represents a problem that is sizable enough to make meaningful conclusions regarding timing and accuracy. This model was analyzed using a program employing four variations of the RISP algorithm

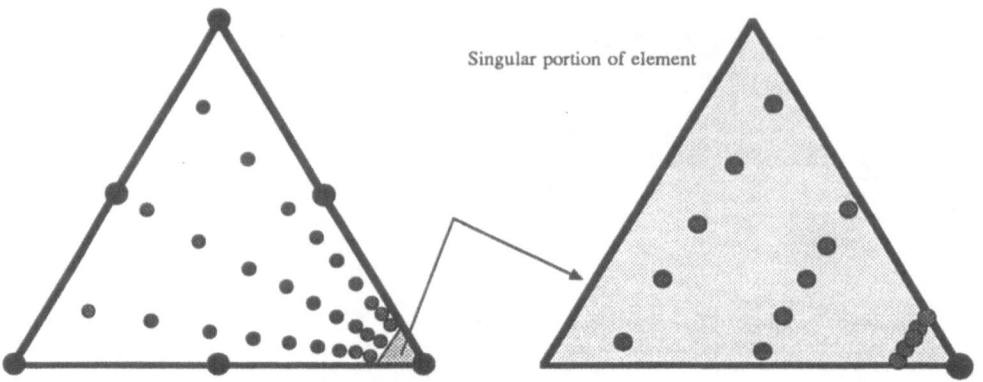

Singular portion of element

Figure 11. Sample point location for inside integration of element with point load at a corner node

described above. To provide timing and accuracy comparisons, the model was also analyzed using an available boundary element code[43] that utilizes the element subdivision approach. All programs were compiled in FORTRAN 77 and executed on the same computer. For all the results reported below, the CPU times quoted are for the integration step only. In all cases, the accuracy of the response predicted by the program using the RISP numerical integration algorithm was the same or marginally better than the accuracy of the results obtained by the program employing the element subdivision approach. The results are shown in Figure 13, for comparison.

Case 1. The 176-element, 354-node model was first run in an available program based on the element subdivision approach[43]. The numerical integration step in this run consumed 5529 seconds of CPU time. This timing can be used an a standard for the purpose of comparison of the degree of efficiency obtained by the other RISP algorithm. Also, the response predicted by this program was used to compare the accuracy of the program using the RISP algorithm.

Case 2. In this first version of the program using the RISP algorithm, the inside integration was performed using a strategy for clustering the intrinsic sample points that involved the use of 3 intrinsic triangles within each of the quadrilateral sub-elements shown in Figure 4. In each of these quadrilateral sub-elements 21 integration sample points were used (7 in each triangle). Furthermore, 5 quadrilateral sub-elements were used in the non-singular portion of the intrinsic element. This version of the inside integration scheme also employed a 5 X 5 integration rule in the triangular sub-regions containing the singularities. This resulted in an inside integration rule with 109 integration sample points (5 X 21 + 25 = 109) for the case when the load point was at a corner node. A CPU time of 1299 seconds was required in this case for the pressurizes cylinder model. This comparison (5529 seconds versus 1299 seconds) reveals the dramatic computational advantage associated with the use of predetermined sets of integration rules compared to the constant generation of optimum rules inside the boundary element program.

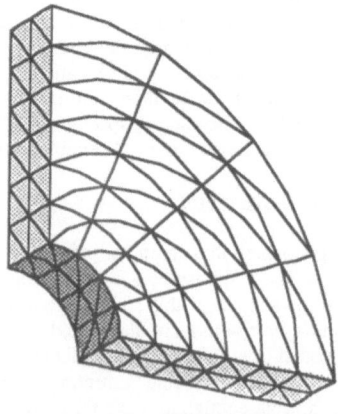

Figure 12 . Pressurized Thick Cylinder BEA Model Used in Case Studies

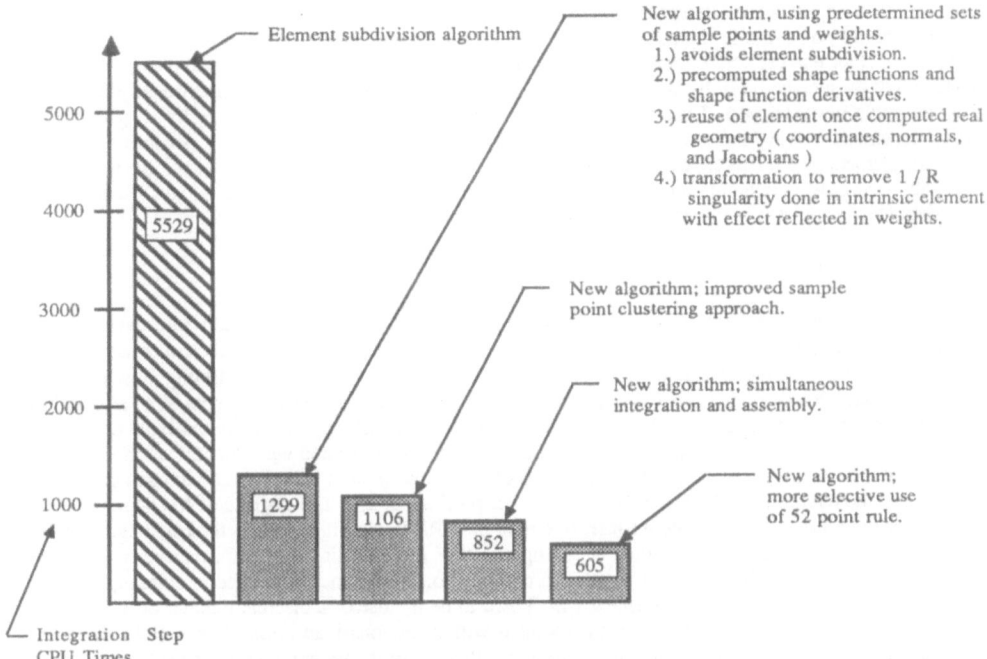

Figure 13. Results of case testing of 3D pressurized cylinder model

Case 3. The inside integration rules illustrated in Figure 10 were incorporated into the boundary element program for this case study. As a result of this improvement to the RISP algorithm, the numerical integration of the boundary element matrices was performed in 1106 CPU seconds for the pressure vessel model under consideration. This rule uses only 42 integration sample points (3 X 9 + 15 = 42) for the case when the load point was at a corner node. It is noted that while the number of sample points used in the inside integrations was reduced by a factor of 42 / 109 (.385), the timing performance was reduced by only a factor of 1106 / 1299 (.851). This result can be explained by the fact that the inside integration step is only done six times for each element and therefore contributes to the overall timing of the overall numerical integration step in a marginal fashion. This comparison also demonstrates the marginal improvements in computational efficiency that one can expect from any further reduction in the number of sample points used in the singular integrations involved in the boundary element method.

Case 4. In the cases discussed above, the boundary element program computed the contribution of an element to the overall boundary element equations and then stored this information on secondary (disk) memory for subsequent use by an assembly subroutine. This operation represents a considerable expenditure of computer resources, regardless of how efficiently it is programmed. A new version of the RISP algorithm was therefore developed that avoids this operation of writing the element information to secondary memory after each element integration. This version of the program assembles the contribution of each element to the overall boundary element matrices immediately after its formation by numerical integration. It thus represents a simultaneous numerical integration and assembly procedure. When the pressure vessel model was run through this version of the program, the CPU time required for the combined numerical integration and assembly step was found to be 852 seconds. This 254 second (1106 - 852) savings in computer time represents the CPU time used in the writing of the element contributions to a scratch file during the numerical integration process. An additional savings of approximately the same magnitude is also accomplished using this strategy in the assembly process because this information need not be read from a file during the assembly process. This total savings of approximately 500 CPU seconds due to avoiding the writing out and reading in of the element [H] and [G] matrices is of the same order of magnitude as the total time required for the numerical integration process in its final form. The significance of this improvement to the algorithm is thus evident.

Case 5. It was suspected that the use of the 52-point outside integration rule may represent an overkill for many boundary element models. Accordingly, a version of the RISP algorithm was developed that restricted

247

the use of the 52-point outside integration rule to cases where it was considered essential. The criteria employed for using the 52-point rule in the discrete set of outside integration rules used by a particular element, was based on whether that element encounters a load point closer than $L/8$, where L is taken to be the smallest dimension of the element. For boundary element models that consist of fairly well-proportioned elements, this situation might never happen. Even for boundary element models that contain some badly-shaped elements, this approach can provide some computational economy in the numerical integration of portions of the model where the elements are relatively well-shaped. For the cylinder problem studied in this paper, the restricted use of the 52-point outside integration rule lead to a further savings of 247 CPU seconds as the CPU time for the numerical integration step in this version of the program amounted to only 605 seconds.

CONCLUSIONS

A discussion of a number of techniques has been presented that produce significant improvements in the computational performance of the BEM. The concepts of multi-zone BEA, sparse blocked equation solving, and zone substructuring (condensing) have been discussed. It has been shown that selecting master degrees of freedom at nodes that lie on boundary element zone interfaces, facilitates the development of an algorithm that allows for both full and condensed boundary element zones to consistently coexist in the same analysis. An example was given that demonstrate the computational efficiency associated with the utilization of the newly developed equation solver for the solution of classes of problems where it is appropriate to form substructures for a part of the overall model while maintaining other portions at their full degrees of freedom. It has also been demonstrated that the Reusable Intrinsic Sample Point, RISP, algorithm retains the beneficial features of earlier numerical integration algorithms based upon either the coordinate transformation or the element subdivision technique, while providing significant computational efficiency compared to other algorithms. A key factor in this performance improvement was shown to be the use of a discrete number of predetermined integration rules. This allows integrations associated with numerous load points to reuse the same set of element geometric data. It was shown that this strategy allowed for the pre-computing and subsequent sharing of an element's shape functions and shape function derivatives for all integrations that use each particular rule. This work clearly demonstrates that numerical integration algorithms for boundary element analysis are the most efficient when the entire number of calculations performed is made smaller, instead of when just the number of integration sample points is reduced as much as possible. In aggregate, these techniques can begin to allow for the effective utilization of Boundary Element Techniques on significantly larger problems.

ACKNOWLEDGEMENT

Portions of the research discussed herein have been supported by grants from the U. S. National Science Foundation (DMC-8708254) to Worcester Polytechnic Institute and also by grants or contracts from the Wyman-Gordon Company of Worcester Massachusetts, General Dynamics, Electric Boat Division in Groton CT, and from the United Technologies Corporation. Any opinions, findings, and conclusions or recommendations expressed in this publication are those of the authors and do not reflect the views of these other organizations.

REFERENCES

1. P. K. Banerjee and R. Butterfield, <u>Boundary Element Methods In Engr. Sci.</u>, McGraw Hill Book Co. UK. London, 1981
2. C. A. Brebbia and S. Walker, <u>Boundary Element Techniques in Engineering</u>, Newnes Butterworths, London and Boston, 1980
3. C. A. Brebbia, J. C. F. Telles, and L. C. Wrobel, <u>Boundary Element Techniques</u>, Springer Verlag, Berlin and New York, 1984
4. J. C. Lachat, and J. O. Watson, "A Second Generation Boundary Integral Program for Three Dimensional Elastic Analysis," in, <u>Boundary Integral Equation Method: Computational Applications in Applied Mechanics</u> (Eds. T.A. Cruse and F. J.Rizzo) Appl. Mech. Div., ASME Vol 11, New York, 1975.
5. J. C. Lachat, "Further Developements of the Boundary Integral Technique for Elasto-statics," Ph.D. Thesis, Southampton Univ., 1975.
6. J. C. Lachet and J. O. Watson, "Progress in The Use of Boundary Integral Equations, Illustrated by Examples," <u>Computational Methodes in Applied Mechanical Engineering</u>, 10, 273-289.
7. J. M. Crotty, "A Block Equation Solver for Large Unsymmetric Matrices Arizing in the Boundary Element Method," <u>International Journal for Numerical Methods in Engineering</u>, Vol 18, pp 997-1017, 1982
8. P. C. Das, "A Disc Based Block Elimination Technique Used for the Solution of Non-Symmetrical Fully Populated Matrix Systems Encountered in the Boundary Element Method," <u>Proc. Int. Symp. On Recent. Dev. In Boundary Element Meth.</u>, Southampton Univ. pp 391-404, 1978

9. R. Bailecki and R. Nahlik, "Linear Equations Solver for Large Block Matrices Arising in Boundary Element Methods," Boundary Elements IX, Vol 1, Computational Mechanics Publications, Springer-Verlag, Southampton and Boston, 1987

10. R. Bailecki, "Nonlinear Equations Solver For Large Equation Sets Arizing When Using BEM In Homogenous Regions of Nonlinear Material, " Boundary Elements IX, Vol 1, Computational Mechanics Publications, Springer-Verlag, Southampton and Boston, 1987

11. G. R. Tomlin, "Numerical Analysis of Continuum Problems in Zoned Anistropic Media," Ph.D. Thesis, Southampton Univ., 1972.

12. R. Butterfield and G. R. Tomlin," Integral Techniques for Solving Zoned Anistropic Continuum Problems," Proceedings, International Conference on Variational Methods in Engineering, Southampton Univ., pp 9/31-9/51, 1971

13. O. V. Chang, "Boundary Elements Applied to Seepage Problems in Zoned Anisotropic Soils," MSc. Thesis, Southampton Univ., 1979.

14. G. Beer, "Implementation of Combined Boundary Element-Finite Element Analysis with Applications In Geomechanics," Chapter 7 in, Developments in Boundary Element Methods, (Eds. P. K. Banerjee and J. O. Watson), Elsevier Applied Science Publishers, London and New York, 1986

15. G. G. W. Mustoe, "A Combination of the Finite Element Method and Boundary Solution Procedures for Continuum Problems," Ph.D. Thesis, University of Wales, University College, Swansea.

16. T. G. Davies, "Linear and Nonlinear Analysis of Pile Groups," Ph.D. Thesis, University of Wales, University College, Cardiff, 1979.

17. H. Jin, K. Runesson, and A. Simuelsson, "Application of the Boundary Element Method to Contact Problems in Elasticity with a Nonclassical Friction Law," Boundary Elements IX, Computational Mechanics Publications, Springer-Verlag, Southampton and Boston, Vol 2 pp 397-415, 1987

18. S. Margenov, K. Georgiev, L. Hadjikov, and M. Novakova,"An Effective Approach for Boundary Element Method Application to Friction Contact Problems," Boundary Elements IX, Vol 1, Computational Mechanics Publications, Springer-Verlag, Southampton and Boston, pp 439-445, 1987.

19. B. Nour-Omid and P. Wriggers, "A Two Level Iteration Method for the Solution of Contact Problems," Computer Methods in Applied Mechanics in Engineering, 54, pp 131-144, North Holland Publisher, 1986.

20. J. H. Kane, "Shape Optimization Utilizing a Boundary Element Formulation," BETECH 86, Proceedings, 1986 Boundary Element Technology Conference, MIT, Computational Mechanics Publications,
Springer-Verlag, Southampton and Boston, 1986

21. J. H. Kane and S. Saigal, "Design Sensitivity Analysis of Solids Using BEM," Journal of Engineering Mechanics, ASCE, to appear, Oct. 1988.

22. S. Saigal and J. H. Kane, "A Boundary Element Shape Optimization System for Aircraft Components," AIAA Journal, Accepted for publication.

23. S. Saigal, R. Aithal and J.H. Kane, "Conforming Boundary Elements in Plane Elasticity for Shape Design Sensitivity," International Journal for Numerical Methods in Engineering, submitted for review.

24. S. Saigal, J. H. Kane and R. Aithal, "Semi-Analytical Structural Sensitivity Formulation Using a Boundary Elements," AIAA Journal, submitted for review.

25. S. Saigal, J. T. Borggaard and J. H. Kane, "Boundary Element Implicit Differentiation Equations for Design Sensitivities of Axisymmetric Structures," International Journal of Solids and Structures, submitted for review.

26. J. H. Kane and S. Saigal, "Design Sensitivity Analysis of Boundary Element Substructures," Second NASA / Airforce Symposium on Recent Experiences in Multi Disciplinary Analysis and Optimization, September, 1988, Accepted for presentation, Proceedings to be published.

27. M. F. Hodous, D. G. Bozek, D. M. Ciarelli, K. J. Ciarelli, R. B. Katnik, and K. A. Kline, 'Vector Processing Applied to Boundary Element Algorithms on the CDC Cyber 205,' Bulletin de la Direction des Etudes et des Recherches, Serie C, No. 1, 1983, pp 87-94.

28. K. A. Kline, N. K. Tsao, and C. B. Friedlander, 'Parallel Processing and the Solution of Boundary Element Equations,' Advanced Topics in Boundary Element Analysis, ASME AMD Publication, (editors T. A. Cruse, A. B. Pifko, and H. Armen), 1985.

29. J. A. Bettess, 'Solution Techniques for Boundary Integral Matrices,' Numerical Methods in Transient and Coupled Problems, Wiley-Interscience, (editors R. W. Lewis, E.Hinton, P. Bettess, and B. A. Schrefler), 1987.

30. J. H. Kane, S. Saigal, and B. L. Kesheva Kumar, 'An Arbitrary Zone Condensing, Noncondensing Forunlation with an Associated Sparse Blocked Equation Solver for Large Scale Boundary Element Analysis,' International Journal for Numerical Methods in Engineering, submitted for review.

31. J. H. Kane and S. Saigal, 'An Arbitrary Zone Condensing Technique for Boundary Element Structural Design Sensitivity Analysis,' AIAA Journal, Submitted for Review.

32. Mustoe, G. G. W., 'Advanced Integration Schemes Over Boundary Elements and Volume Cells for Two- and Three-dimensional Non-linear Analysis,' Eds., Banerjee, P. K., and Mukherjee, S., Developments in Boundary Element Methods -- Vol. III, Applied Science Pub. Ltd., England, (1982)

33. Watson, J. O., 'Advanced Implementation of the Boundary Element Method for Two- and Three-dimensional Elastostatics,' Eds., Banerjee, P. K. and Butterfield, R., Developments in Boundary Element Methods -- Vol. I, Applied Science Pub. Ltd., England, (1979)

34. Telles, J. C. F., 'A Self-Adaptive Coordinate Transformation for Efficient Numerical Evaluation of General Boundary Element Integrals,' International Journal for Numerical Methods in Engineering, 24, 959-973, (1987)

35. Cruse, T. A., 'Numerical Solutions in Three Dimensional Elastostatics,' International Journal of Solids and Structures, 5, 1259-1274, (1969)

36. Aliabadi, M. H., Hall, W. S., and Phemister, T. G., 'Taylor Expansions for Singular Kernels in the Boundary Element Method,' International Journal for Numerical Methods in Engineering, 21, 2221-2236, (1985)

37. Lean, M. H. and Wexler, A., ' Accurate Numerical Integration of Singular Boundary Element Kernels Over Boundaries With Curvature,' International Journal for Numerical Methods in Engineering, 21, 211-228, (1985)

38. Li, B. H., Han, G. M., and Mang, H. A., ' A New Method for Evaluating Singular Integrals in Stress Analysis of Solids by the Direct Boundary Element Method,' International Journal for Numerical Methods in Engineering, 21, 2071-2098, (1985)

39. Schwab, C. H., and Wendland, '3D BEM Numerical Integration,' in Proc. 7'th Conference on B.E.M., Eds. Brebbia, C. A., and Maier, G., Springer-Verlag, Berlin, (1985)

40. Cook, R. D., Concepts and Applications of Finite Element Analysis, Second Ed., J. Wiley, New York, 201-204, (1981)

41. Bathe, K. J., Finite Element Procedures in Engineering Analysis, Prentice-Hall, New Jersey, page 280, (1982)

42. Timoshenko, S. P., and Goodier, J. N., Theory of Elasticity ,3'ed., McGrall-Hill, New York, (1970)

43. Lindholm, U. S., 'Used's Manual For CRX3D a 3-D Fracture Mechanics Program,' Southwest Research Institute, San Antonio, Texas, (1986)

44. Kane, J. H. and Saigal, S., 'Reusable Intrinsic Sample Point (RISP) Algorithm for the Efficient Numerical Integration of Three Dimensional Curved Boundary Elements', International Journal for Numerical Methods in Engineering, accepted for publication, to appear.

DOMAIN DECOMPOSITION TECHNIQUES FOR LARGE SPARSE NONSYMMETRIC SYSTEMS ARISING FROM ELLIPTIC PROBLEMS WITH FIRST-ORDER TERMS

David E. Keyes*,William D. Gropp** and Ali Ecder*

Research Center for Scientific Computation
Yale University
New Haven,CT 06520

ABSTRACT

Parallel block-preconditioned domain-decomposed Krylov methods for sparse linear systems are described and illustrated on two-dimensional model problems of algebraic dimension up to 65,025. Four convective-diffusive transport problems typical of implicit upwind finite-difference discretizations of heat and mass transfer applications (pure conduction, a plug flow, a jet flow, and a recirculating flow) are tested for practicality of parallel solution under the domain decomposition paradigm. The discrete operators corresponding to the latter two lack constant coefficients and symmetry, and there is little iterative convergence theory to guide their solution, but much practical progress can be made. We describe techniques depending only on the sparsity structure and approximate diagonal dominance of the linear operator and thus of broad applicability. Results of tests run on an Encore Multimax with up to 16 processors demonstrate their utility in the coarse-granularity parallelization of hydrocodes.

INTRODUCTION

Domain decomposition is a natural route to parallel computing for partial differential equation solvers. In the most general form of this procedure, a solver that is presumed satisfactory as a serial code is replicated on a collection of parallel processors to each of which is assigned a subdomain of the domain of definition of the PDE. The boundary conditions required by the subdomain solver are replaced by interface conditions which enforce the requisite degree of continuity of the solution. The price of this decomposition is periodic coordination between processors to compute global parameters and maintain consistency of the interfacial data.

If the subdomain problems are solved exactly (in terms of the interfacial data) to obtain a reduced system, domain decomposition is equivalent to the block Gaussian elimination process known as "static condensation" in the finite element literature. The interface equations then contain all of the global coupling of the original problem, and form an inherently sequential system. Though complex in structure, they involve a lower-dimensional subset of the unknowns

*Department of Mechanical Engineering. The work of these authors was supported in part by the National Science Foundation under contract number EET-8707109.
**Department of Computer Science. The work of this author was supported in part by the Office of Naval Research under contract N00014-86-K-0310 and the National Science Foundation under contract number DCR 8521451.

of the original problem. With the development of an approximate inverse for the interfacial system, a preconditioned iterative scheme of low dimension can be constructed (see, *e.g.*, [8, 11, 16, 17, 31]). If, instead, the subdomain problems are solved only approximately, a preconditioned iterative scheme involving all of the unknowns of the problem can be employed (see [3]). The solution (or preconditioning) of the full collection of the subdomain problems is often cheaper (in terms of total operation count per iteration) than the solution (or preconditioning) of the full domain problem. This advantage can be traded off against the possibly greater number of iterations needed due to a poor approximation of the interface equations. However, as emphasized in [4], the divide-and-conquer approach of domain decomposition can be beneficial even sequentially – it is often possible to reduce the condition number of the overall iteration (relative to an undecomposed approach), provided the approximate inverse for the interfacial system is of a level of accuracy compatible with that of the subdomains, and if sufficient global coupling is embedded in the preconditioner. For general operators, the cost of deriving the exact lower-dimensional systems can greatly exceed the cost of direct banded Gaussian elimination on the original system. It follows that the success of domain decomposition hinges on the ability to efficiently approximate the lower-dimensional operators, by taking advantage of either known or "probed" structure.

Most published work to date on domain decomposition algorithms has concentrated on self-adjoint scalar elliptic equations, and several optimal algorithms are now known for this case, in the sense that the number of iterations required to solve the discretized PDE does not grow asymptotically with the gridpoint density or the number of subdomains as these quantities are refined in proportion (see, *e.g.*, [3, 32]). These algorithms typically employ the conjugate gradient method, and are distinguished from each other primarily by the selection of solvers or preconditioners for the decoupled systems of equations for the subdomain interiors and for the coupled interface equation system. For certain constant coefficient problems, including non-self-adjoint cases, exact preconditioners can be obtained by means of Fourier analysis so that the iterations converge in a single step [5, 7]. However, problems involving convective-diffusive equations (especially *several* such equations coupled to each other by source terms) have received a disproportionately small share of attention in the domain decomposition context. The solution of such problems is an important computational kernel in implicit methods (for instance, Newton-like methods and linearized implicit time-stepping methods) commonly used for systems of nonlinear PDEs arising in science and engineering, and is often CPU-bound or memory-bound or both on the fastest and largest serial computers available. Furthermore, it is often the only computationally intensive part of such codes whose efficient parallelization is not straightforward, particularly when the distribution of data throughout the computer's memory hierarchy cannot be dictated exclusively by linear algebra considerations. (See [19] for a typical complexity analysis.)

The present contribution is a companion paper to [21] and extends the range of examples in that paper to flows which are dominated by *non*-unidirectional convection. In [21] approximate block-diagonal inverses for the interfacial system are constructed by "probing" the Schur complement corresponding to the interface unknowns. The low-bandwidth diagonal blocks are required to produce the same matrix-vector products as practically computable approximations to the true (dense) Schur complement when acting on a given set of trial vectors, and are obtainable by solving independent problems on subdomains. This technique exploits only the sparsity structure and clustering of large magnitudes around the diagonal of the actual operator, and no other properties like symmetry or constant coefficients. As a result, it applies to quite general systems, though it performs suboptimally on some special problems. Some experimental results illustrate the technique. They are preceded below by a brief description of the overall computational procedure and its parallel implementation.

ALGORITHMIC DESCRIPTION

Iterative substructuring by domain decomposition should be regarded as a framework into which various combinations of well-known algorithmic modules can be inserted. It consists of an iterative procedure together with preconditioners for the subdomain and interface systems. The combination of a generalized minimum residual (GMR) iterative scheme with a modified

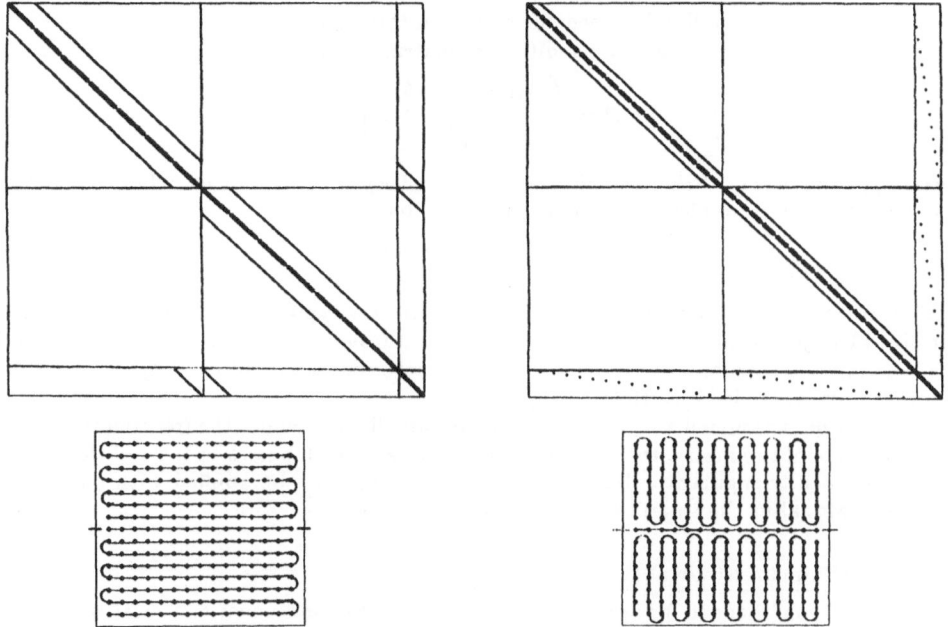

Figure 1. Sparsity patterns for two orderings of a decomposition of a rectangular region into two strips with a common edge, based on a 5-point finite-difference template. Within each subdomain the gridpoints are ordered as shown aligned with the direction of fastest ordering on the left; two strips, one edge, normal to the direction of fastest ordering.

incomplete LU (MILU) preconditioning on the subdomains and a low-bandwith interfacial probe (IP) preconditioning is a useful general purpose one. The combination of GMR with MILU is a robust serial technique for upwinded scalar convective-diffusive operators. Other choices for the iterative method (Chebyshev, block-SOR, and *restarted* GMR) other varieties of incomplete factorization subdomain preconditioning are applied to similar problems in [19] and in [14], respectively.

Unreduced Domain Decomposition

In this paper, domain decomposition enters our considerations at the level of the solution of a system of linear equations. In applications, this system might arise from the linearization of a nonlinear process, which gives the entire domain decomposition procedure the status of an inner iteration. To avoid any deeper nesting of iterations, we iterate simultaneously on all of the unknowns in the linear system, meaning that the subdomain problems are not individually iterated to convergence before their values are used to update the right-hand sides of the equations for the interfacial unknowns.

This paradigm for domain decomposition is most easily illustrated in the decomposition into two strips of a rectangular region overlaid by a tensor-product grid. The single cut follows a line of gridpoints, which are ordered separately. For a 5-point operator on a grid with 16 interior subintervals in each direction (with Dirichlet boundary conditions eliminated) the resulting sparsity pattern for the operator A is indicated graphically in Fig. 1, and in matrix notation as follows:

$$A = \begin{pmatrix} A_{11} & 0 & A_{13} \\ 0 & A_{22} & A_{23} \\ A_{31} & A_{32} & A_{33} \end{pmatrix}. \tag{1}$$

Here, A_{33} renders the coupling between the points on the interface itself.

The conformally partitioned preconditioning matrix we propose for A is

$$B = \begin{pmatrix} \tilde{A}_{11} & 0 & \tilde{A}_{13} \\ 0 & \tilde{A}_{22} & \tilde{A}_{23} \\ 0 & 0 & \tilde{C} \end{pmatrix}, \tag{2}$$

where $\tilde{C}$ approximates the Schur complement of A_{11} and A_{22} in A. The exact Schur complement, C, may be obtained from block-Gaussian elimination on A as:

$$C \equiv A_{33} - A_{31}A_{11}^{-1}A_{13} - A_{32}A_{22}^{-1}A_{23}. \tag{3}$$

The tilde-notation in the definition of B accommodates the replacement, if convenient, of the exact A_{ij} with approximations thereto. We assume throughout that the A_{ii} are invertible. (This is certainly a reasonable requirement for a discrete convective-diffusive operator.) Under this assumption, C is also invertible [9].

It has not been assumed above that A is symmetric. This provides the freedom to consider, without sacrifice of symmetry, the nonsymmetric B in (2), which possesses instead the valuable property of block triangularity. Note that the inverse of B can be applied with one solve in each subdomain, after the small system involving $\tilde{C}$ is solved.

Generalized Minimum Residual (GMR) Iteration

Any algorithm intended for use in fluid dynamical applications must be robust with respect both to asymmetry, to allow for the presence of convective terms, and (in some applications) indefiniteness, to allow for the presence of linearized source terms whose coefficients oppose the algebraic sign of the diagonal term of the discrete convective-diffusive operator. The generalized minimum residual method (GMR)* [27] is well-suited for such cases. Adaptive Chebyshev iteration [22] is an alternative suitable for cases in which the indefiniteness can be controlled; it enjoys an advantage over GMR on parallel machines with very large communication latency because it requires no inner products between adaptive steps.

Given a system of equations, $Mx = f$, M nonsingular, and an initial iterate, x_0, with initial residual, $r_0 = f - Mx_0$, GMR computes the solution x from finding $z \in K_m$ such that

$$(r_0 - Mz, v) = 0 ,$$

for all $v \in L_m$, and setting $x = x_0 + z$, where K_m and L_m are Krylov spaces based on r_0:

$$K_m \equiv \text{span}\{r_0, Mr_0, \ldots, M^{m-1}r_0\}, \quad L_m \equiv \text{span}\{Mr_0, M^2r_0, \ldots, M^m r_0\}.$$

The solution computed after m steps of GMR minimizes $\|r\|_2$ in the affine space $x_0 + K_m$. In a practical implementation, an orthogonal basis for K_m is built up by means of a Gram-Schmidt or Householder process, which obviates the necessity of working with the normal equations. Suitable implementations have been given in [27] and [29], of which we use the former. Among the desirable properties of GMR are: (1) the only reference to M is in form of matrix-vector products, (2) it cannot break down (in exact arithmetic) short of delivering the solution even for nonsymmetric systems with indefinite symmetric part, (3) it requires less storage and fewer operations per step than the mathematically equivalent GCR and ORTHODIR algorithms, and (4) the 2-norm of the residual is non-increasing and can be monitored without constructing intermediate solution iterates. The main disadvantage of GMR is the lack of a bounded recurrence relation, which causes the operation count and storage requirements to grow quadratically and linearly, respectively, in the iteration index. In many applications, restarting GMR after a predetermined number of steps is amelioratory, but restarted GMR can also fail through stagnation.

---

*abbreviated GMRES by its developers

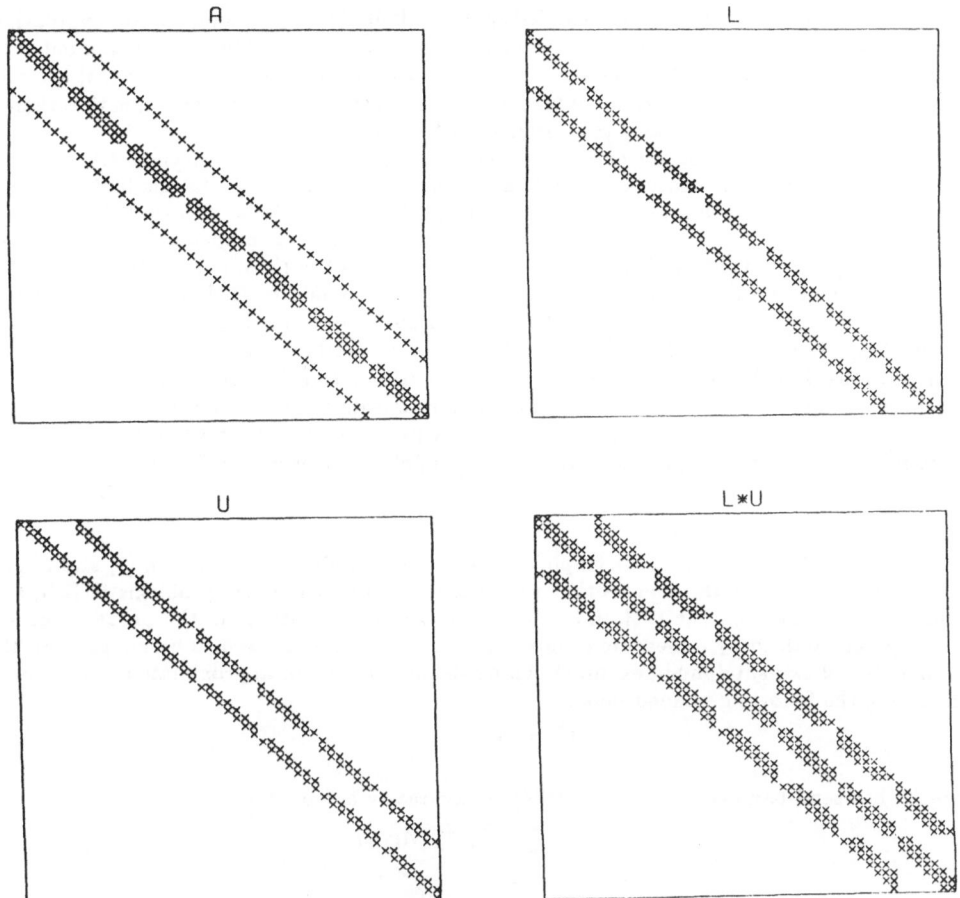

Figure 2. Sparsity patterns illustrating the MILU(1,α) decomposition: the original matrix A; the lower factor L and upper factor U, admitting one diagonal of fill each, just inside edge of the band; and the product LU, showing a second diagonal of fill. The elements of this second diagonal are weighted by α and lumped to the main diagonal.

Modified Incomplete LU (MILU) Subdomain Preconditioning

GMR is often uneconomical when acting unaided on a general reaction-convection-diffusion operator. In an effort to control the work and storage required by GMR when A has a widely distributed spectrum, we precondition the iterations by taking M in the formulae above to be AB^{-1}, for the A and B given above. This is "right" preconditioning, which first solves $M\tilde{x} = f$ for $\tilde{x}$, then $Bx = \tilde{x}$ for x. We adopt right over left preconditioning because in the latter the matrix B enters into the GMR residual convergence criterion in a direct way, making convergence comparisons between different preconditioning techniques difficult.

Approximate factorizations of the original matrix into triangular matrices, such as incomplete LU-decomposition (ILU) [23] or modified ILU (MILU) [13], though often disappointing as splitting matrices, are very useful general purpose preconditioners for Krylov and Chebyshev acceleration methods. Such factorizations can be bottlenecks in parallel implementations, however, because of the sequential nature of triangular factorizations and solves. Though wavefront-based or red-black reorderings of the standard sequential operations can alleviate this problem in the context of sparse banded matrices [26], domain-decomposition approaches side-step it altogether by applying the incomplete factorizations within subdomains only. In this connection, we mention the obvious fact that wavefront or other orderings [15] are still useful in the domain decomposition context for producing finer-grained parallelization than a one-to-one mapping of subdomains to processors will allow.

In the present examples we employ a technique appropriate to nonsymmetric systems derived from 5-point discretizations, namely a Crout MILU($1,\alpha$), where the one indicates that one diagonal of fill-in outside the original sparsity pattern of A is allowed (see [30]), and the parameter α indicates the fraction of the remaining fill-in which is lumped onto the diagonal of the MILU approximation. (This parameter is distinct from the parameter introduced in [13] to overcompensate for the diagonal entries.) $\alpha = 0$ corresponds to ILU, which discards the residual fill. $\alpha = 1$ corresponds to pure MILU. It is now appreciated that α_{opt} generally lies between these extremes. For the type of problems we consider herein α_{opt} is less than unity by an amount roughly inversely proportional to the resolution parameter of the grid, approaching unity for fine grids. The usefulness of such a parameter was illustrated on a variety of symmetric systems in [1] and [2]. Figure 2 displays the sparsity patterns of A, L and U (allowing for one level of fill in each), and the product LU for a five point operator on 7×7 grid.

Interface Probe Preconditioning

For the interface equations, we employ the probing technique suggested in [6] and implemented and compared with several other preconditioners on symmetric problems in [20], and on nonsymmetric problems in [21]. Like the subdomain preconditioning, the probe preconditioner is parameterized by a measure of how many diagonals are allowed to be nonzero. In the two-dimensional two-subdomain example (1), we define a family of approximate inverses for C by factoring the following banded matrix:

$$\tilde{C}_k = \tilde{A}_{33} - E_k ,$$

where E_k has semi-bandwidth k and satisfies as accurately as possible

$$E_k v_l = \sum_i (\tilde{A}_{3i} \tilde{A}_{ii}^{-1} \tilde{A}_{i3}) v_l \tag{4}$$

for a set of vectors v_l, $l = 1, 2, \ldots , L$.

For a nonsymmetric scalar system of equations, we set $L = 2k+1$ and use for the respective E_k the vectors:

$$
\begin{aligned}
E_0 &: v_1 = [1, 1, 1, 1, 1, 1, \ldots]^T \\
E_1 &: v_1 = [1, 0, 0, 1, 0, 0, \ldots]^T \\
v_2 &= [0, 1, 0, 0, 1, 0, \ldots]^T \\
v_3 &= [0, 0, 1, 0, 0, 1, \ldots]^T \\
&\vdots
\end{aligned}
\tag{5}
$$

There are $(2k+1)n - k(k+1)$ distinct elements in E_k of dimension n, and $(2k+1)n$ scalar equations in (4). Therefore, only the $k = 0$ (simple row-sum preserving) case is well defined. $(2k+1)n - k(k+1)$ of the equations in (4) explicitly assign individual elements of E_k. The remaining $k(k+1)$ equations, for $k > 0$, involve none of the elements of E_k at all, but require a certain sum of elements from the matrix on the right-hand side of (4), each of which is at least $k+1$ diagonals away from the main diagonal, to vanish. This overdetermination is inconsistent, but not detrimental for a preconditioner, particularly if the Schur complement matrix being approximated has terms which decay rapidly away from the diagonal. It costs $2k+1$ solves on each subdomain to compute the right-hand side of (4).

Setting $k = n - 1$ determines the Schur complement exactly (assuming that the tilde-quantities in (4) are identical to their non-tilde counterparts), in which case the overall algorithm converges in one iteration. This would be, in effect, the approach used by Przemieniecki in [24], and requires as many subdomain solves as there are degrees of freedom on the interface. As a means of obtaining a specified finite level of precision in the final result (commensurate, for example, with the discretization error), taking k to be $O(n)$ is inefficient, and particularly so if the subdomain solves are themselves not exact so that more than one iteration is needed. Between the extremes of many cheap iterations and few expensive iterations determined by the (possibly different) index k in each of MILU(k,α) and IP(k) will be an optimal trade-off.

Probing Multiple Interfaces

The description above applies to a simple interface between two subdomains, as depicted in Fig. 1. Generalizations to the compound interfaces and finer decompositions can take a variety of forms. Their development is aided by a heuristic approach to approximating the compound-interface generalization of the Schur complement (3), recognizing its inverse to be, essentially, a discrete Green's function.

We note that the element C_{ij} represents the influence of the data at interfacial node j on the discrete residual at interfacial node i [17]. Varying each interface point separately in turn would enable filling in C column-by-column. In diffusive problems, we would expect this internodal influence to decay rapidly with physical separation. In mixed convective-diffusive problems, we would expect dependencies of even shorter range on the values at "downwind" nodes, and of somewhat longer range on the values at "upwind" nodes (see [5] for exact expressions for a single interface). Depending upon the magnitude of these influences we might want, in the interest of economy and efficient parallelism, to set large off-diagonal blocks of C corresponding to sufficiently distantly coupled degrees of freedom to zero, and to determine the remaining (assumed non-neglible) blocks by varying large numbers of source points simultaneously. This is analogous to the Curtis-Powell-Reid [10] technique for efficient sparse Jacobian estimation using vector function evaluations, except that we are prepared, in general, to accept much looser restrictions on which columns may be treated as corresponding to unrelated degrees of freedom and thus be evaluated simultaneously.

In the extreme limit of $k = 0$, we attempt to probe all columns of C simultaneously. As k grows, the resources for resolving more of the structure of C can be invested in different ways. This is done by the selection of the v_l in (4). The v_l listed are appropriate in a purely diffusive problem with a spatially uniform diffusion coefficient and isolated interfaces (assuming that all the nodes on a given interface are ordered consecutively). By spreading out the active source points as evenly as possible, these v_l put a premium on resolving the influence of nearest neighbors *along* an interface.

The extension of IP preconditioning to the multiple-interface case can be done (as herein) by reducing the evaluation of a block diagonal $\tilde{C}$ to a series of independent interface problems, or by allowing $\tilde{C}$ to include coupling between different interfaces. The former (block diagonal) preserves in B the simple hierarchy of the two-subdomain case: each processor first solves a problem associated with one of the interfaces of its subdomain (modulo edge effects), followed by a subdomain interior solve. The latter (allowing some off-diagonal blocks which couple the interfaces) constitutes a more sequential algorithm, since interfaces belonging to different processors must complete their interactions before subdomain interior solves can be carried out. However, some form of global communication in the preconditioner phase is necessary in order to overcome the dependence of the overall iteration count on the number of subdomains employed. This becomes crucial at high decomposition granularity. The algorithms for symmetric operators proposed and analyzed in [3] and [12, 32], for instance, converge in a number of iterations which is either independent of the number of subdomains, or dependent only on the ratio of the mesh spacing to the number of subdomains.

PARALLEL IMPLEMENTATION

There are potentially three penalties to be paid in distributing the GMR/ILU/IP solution algorithm over an array of independent processors: synchronization overhead, communication overhead, and degradation of convergence. These penalties are measured indirectly through the speedup and efficiency figures-of-merit of a parallel implementation. The speedup is the ratio of the uniprocessor execution time of a given algorithm to that of the multiprocessor execution of the same algorithm. The efficiency is usually defined as the speedup divided by the number of processors. For some algorithms, these definitions are unnatural in the sense that one would not use the same algorithm in both uniprocessor and multiprocessor environments in practice. (Often *better* uniprocessor algorithms exist, so the parallel efficiency as defined above is inflated relative to its advantage.) We adopt measures in which the execution times are obtained from the most *natural* algorithm for each environment, namely, given p processors we employ exactly p subdomains.

The synchronization penalty arises if the processors have dependencies which force them to wait for data which are not yet available. Even if the processors are programmed homogeneously this can happen at convergence checkpoints, for instance, if they have unequal amounts of work to do. If the gridpoints are allocated to the processors as evenly as possible within shape and contiguity constraints synchronization delays can be made relatively unimportant.

The communication penalty is the time spent gaining access to shared data even after it becomes available. The significance of this penalty depends on the amount of data to be shared, on its routing between memory and processors, on the amount of arithmetic which the algorithm must perform between fetches and writes, and on the communication-to-computation speed ratios of the hardware in question.

In an effort to increase the number and length of the independent threads which comprise a parallel computation, global coupling may be reduced in ways that degrade the convergence rate of the algorithm. In the context of domain decomposition, this penalty may arise as the ratio of interface to interior degrees of freedom is increased in refinements of the decomposition, since the approximations required to form diagonal preconditioner blocks for the former are often more severe.

In view of the above considerations as well as programming convenience, our parallel implementation consists of decomposing the logical tensor product computational domain into logically congruent contiguous strips, mapping these subdomains onto a network of processors, and programming the processors homogeneously (apart from edge effects at the true domain boundaries). In this paper, our principal interest is in convergence rates, not architectural advantages, so we report on a bus-connected shared memory machine only: an Encore Multimax 320. The dependent variable arrays are placed in the shared memory and each processor is confined to roam over subranges of the array indices. The timings in the tables to follow include the entire preconditioned GMR process, but exclude the generation of the coefficients of the original operator and of the preconditioner blocks. The preconditioner coefficient generation has been efficiently parallelized, but we omit it from the timings because in nonlinear applications the preconditioning of the Jacobian matrix, as well as the Jacobian itself, may often be computed an order of magnitude less frequently than the solution of linear systems relying on these operators. To reduce the number of synchronization points inside each GMR iteration, a small QR factorization from which the coefficients of the Krylov basis vectors in the solution vector are derived is carried out redundantly in each processor. An analogous consideration led to the redundant solution of the equations for the vertex degrees of freedom in the parallel domain decomposition method described in [18].

RESULTS FOR MODEL PROBLEMS

This section contains numerical results that display a few of the possibilities of probe preconditioning, and, more generally, of the GMR/ILU/IP algorithm. In the solution of linear systems arising from finite-differenced systems of conservation equations of the form

$$\frac{\partial u_i}{\partial t} + \vec{c}(\vec{u}) \cdot \nabla u_i - \nabla \cdot D_i(\vec{u}) \nabla u_i = f_i(\vec{u}), \tag{6}$$

for $i = 1, \ldots, r$, the cases $|\vec{c}| > 0$ and $r > 1$ are of particular interest. The $r > 1$ case has been discussed and illustrated in [21], and we focus instead on different representative $\vec{c}$ here, discarding the unsteady term, and taking D to be constant in the process.

The ratio of $|\vec{c}|$ to D has the physical dimension of reciprocal length, and the dimensionless combination formed when this is multiplied by a typical mesh cell diameter is known as the cell

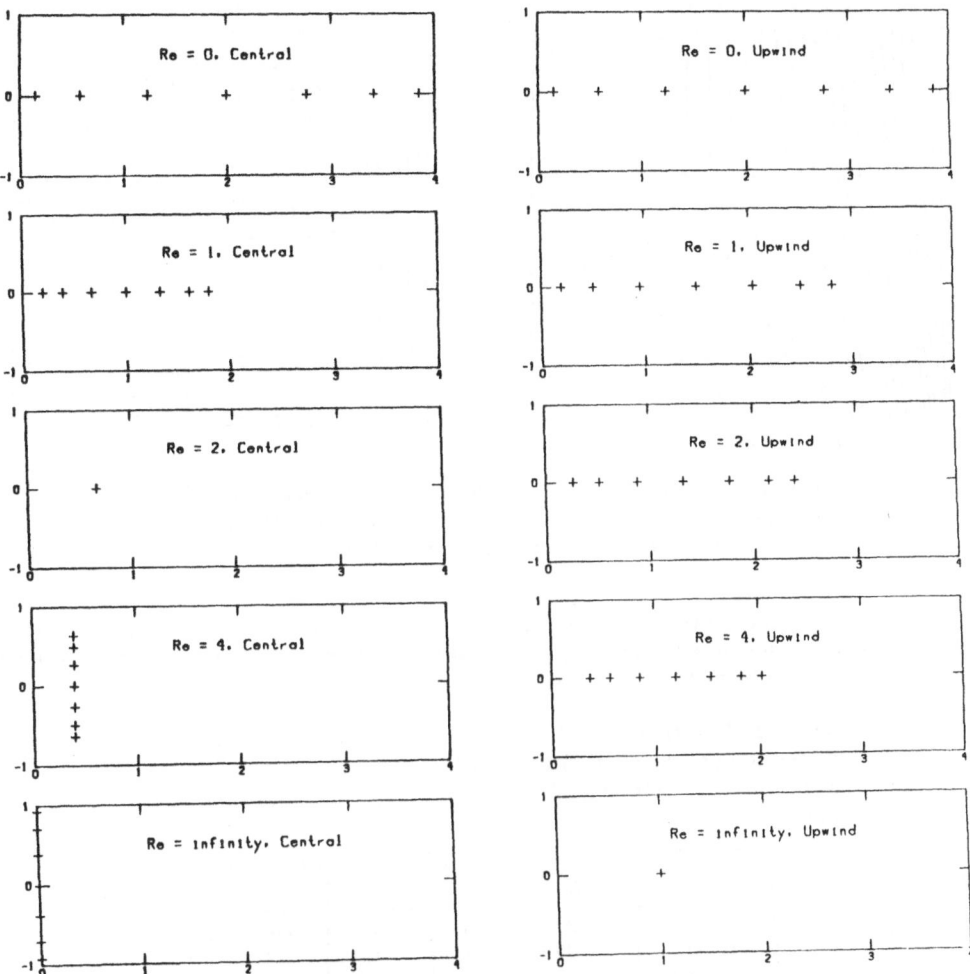

Figure 3. Spectra of a discrete 1-D convection-diffusion operator at five cell Reynolds numbers ranging from 0 to ∞: central differencing on the left, and upwind differencing on the right.

Reynolds number, Re_c. In the limit of unbounded mesh refinement and bounded velocity $\vec{c}$, the nonsymmetric convective term in the discrete operator will be dominated by the symmetric positive definite diffusive term, and FFT-implementable domain-decomposed Poisson solvers (see, *e.g.*, [7]) will be effective preconditioners. However, much practical CFD computation occurs in the opposite (singularly perturbed) limit, especially in the early, coarse-grid stages of an adaptive grid calculation or in defect-correction methods.

In contrast to published studies of iterative methods on convectively-dominated model problems which are based on central differences, we employ a simple bi-directional upwind-differenced discretization of (6). First-order upwinding for a constant $\vec{c}$ field keeps the spectrum of the discrete operator real as Re_c becomes large, in spite of the asymmetry. (We have not yet investigated any higher-order upwinding schemes.) For spatially varying $\vec{c}$, the upwinded spectrum is not necessarily real. We provide Fig. 3 to illustrate the spectral contrast between central and upwind discretizations at five different values of Re_c. Pictured are the seven eigenvalues of the constant coefficient operator $c \partial/\partial x - D \partial^2/\partial x^2$ discretized on grid of eight uniform subintervals, assuming Dirichlet boundary conditions. For $Re_c \leq 2$ both spectra are

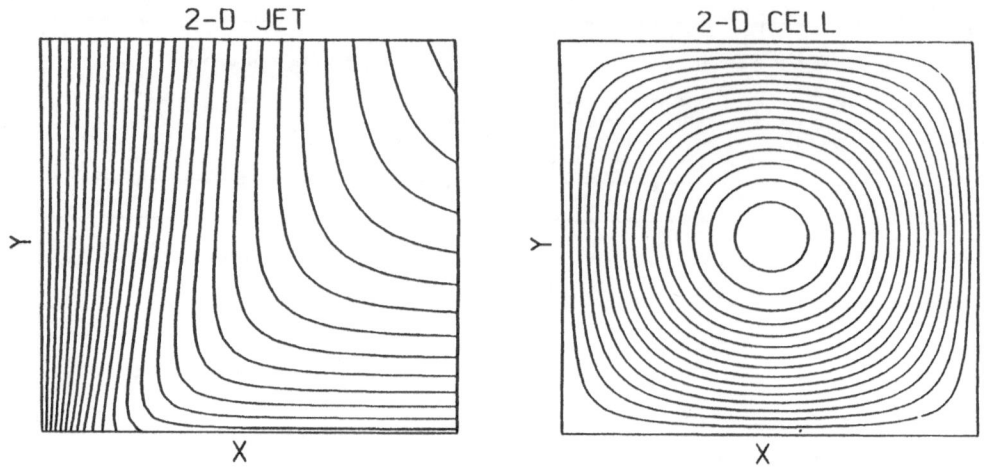

Figure 4. Streamline contours for the jet flow (left) and cell flow (right) problems. The jet is symmetric about the y-axis with its (point-source) orifice located at $(0, -\frac{1}{10})$.

real, but the eigenvalues of the centrally-differenced operator have non-zero imaginary parts at higher Re_c, and become purely imaginary asymptotically, while the eigenvalues of the upwinded operator asymptotically coalesce to unity.

Table 1. Iteration count I, CPU time T (sec.), and efficiency e for the pure conduction problem as a function of number of processors p and spatial resolution n, with total discrete dimension N.

			Conduction/MILU$(1, 1-h)$		
p	n	N	I	T	e
1	15	225	7	2.8	1.00
2	15	225	9	1.9	.75
4	15	225	9	1.0	.67
1	31	961	11	16.8	1.00
2	31	961	13	10.3	.82
4	31	961	13	5.5	.77
8	31	961	15	3.3	.63
1	63	3,969	15	93.5	1.00
2	63	3,969	19	61.7	.76
4	63	3,969	18	30.3	.77
8	63	3,969	21	17.9	.65
16	63	3,969	25	12.1	.48
1	127	16,129	21	546.	1.00
2	127	16,129	29	408.	.67
4	127	16,129	27	188.	.73
8	127	16,129	30	112.	.61
16	127	16,129	36	77.0	.44

The four model problems we consider derive from two-dimensional convective-diffusive systems on unit square domains: (1) pure diffusion, in which $\vec{c} = 0$ (as a benchmark); (2) plug flow, in which $\vec{c}$ is constant and oriented along one of the coordinate directions (the one normal to the direction of fastest ordering of the unknowns); (3) a jet flow, in which

Table 2. Iteration count I, CPU time T (sec.), and efficiency e for the plug flow at $\mathrm{Re}_c = 2.0$ as a function of number of processors p and spatial resolution n, with total discrete dimension N.

p	n	N	Plug Flow/ILU(1,0.0)			Plug Flow/MILU(1,1 − h)		
			I	T	e	I	T	e
1	15	225	8	2.9	1.00	7	2.8	1.00
2	15	225	9	1.7	.86	8	1.7	.82
4	15	225	10	1.1	.68	10	1.1	.62
1	31	961	11	14.8	1.00	10	15.4	1.00
2	31	961	12	8.8	.84	11	8.9	.87
4	31	961	13	5.0	.74	12	5.0	.77
8	31	961	16	3.4	.55	16	3.5	.55
1	63	3,969	15	85.2	1.00	13	81.4	1.00
2	63	3,969	16	47.5	.90	14	45.3	.90
4	63	3,969	17	26.3	.81	15	25.0	.81
8	63	3,969	20	16.6	.64	19	16.5	.62
16	63	3,969	26	12.3	.43	25	12.1	.42
1	127	16,129	21	518.	1.00	17	437.	1.00
2	127	16,129	22	280.	.93	18	239.	.91
4	127	16,129	23	149.	.87	20	138.	.79
8	127	16,129	26	97.5	.66	23	80.0	.68
16	127	16,129	31	60.0	.53	30	60.8	.45
1	255	65,025	32	3490.	1.00	23	2490.	1.00
2	255	65,025	32	1790.	.97	24	1310.	.95
4	255	65,025	34	973.	.90	26	745.	.84
8	255	65,025	36	537.	.81	29	436.	.71
16	255	65,025	41	357.	.61	36	311.	.50

$\vec{c}$ is the incompressible flow field for discharge through a slit into a quiescent reservoir (with entrainment) derived approximately by analytical means in [28]; and (4) a recirculating cell flow, given by $(\sin \pi x \, \cos \pi y, -\cos \pi x \, \sin \pi y)$. The jet and cell flows are pictured in Fig. 4. The boundary degrees of freedom are not eliminated *a priori* and are Dirichlet throughout, except for homogeneous Neumann (outflow) conditions at $y = 1$ for cases (2) and (3), a homogenous Neumann (symmetry) condition at $x = 0$ for case (3), and homogeneous Neumann (adiabatic) conditions at $y = 0, 1$ for case (4). A uniform grid is employed, and the Neumann conditions are discretized to first-order accuracy. All of the cases except the purely diffusive one are studied herein for a velocity field $\vec{c}$ scaled so that the maximum cell Reynolds number anywhere in the domain is 2.

Convergence results are presented in Tables 1 through 4. Except in Table 2, MILU$(1,1 - h)$ is used exclusively as the subdomain preconditioner (where $h = 1/(n - 1)$) and each problem is solved at four resolutions consisting of $n = 2^l - 1$ gridpoints in each direction for $l = 4, 5, 6, 7$. (These somewhat unconventional values for the resolution parameter perfectly balance the load on the subdomain interior problems in the parallel decompositions.)

For the pure conduction problem, Table 1 reveals an iteration count penalty for inserting the first interface (going from $p = 1$ to $p = 2$), which generally does not grow in severity until the penultimate dissection, which renders the individual subdomains just 3 discrete unknowns "wide" (or 7 in the finest grid case). Until this second high subdomain aspect ratio penalty sets in, the parallel efficiency sits in the 60-80% range and has a tendency to degrade at constant n as p increases. However, efficiency improves as some locally superlinear speedups are observed, as in going from 2 to 4 processors at $n = 63$ and $n = 127$.

Results for the plug flow problems are reported in Table 2, which extends the resolution comparison to $l = 8$ (a time-consuming problem for a single Encore Multimax processor), and also contrasts performance of MILU$(1,1 - h)$ to ILU(1) to illustrate the utility of the α

parameter, although no attempt is made to optimize α to the particular operator or grid sizes considered. Speedups of 8 to 10 are obtained on the largest problems using 16 processors. Here, efficiency always degrades at constant n as p increases, and improves at constant p as n increases. Note that MILU$(1,1-h)$ is superior to ILU(1) at all granularities of the decomposition. However, its efficiency at high p is always slightly less than ILU(1). This shows the deceptive nature of efficiency as an exclusive criterion for judging parallel performance. MILU's lesser efficiency i due to the fact that it improves the lower p cases more than it improves the higher p cases

Table 3. Iteration count I, CPU time T (sec.), and efficiency e for the jet flow at maximum Re$_c$ = 2.0 as a function of number of processors p and spatial resolution n, with total discrete dimension N.

p	n	N	Jet Flow/MILU$(1,1-h)$		
			I	T	e
1	15	225	10	3.7	1.00
2	15	225	11	2.2	.84
4	15	225	12	1.3	.71
1	31	961	15	22.5	1.00
2	31	961	17	13.6	.83
4	31	961	17	7.0	.80
8	31	961	23	5.3	.53
1	63	3,969	23	148.	1.00
2	63	3,969	24	78.2	.95
4	63	3,969	24	40.7	.91
8	63	3,969	27	24.4	.76
16	63	3,969	42	25.3	.37
1	127	16,129	32	896.	1.00
2	127	16,129	32	453.	.99
4	127	16,129	32	235.	.95
8	127	16,129	32	135.	.83
16	127	16,129	38	89.9	.62

The jet flow results in Table 3 are noteworthy for the several successive domain dissections which are *not* accompanied by an iteration penalty, and sharp high subdomain aspect ratio penalties for the larger systems.

The cell flow results show the problems encountered by limited-fill preconditioners in the presence of recirculation; the iteration counts even for the undecomposed cases grow with n at a rate which is clearly faster than the $O(\sqrt{n})$ observed for the previous three cases and expected on theoretical grounds for the purely diffusive case. In general, we would recommend a stronger preconditioner than MILU$(1,\alpha)$ for the subdomain problems. The parallel efficiencies as p grows are also somewhat disappointing, especially the failure of the largest case to converge within even n iterations. The quadratic growth of Gram-Schmidt work in iteration index for non-restarted GMRES makes this unacceptable. The cyclical connectivity of the cell operator, which can be visualized on purely physical grounds, calls for a preconditioner with greater interface coupling as well as stronger subdomain preconditioning. However, domain decomposition still appeals as a parallelization paradigm at modest granularities, even for this problem.

We conclude this section by showing in Fig. 5 some estimated spectra for the preconditioned operators $B^{-1}A$ of the 63×63 cases of each of the four problems. Two different preconditioners are used: the single subdomain MILU and the eight-subdomain MILU/IP. These spectra were obtained (via Arnoldi's method, see *e.g.*, [25]) from the eigendecomposition of the upper Hessenberg matrices obtained in the first 20 iterations of each of the test problems. Residual calculations show that the eigenvalues with the extreme magnitudes (large and small) are fairly accurate even with just 20 iterations, but the interior portions of the spectra are estimated crudely. Although the estimated unpreconditioned spectra (not shown) are real or nearly so

Table 4. Iteration count I, CPU time T (sec.), and efficiency e for the cell flow at maximum $\mathrm{Re}_c = 2.0$ as a function of number of processors p and spatial resolution n, with total discrete dimension N.

			Cell Flow/MILU$(1,1-h)$		
p	n	N	I	T	e
1	15	225	16	5.8	1.00
2	15	225	18	3.6	.80
4	15	225	19	2.1	.69
1	31	961	29	46.3	1.00
2	31	961	31	26.4	.88
4	31	961	34	15.6	.74
8	31	961	37	9.1	.63
1	63	3,969	50	391.	1.00
2	63	3,969	53	216.	.91
4	63	3,969	59	129.	.76
8	63	3,969	71	85.3	.57
16	63	3,969	87	63.4	.39
1	127	16,129	88	3710.	1.00
2	127	16,129	95	2110.	.88
4	127	16,129	101	1180.	.79
8	127	16,129	123	810.	.57
16	127	16,129	> 150	-	-

(with at most a pair of complex eigenvalues of large real and small imaginary part, and a ratio of the largest to smallest eigenvalue magnitudes on the order of n^2), the MILU-preconditioned spectra are generally complex. The pure conduction problem is an exception, but even here the eight-subdomain preconditioner has a complex spectrum. Though some of the multi-domain spectra are more compact, their hulls lie closer to the origin, and the multi-domain problems do not converge as well as the single-domain problems, in general. The reason for this is well understood in the symmetric positive definite context and well intuited in the present one: nearest neighbors-only coupling in the preconditioner allows for only gradual propagation of signals across the domain. Upwinded convection tends to parabolize the operator, but the elliptic terms are still present.

CONCLUSIONS

Included in the framework of this paper is a family of schemes which lie between the extremes of decoupled block-diagonal preconditioned iteration and domain-decomposed direct elimination, governed by bandwidth parameters for subdomain and interface preconditioning blocks. Though surpassed in efficiency by known methods in several contexts, they provide a means for the parallel solution of rather general linear systems, at least in the coarse-to-medium granularity range, offering typically 8-fold speedups on 16-processor arrays. One near-term motivation for techniques which perform acceptably over a range of granularities is that jobs which use nearly all of the memory of a multiprocessor supercomputer, such as the Cray-XMP or the ETA-10, effectively tie up most of the processors sharing that memory, whether their cycles are actually employed or not. The ability to keep the processing force in rough proportion to memory usage over a range of problem sizes can be an economic advantage. Independent of economic considerations, there is obviously a premium on the ability to reduce wall-clock time by adding processors in "production mode" in many applications. Until the elusive day when native software facilities make parallelization automatic, domain decomposition will have a significant role to play in meeting these objectives.

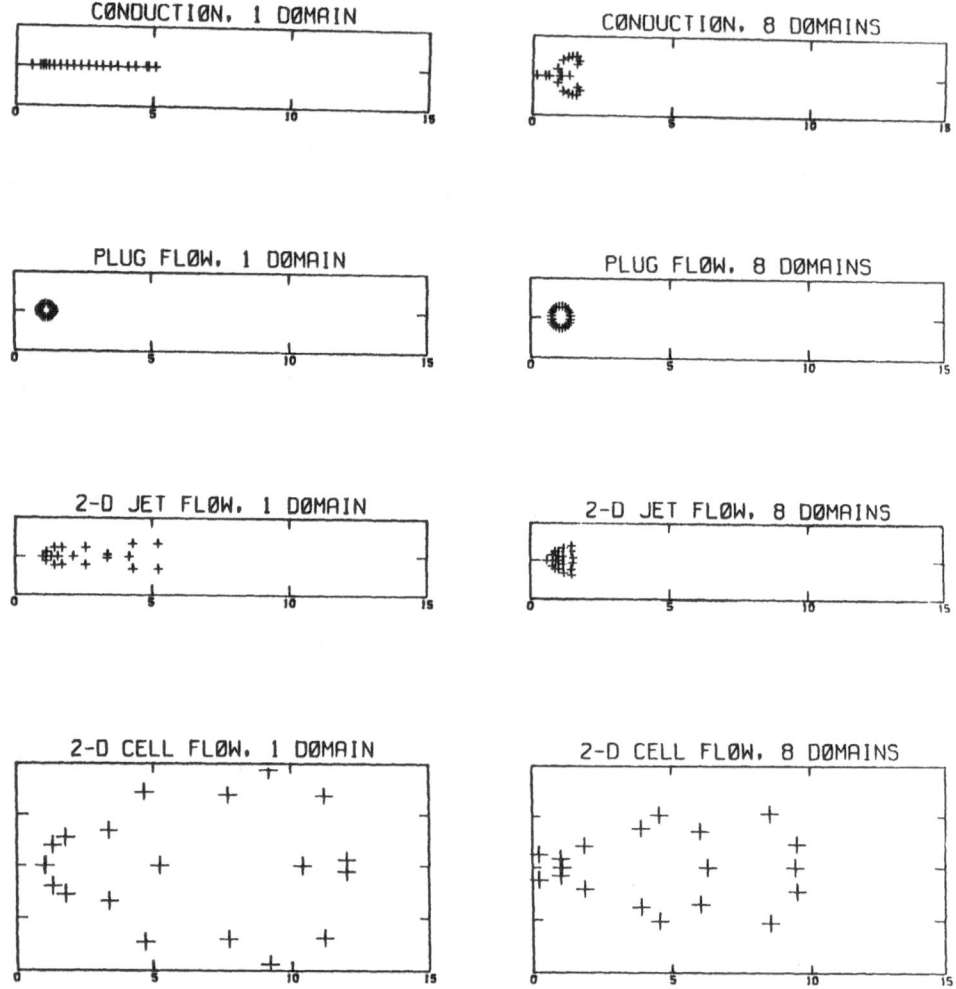

Figure 5. Spectra of the preconditioned discrete operators for each of the four model problems: single-domain spectra on the left, multiple-domain spectra on the right.

REFERENCES

[1] C. Ashcraft, *Domain Decoupled Incomplete Factorizations*, Technical Report ETA-TR-49, Boeing Computer Services, April 1987.

[2] O. Axelsson & G. Lindskog, *On the Eigenvalue Distribution of a Class of Preconditioning Methods*, Numer. Math., 48 (1986), pp. 479–498.

[3] J. H. Bramble, J. E. Pasciak & A. H. Schatz, *The Construction of Preconditioners for Elliptic Problems by Substructuring, I*, Math. Comp., 47 (1986), pp. 103–134.

[4] T. F. Chan & D. Goovaerts, *Domain Decomposition Beneficial Even Sequentially*, Technical Report 88-18, UCLA Comp. and App. Math., June 1988.

[5] T. F. Chan & T. Y. Hou, *Domain Decomposition Preconditioners for General Second Order Elliptic Operators*, Technical Report 88-16, UCLA Comp. and App. Math., June 1988.

[6] T. F. Chan & D. Resasco, *A Survey of Preconditioners for Domain Decomposition*, Technical Report 414, Computer Science Dept., Yale University, September 1985. In Proceedings of the IV Coloquio de Matemáticas del CINVESTAV, Workshop in Numerical Analysis and its applications, Taxco, Mexico, Aug. 18-24, 1985.

[7] ———, *A Domain-Decomposed Fast Poisson Solver on a Rectangle*, SIAM J. Sci. Stat. Comp., 8 (1987), pp. s14-s26.

[8] ———, *A Framework for the Analysis and Construction of Domain Decomposition Preconditioners*, R. Glowinski, G. H. Golub, G. A. Meurant & J. Periaux (eds.), *First International Symposium on Domain Decomposition Methods for Partial Differential Equations*, SIAM, 1988, pp. 217–230.

[9] R. W. Cottle, *Manifestations of the Schur Complement*, Lin. Alg. Appl., 8 (1974), pp. 189–211.

[10] A. R. Curtis, M. J. Powell & J. K. Reid, *On the Estimation of Sparse Jacobian Matrices*, J. Inst. Math. Appl., 13 (1974), pp. 117–119.

[11] M. Dryja & W. Proskurowski, *Capacitance Matrix Method using Strips with Alternating Neumann and Dirichlet Boundary Conditions*, Appl. Numer. Math., 1 (1985).

[12] M. Dryja & O. B. Widlund, *An Additive Variant of the Schwarz Alternating Method for the Case of Many Subregions*, Technical Report 339, Courant Institute of Mathematical Sciences, NYU, December 1987.

[13] T. Dupont, R. Kendall & H. H. Rachford, *An Approximate Factorization Procedure for Solving Self-Adjoint Elliptic Difference Equations*, SIAM J. Num. Anal., 5 (1968), pp. 559–573.

[14] A. Ecder & D. E. Keyes, *A Parametric Study of the Spectra and Convergence Rates of Preconditioned Generalized Minimum Residual Schemes for Convective-Diffusive Problems*, 1988. (in preparation).

[15] H. C. Elman & E. Agrón, *Ordering Techniques for the Preconditioned Conjugate Gradient Method on Parallel Computers*, Comp. Phys. Comm. (1989, to appear).

[16] R. Glowinski & M. F. Wheeler, Domain Decomposition and Mixed Finite Element Methods for Elliptic Problems, R. Glowinski, G. H. Golub, G. A. Meurant & J. Periaux (eds.), *First International Symposium on Domain Decomposition Methods for Partial Differential Equations*, SIAM, 1988, pp. 144–172.

[17] G. H. Golub & D. Mayers, The Use of Preconditioning over Irregular Regions, R. Glowinski & J. L. Lions (eds.), *Computing Methods in Applied Sciences and Engineering VI*, North Holland, 1984, pp. 3–14.

[18] W. D. Gropp & D. E. Keyes, *Complexity of Parallel Implementation of Domain Decomposition Techniques for Elliptic Partial Differential Equations*, SIAM J. Sci. Stat. Comp., 9 (1988), pp. 312–326.

[19] D. E. Keyes, *Domain Decomposition Methods for the Parallel Computation of Reacting Flows*, Comput. Phys. Comm. (1989, to appear).

[20] D. E. Keyes & W. D. Gropp, *A Comparison of Domain Decomposition Techniques for Elliptic Partial Differential Equations and their Parallel Implementation*, SIAM J. Sci. Stat. Comp., 8 (1987), pp. s166-s202.

[21] ————, *Domain Decomposition Techniques for the Parallel Solution of Nonsymmetric Systems of Elliptic PDEs*, Appl. Num. Meths. (1989, to appear).

[22] T. A. Manteuffel, *The Tchebychev Iteration for Nonsymmetric Linear Systems*, Numer. Math., 28 (1977), pp. 307–327.

[23] J. A. Meierink & H. A. Van der Vorst, *Guidelines for the Usage of Incomplete Decompositions in Solving Sets of Linear Equations as they Occur in Practical Problems*, J. Comp. Phys., 44 (1981), pp. 134–155.

[24] J. S. Przemieniecki, *Matrix Structural Analysis of Substructures*, AIAA J., 1 (1963), pp. 138–147.

[25] Y. Saad, *Chebyshev Acceleration Techniques for Solving Nonsymmetric Eigenvalue Problems*, Math. Comp., 42 (1984), pp. 567–588.

[26] Y. Saad & M. H. Schultz, *Parallel Implementations of Preconditioned Conjugate Gradient Methods*, Technical Report YALEU/DCS/RR–425, Computer Science Dept., Yale University, October 1985.

[27] Y. Saad & M. H. Schultz, *GMRES: A Generalized Minimum Residual Algorithm for Solving Nonsymmetric Linear Systems*, SIAM J. Sci. Stat. Comp., 7 (1986), pp. 856–869.

[28] H. Schlichting, *Boundary Layer Theory*, 7th ed., McGraw-Hill, New York, 1979.

[29] H. F. Walker, *Implementation of the GMRES Method Using Householder Transformations*, SIAM J. Sci. Stat. Comp., 9 (1988), pp. 152–163.

[30] J. W. Watts, III, *A Conjugate Gradient-Truncated Direct Method for the Iterative Solution of the Reservoir Simulation Pressure Equation*, Soc. Petrol. Engin. J., 21 (1981), pp. 345–353.

[31] O. B. Widlund, Iterative Substructuring Methods: Algorithms and Theory for Elliptic Problems in the Plane, R. Glowinski, G. H. Golub, G. A. Meurant & J. Periaux (eds.), *First International Symposium on Domain Decomposition Methods for Partial Differential Equations*, SIAM, 1988, pp. 113–128.

[32] ————, *Some Domain Decomposition and Iterative Refinement Algorithms for Elliptic Finite Element Problems*, Technical Report 386, Courant Institute of Mathematical Sciences, NYU, June 1988.

PARALLEL COMPUTATION IN MECHANICS

Gary M. Johnson

San Diego Supercomputer Center
10100 John Jay Hopkins Drive
San Diego, California   92121

ABSTRACT

The success of numerical simulation as an independent approach to the solution of engineering problems requires computing capability far exceeding that which is presently available. In this paper, the computing requirements posed by challenging problems in mechanics are examined and contrasted with contemporary supercomputer resources. Of the means available to help fill the gap between the demands of scientific computation and the performance level of present-generation supercomputer systems, parallel processing appears to have the greatest potential for near-term success. Typical parallel computer architectures are reviewed and categorized. Philosophies of parallel processing are distinguished by the number and size of the parallel tasks which they employ. Selected engineering problems are examined for parallelism inherent at the physical level. Typical algorithms and their mappings onto parallel architectures are discussed. Computational examples are presented to document the performance of scientific applications on present-generation parallel processors. Projections are made concerning future algorithms and machine architectures.

INTRODUCTION

Computation has begun to emerge from its supporting role and establish itself as an independent approach to engineering. Although the origins of computational engineering lie further in the past, the relatively recent introduction of the supercomputer has enabled it to assume the same status enjoyed by the traditional theoretical and experimental approaches. At present, this elevated status is based in part on impressive applications which have yielded significant economic impact in selected areas, such as aircraft design, petroleum recovery, and structural analysis and in part on the expectation that its future holds both broader application and greater impact.

The greatest potential for increasing the economic impact of computational science lies in the solution of large-scale, challenging problems which are not amenable to alternative approaches. Wilson[1], who refers to such problems as "Grand Challenges", opines that for at least two of these challenges, those concerning electronic structure and turbulence, a significant breakthrough would have the same impact as the discovery of a law of nature.

Where it is possible to estimate the computing requirements posed by large-scale engineering problems, a universal conclusion emerges. While the methods and machines presently available have served to demonstrate the potential of computational engineering, they are inadequate to the task of realizing that potential. Large improvements are necessary in both computational methods and the computers which run them.

Because the technologies of scalar and vector computing have been heavily exploited, further refinements are unlikely to yield large increases in performance. Consequently, attention has shifted to parallel processing. The nature of parallel computing is such that its successful use presently requires detailed understanding of the interactions among large-scale applications, parallel algorithms, and the architectures of parallel-processing devices. While a goal of parallel computing is to make such considerations transparent, in the short term, interdisciplinary teams of applications specialists, computational mathematicians, and computer scientists constitute the best hope for progress.

ENGINEERING REQUIREMENTS

As researchers have now begun to attack large-scale problems in a number of disciplines, it is possible to make rough estimates of the computer speed and memory requirements posed by a fairly broad sampling from the spectrum of engineering. Figures 1 through 6 present such requirements and contrast them with the capabilities of past-, present-, and prospective future-generation computers. These figures are modeled on the work of Peterson and Arnold[2], which may be consulted for further details on some of them. Some general observations are possible. In all the areas shown, the requirements exceed present computer capability by wide margins. In many areas, several orders of magnitude improvement in both speed and memory are necessary simply to satisfy the needs of problems of contemporary interest. Present-generation supercomputers have 4 to 8 processors. Those available by 1992 are expected to have as many as 64 processors. The figures show that even this level of parallelism is inadequate. Thus, it is to be assumed that massively-parallel supercomputers will be required in order to carry computational engineering forward into the next century.

COMPUTATIONAL RESOURCES

Assuming that adequate memory is available, the main issue is reducing the solution time required by challenging problems to tractable levels. Minimizing solution time means improving the performance of both algorithms and machines.

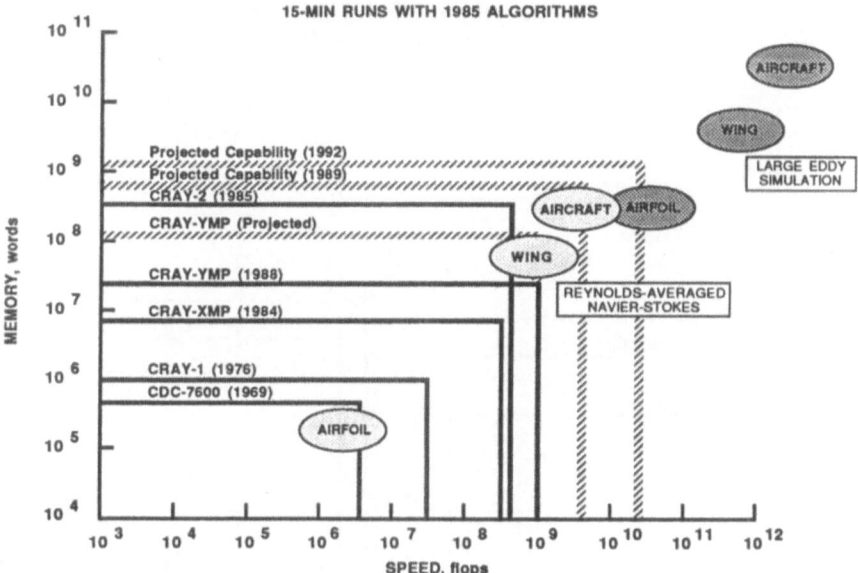

Figure 1.

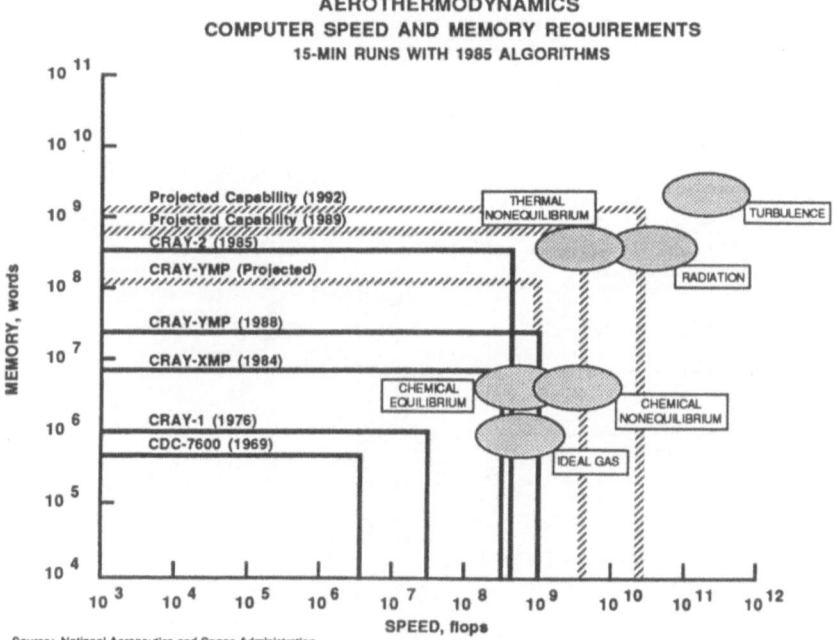

Figure 2.

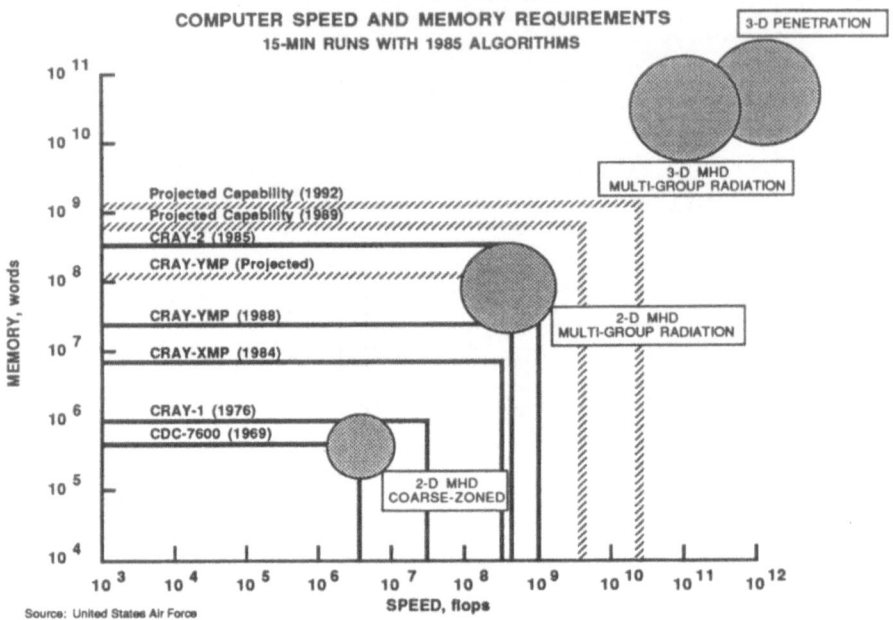

Figure 3.

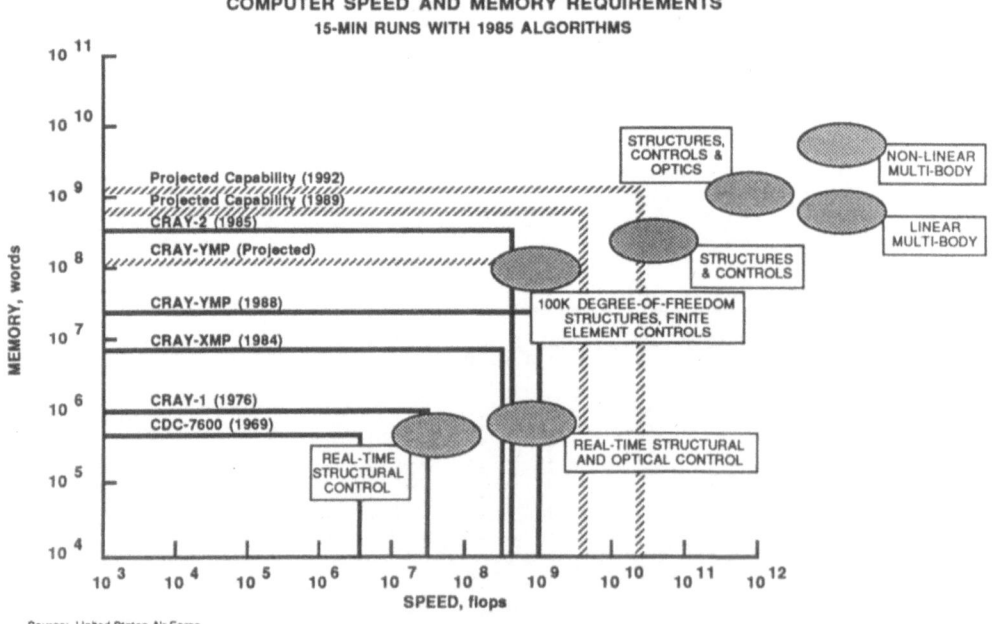

Figure 4.

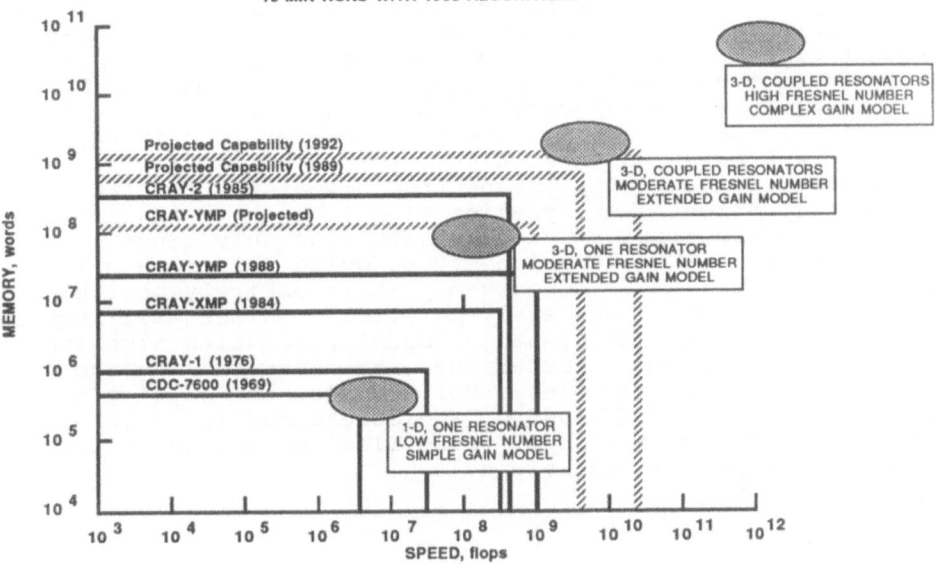

Figure 5.

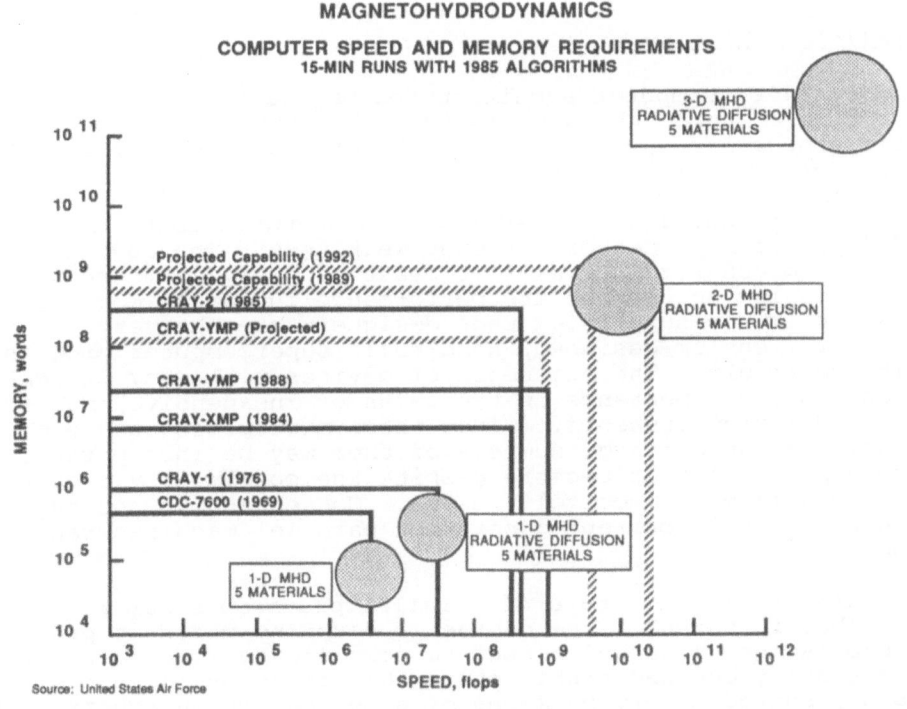

Figure 6.

Historically, both algorithms and machines have improved at an impressive pace. Peterson and Arnold cite an example drawn from computational aerodynamics where, over a 15-year period, a net improvement of 10^5 was attained. This improvement was the product of compounding a 10^3 improvement in algorithm performance on a given machine with a 10^2 improvement in machine performance for a given algorithm. While such compounding is valid for scalar machines, care must be exercised in applying it to more sophisticated architectures. The major architectural innovation during the 15-year period cited above was vectorization. Aerodynamics algorithms have adapted fairly easily to vectorization, usually requiring only minor restructuring. The situation may be quite different where massive parallelism is concerned. It is entirely possible that the efficient use of massively-parallel machines will require the adoption of new mathematical models, solution algorithms, and computer programs, rather than the restructuring and adaptation of existing ones. For example, the lattice gas approach[3], which is based on a discrete model, could displace the continuum description of fluid flow and its attendant partial differential equations.

PARALLEL PROCESSING

The term parallel processing, as used in this paper, refers to the use of multiple instruction, multiple data (MIMD) machines. Such machines allow several different instructions to concurrently operate on several distinct sets of operands. SIMD (single instruction, multiple data) machines, such as array and vector processors, which permit only one instruction to simultaneously operate on multiple sets of operands, are not further considered.

Parallel Architectures

Parallel computer architectures may be classified according to three characteristics: The number and power of their processing units; the location of their memory; and the nature of their interconnection scheme.

Current parallel-processing architectures contain either a few powerful processors or many weak ones. The cost of a parallel processor increases at least linearly with the number of processing units, but its performance increases more slowly. Thus, it is not surprising that designers use the very expensive, custom-designed, hand-built supercomputer processing units sparingly. The existence of devices with many processors is due to the appearance in the 1970s of inexpensive, mass-produced microprocessors. Since these microprocessors are relatively cheap, large numbers of them may be incorporated into a parallel architecture despite the possibility that some of them will be effectively wasted. Consequently, the number of processors in current-generation parallel machines ranges from 2 to 2^{10}.

The memory structure of parallel processors may be described in terms of the degree of coupling which it provides to processing units which need to communicate with each other. In a tightly-coupled architecture the processors access a global, shared memory by means of some connection scheme. The processors communicate by reading from and writing to this

272

memory. In a loosely-coupled, distributed-memory machine, each processor has its own local memory and communicates with other processors by sending messages through the interconnect. It is not unusual for real machines to have memory organizations which combine both of these structures. Present-generation parallel processors have memories ranging as high as 256 million 64-bit words.

Parallel processors differ widely in the way in which communications among processors and memory are handled. In general, either a data bus or a switching network is used. Architectures which use a data bus provide one common path which connects all of the processors and memories. As all communications take place over the bus, system performance will degrade when the bus is saturated by processor requests for access. Thus, the rate at which the bus can transfer data is a critical parameter for this sort of multiprocessor.

The use of switching networks in parallel architectures represents an alternative communications scheme which does not suffer from saturation the way that busses do. The most general means of connecting processors and memory is with a full crossbar switch. In such a network, every processor is directly connected to every memory module. As full crossbar switches are quite expensive when large numbers of processors and memory modules are involved, designers often resort to less complete switching schemes. Other switching networks are either based on many small interconnected crossbar networks, or on incomplete crossbar networks. Typical data transmission rates for parallel processor interconnection schemes range from 10 megabits to 100 gigabits per second.

Parallel Algorithms

The main issue in parallel processing applications is designing algorithms which, while minimizing the number of operations per solution, also make efficient use of the parallel architecture in order to maximize machine speed and thus minimize the time to solve problems. The alternative of building a custom parallel processor designed specifically to solve a particular problem quickly is technically feasible but is presently economically unattractive for most applications and will not be discussed here.

The factor by which execution time is reduced when an application is run on a parallel processor rather than on a single-processor machine is called speedup. Parallel-processing speedup, S, may be modeled [4] as

$$S = \frac{1}{R + (1-R)\left[\frac{Y}{KG} + f(\frac{K}{N})(\frac{1}{K} + \frac{O}{KG})\right]}$$

where:

G	=	task granularity
K	=	number of tasks between synchronizations
N	=	number of processors
O	=	task overhead
R	=	residual sequential code
Y	=	synchronization overhead

$$f\left(\frac{K}{N}\right) = \begin{cases} \dfrac{K}{N} & ; \ K \text{ divisible by } N \\ \dfrac{K}{N}+1; & \text{otherwise} \end{cases}$$

Insight may be gained by examining a few special cases.

Task overhead is the additional time required to complete a task because it is being executed on a parallel processor. Synchronization overhead is the additional time required to properly pace the concurrent execution of tasks. If we assume that both of these effects are trivial and that the number of tasks is divisible by the number of processors, then we may write

$$S = \frac{1}{R + \dfrac{1-R}{N}}$$

This expression demonstrates the effect which the residual percentage of code which must be executed sequentially, R, has on speedup. This effect is also illustrated in Figure 7, where curves of constant R are plotted as functions of the speedup and number of processors. In the figure, the heavy line indicates the point at which each curve reaches ninety percent of its asymptotic value. This asymptote is easily seen to be the reciprocal of R. The diagram shows that residual sequential code places a rather significant constraint on the effectiveness with which any particular algorithm can make use of large numbers of processors.

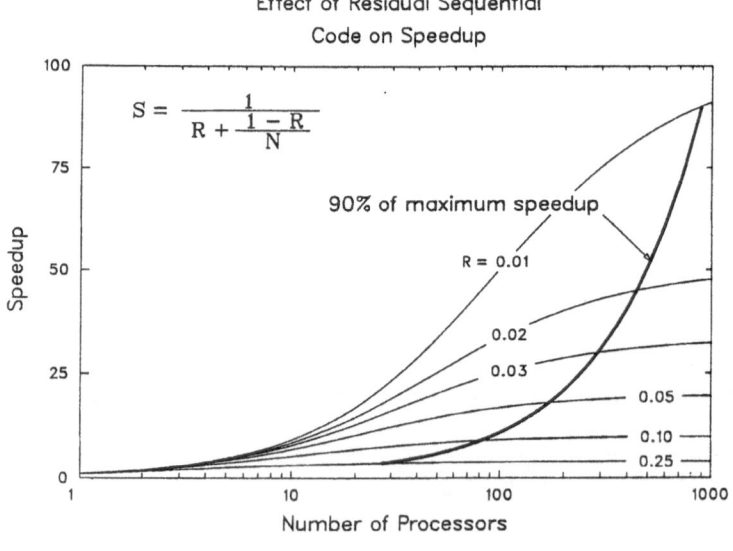

Figure 7.

For the following examples, we assume that the residual sequential code is negligible. If we then fix the number of tasks and allow the number of processors to become arbitrarily large, the speedup approaches

$$S = \frac{K}{1 + \dfrac{Y+O}{G}}$$

Observe that the speedup is independent of the number of processors, so that the parallel-processing efficiency, S divided by N, goes to zero in this limit. This case is of more than theoretical interest because it epitomizes the difficulty of mapping a fixed-size application onto ever larger numbers of processors.

If we assume one task per processor, the speedup reduces to

$$S = \frac{N}{1 + \dfrac{Y+O}{G}}$$

If we then define a relative overhead, Q, such that

$$Q = \frac{Y+O}{G}$$

we see that, for this case, the parallel processing efficiency is $1/(1+Q)$. Thus, if the relative overhead can be made small and held constant as N increases, a strategy of assigning one task per processor and then increasing the problem size as the number of processors grows should result in good speedup and high efficiency.

If the number of processors is fixed while the number of tasks grows without bound, the limiting speedup is

$$S = \frac{N}{1 + \dfrac{O}{G}}$$

This case is similar to the previous one, with the added simplification that the effect of synchronization overhead has disappeared. This may be a deceptively simple example in that an arbitrarily large number of tasks on a fixed-size machine may incur very significant task overhead.

These examples suggest that a successful parallel algorithm will minimize both residual sequential code and relative overhead. This exercise may be difficult. Contriving a parallel implementation of an inherently sequential computation could result in high task overhead. Increasing task granularity to reduce relative overhead will place an upper bound on the number of processors onto which a fixed-size problem may be mapped.

The need to extract high performance levels from parallel processors causes algorithms and architectures to be more interdependent than has previously been the case in computational science. Some numerical methods are more amenable to parallelism than others. If we consider, for example, the numerical solution of partial differential equations, the applicable methods may be classified as either explicit or implicit. Explicit methods use local information and do not involve coupling of locations during the computational step. Thus, explicit methods map well onto parallel machines, allowing as many tasks as there are points in the computational domain. However, the use of only local information can lead to problems. Poor convergence rate (i.e., a high operations count per solution) may be one such problem. Small task granularity is another one.

Implicit methods use all of the information within some block structure and also couple the locations within this structure during the computational step. Generally, this leads to improved convergence rates and larger task granularity. However, options for mapping such methods onto parallel machines are limited by the structure of the implicit blocks. Problem partitions which do not respect this structure may lead to diminished convergence rates or increased overhead.

Parallelism in Mechanics

In spite of all the obstacles to the effective use of parallel computing, our senses tell us that the physical universe contains a great deal of parallelism to be exploited and steady progress is being made in finding and using this parallelism. The main issues are the granularity and partitioning of problems.

A partitioning strategy used to define tasks for simultaneous execution should be based on the parallelism inherent in the physical problem to be solved. Two forms of parallelism may be readily identified: spatial and functional. Alternatively, these might be called domain decomposition and problem decomposition, respectively.

Spatial parallelism (including the possibility of time as a dimension) occurs when the effects of a perturbation are felt principally in its immediate vicinity. Spatial subdomains may then be identified which interact only weakly with each other. Such subdomains may be assigned to different processors for updating. Information must occasionally be exchanged among these tasks in order to respect their spatial interaction. Hyperbolic and hyperbolic-parabolic systems contain abundant opportunities for spatial partitioning.

Functional parallelism occurs when the physical phenomena which comprise the various functions of the model are associated with significantly different time or length scales. It is common (but not necessary) in such a case for the phenomena to be represented by different sets of model equations. Examples of functional parallelism drawn from fluid dynamics are viscous effects confined to a boundary layer or wake and chemical reactions taking place in a multicomponent flow. The computation of such functionally-distinct phenomena may generally be assigned to different processors, subject to occasional communication.

Performance Examples

Due mainly to a significantly increased level of interest in parallel processing during the past year, enough performance data has accumulated to provide a reasonable basis for analysis. The examples discussed here are drawn from sources listed in Reference 5. As the purpose of this paper is to be illustrative rather than exhaustive, the interested reader is referred to the source material for more comprehensive detail.

The performance data is analyzed from three perspectives: speedup, efficiency, and floating-point speed. As all performance measures are not available for every datum, the results should be viewed as ensembles. Also, note that no analysis of the effectiveness of particular algorithmic approaches is attempted, because of the general unavailability of such information.

Parallel-processing speedup examples are presented in Figure 8. Part 8a shows all the examples while the data for shared-memory and distributed-memory machines are segregated in Parts 8b and 8c, respectively. The results are encouraging. The best data indicate that something close to a linear increase in speedup can be maintained, at least through 1000 processors. The results for both shared- and distributed-memory machines exhibit a fair amount of scatter. In some cases this is probably due to naive use, since these machines have only recently become widely available. In other cases poor performance may be the result of inadequate adaptation of

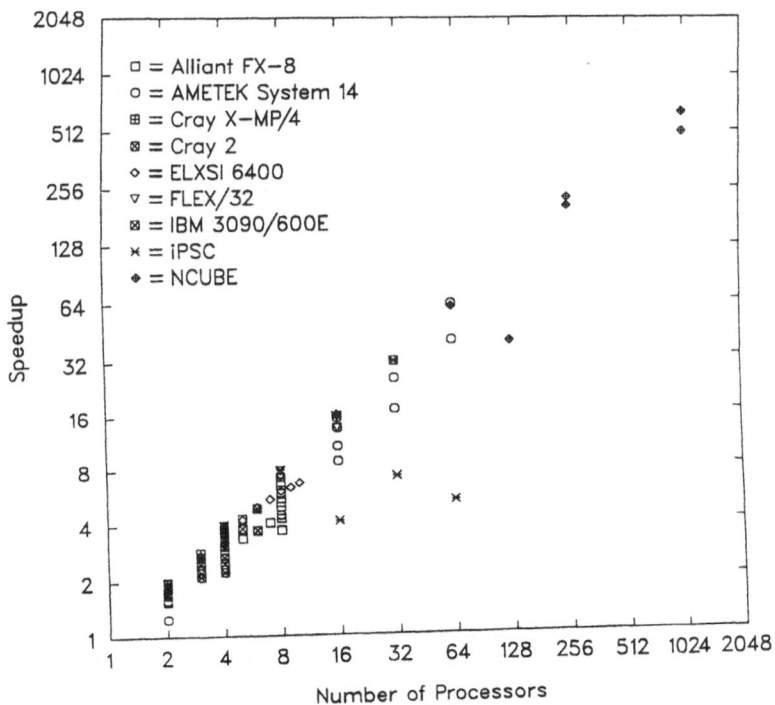

Figure 8a. Parallel-Processing Speedup

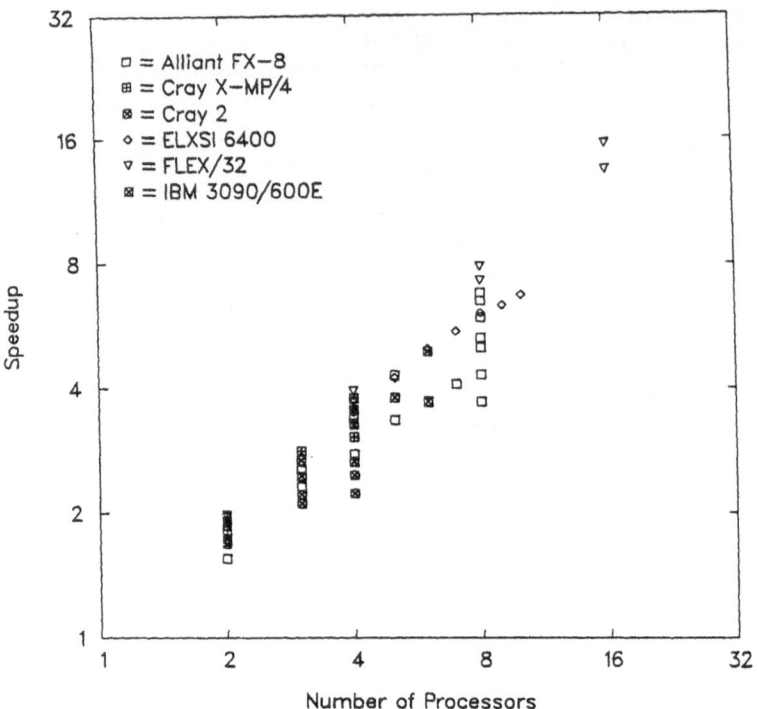

Figure 8b.   Speedup of Shared-Memory Machines

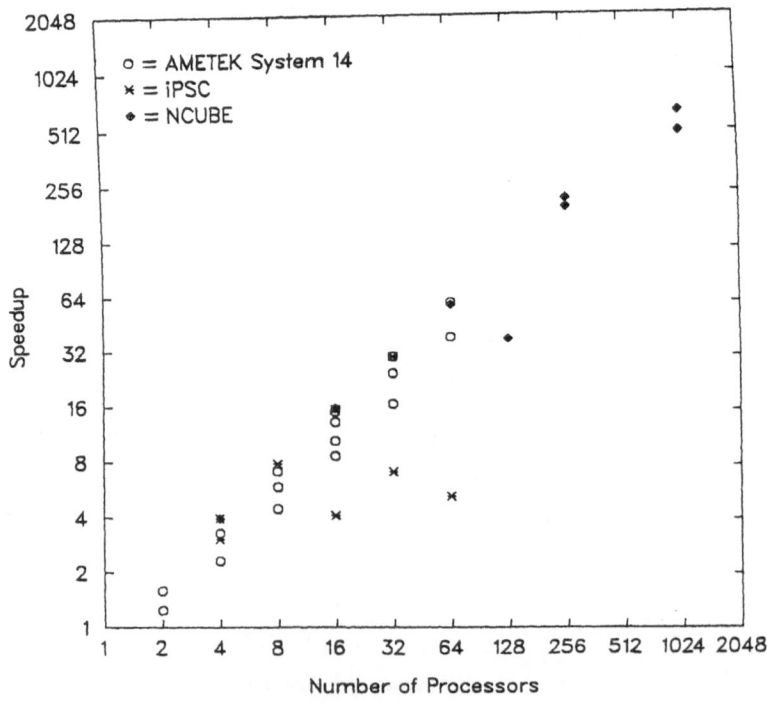

Figure 8c.   Speedup of Distributed-Memory Machines

single-processor architectures to the parallel-processing environment.

Figure 9 illustrates the efficiency examples. Here we observe that the trend is for shared-memory machines with a small number of processors to be significantly more efficient than distributed-memory machines with large numbers of processors. This indicates that either distributed-memory machines need to be made much more efficient or the importance of efficiency as a measure of their performance needs to be reconsidered.

The floating-point speed examples are shown in Figure 10. Here, machines sort out according to the speed of the individual processing units from which they are built. The shared-memory machines all have much better performance than the distributed-memory ones, although the variation in performance within the shared-memory group is quite broad. Simple linear extrapolation of the best results in each memory category indicates that shared-memory machines can sustain a gigaflop with 8 processors, while distributed-memory machines will require about 8,000 processors to do the same.

It should be emphasized that parallel processing is a young and rapidly-evolving part of computational engineering. The examples presented here illustrate trends only to the extent that performance data is available.

PROJECTIONS

Computational engineering and science will continue to pace the development of high-performance computing. The speed requirements posed by challenging applications will progress through the gigaflop range and quickly enter the teraflop realm. The massive computing power required by disciplines now making heavy use of numerical simulation will significantly alter the nature of future hardware and software and will create a strong incentive for teaming among specialists.

Hardware

At present, Cray Research is the principal supplier to the world supercomputer market. At least in the short term, this is unlikely to change. The long-term viability of any supercomputer vendor will depend on its ability to rapidly translate technological innovations into high-performance commercial products, without being impeded by its installed base of obsolescent hardware. Cray's recent products are the X-MP and Cray-2. The largest X-MP configuration has 4 processors, 16 megawords of primary memory and a nominal peak performance of 840 megaflops. The Cray-2 has 4 processors, 256 megawords of memory and a 2 gigaflop nominal peak performance. Within the immediate future, these machines will be superceded by ones with 8 to 16 processors, primary memories in the range of 32 megawords to 2 gigawords, and peak performances of 3 to 16 gigaflops.

In the longer term (mid-1990s), one should expect machines in the 32 to 64 processor range with 10 to 50 gigaword memories and peak performances on the order of 25 to 100 gigaflops. This estimate assumes that the machines are homogeneous, that

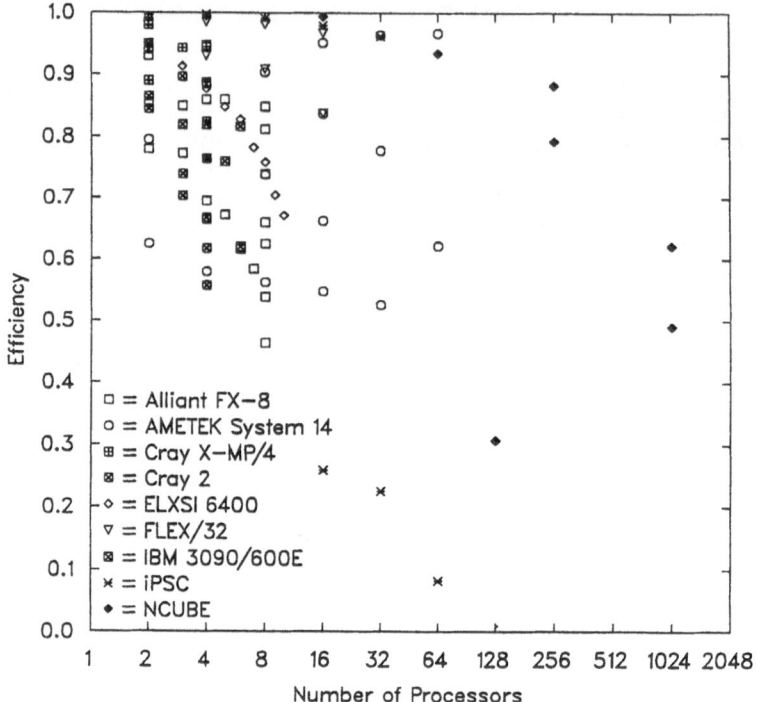

Figure 9a.   Parallel-Processing Efficiency

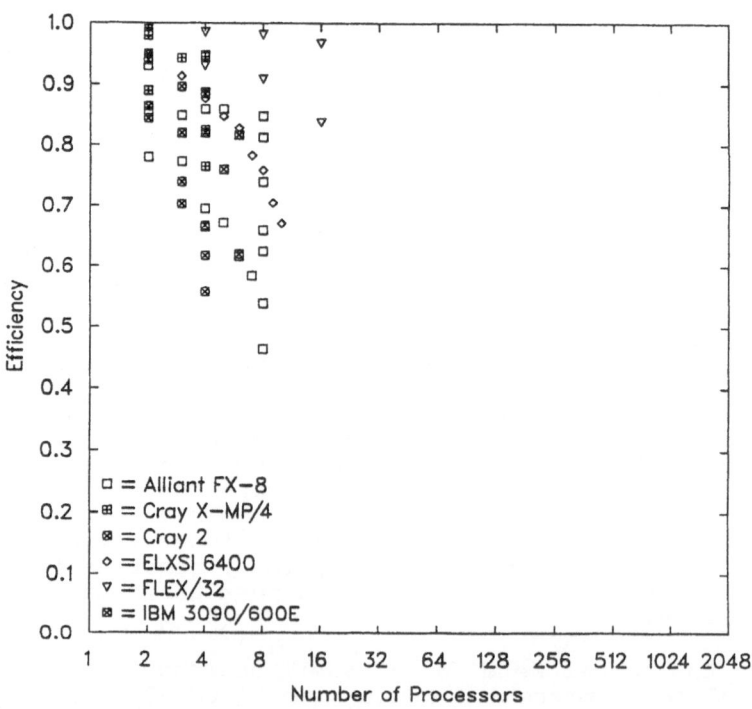

Figure 9b.   Efficiency of Shared-Memory Machines

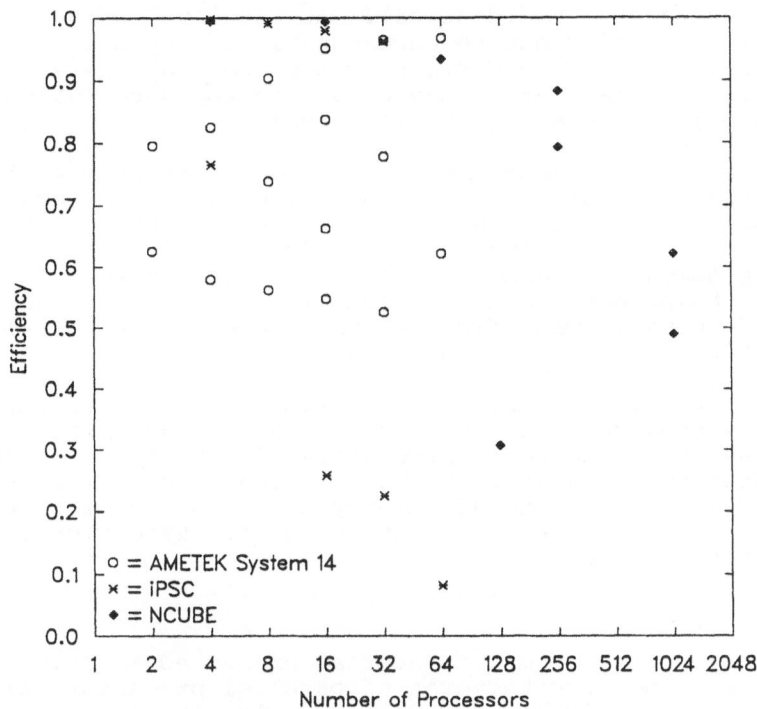

Figure 9c. Efficiency of Distributed-Memory Machines

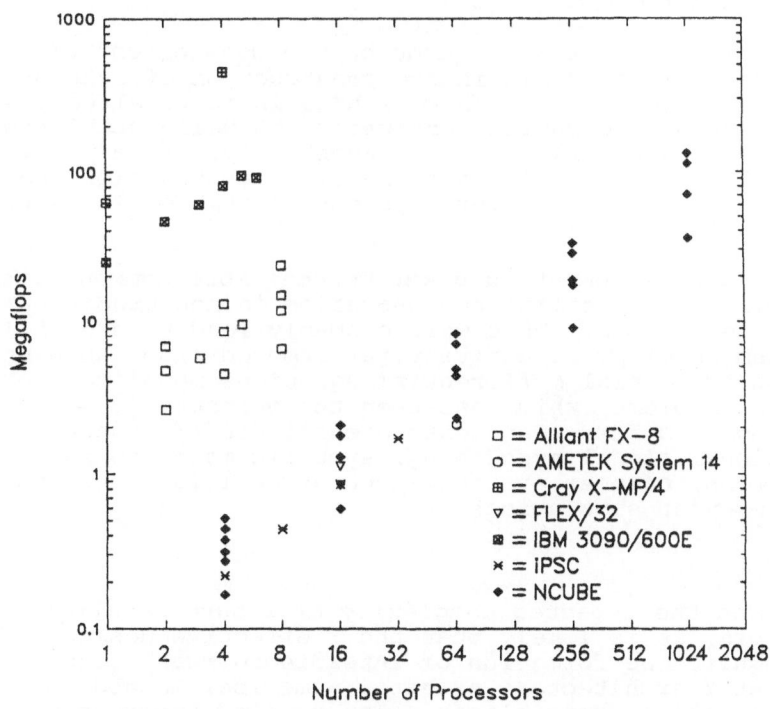

Figure 10. Floating-Point Speed

is, all processors are identical. At the level of 64
processors, it would not be unreasonable to begin introducing
special-purpose CPUs thus creating heterogeneous parallel
processors. In fact, this may be a good approach for pushing
peak performance into the teraflop realm.

The superminicomputer marketplace is presently much more
chaotic than the supercomputer one. The advent of cheap micro-
processor chip sets has substantially lowered the capital
requirements for entering the computer business. Consequently,
there are dozens of vendors vying for market share. It is
difficult to predict which of them will thrive. Indeed,
success depends on many factors other than the technical merits
of a particular parallel-processing architecture.

The distributed-memory superminis have begun to exhibit
remarkably good speed-up performance in the 1000-processor
range. Assuming that microprocessors with better floating-
point speed become available and that the issue of relatively
poor parallel-processing efficiency can be dealt with, these
machines may provide an alternative to more traditional high-
performance architectures.

Software

It should be clear from the material presented in this paper
that the machine architectures of parallel processors are not
transparent to their users. Consequently, it is to be expected
that the next decade will see major efforts on the part of
algorithm designers to develop novel numerical methods which
are intrinsically parallel. This task may prove to be quite
difficult, given the entrenched position of "scalar thinking"
in algorithm design.

Progress will also be made on the development of
programming tools to aid in the construction of systems of
concurrent processes. Such work has, in fact, already started
at the level of concurrent extensions to well-established lan-
guages such as Fortran [6, 7]. Eventually, these programming
tools will operate on a higher level of abstraction and will
probably be driven by expert systems in combination with human
intervention.

As massive computing power becomes more common, its
availability will stimulate innovation in non-traditional
applications areas. This will probably lead to a high level of
algorithm development activity far removed from the numerical
solution of partial differential equations modeling continuum
dynamics problems, which has been the mainstay of scientific
computing thus far. Candidate areas include: Monte-Carlo
simulations, signal processing, symbolic mathematical
computation, complex systems (such as cellular automata), and
discrete event simulations.

Teaming

Given the expected complexity of future parallel
processors, it is likely that their effective design and use
will require the formation of interdisciplinary teams composed
of computer architecture experts, numerical methods developers,
and applications specialists. The teaming approach will
probably appear first in situations of critical economic or

national security importance. Candidates include the design of future-generation supercomputers and the development of space battle management software. Once the utility of this teaming approach to high-performance computing has been demonstrated, it will quickly become the norm.

CONCLUSIONS

Large increases in computing power are necessary for the continued advance of computational mechanics. As scalar and vector computer architectures are rapidly approaching their technological limits, extensive use of parallel processing will be necessary. Parallel processing devices can be quite complicated and their architectures are not transparent to the user. Software innovations may make parallel processing easier in the future, but no significant short-term relief should be expected. Algorithms must be tuned to specific parallel-processing architectures if high levels of performance are to be attained. This, in many cases, is not a straightforward exercise. Fortunately, applications in engineering appear to have a high degree of inherent parallelism.

ACKNOWLEDGEMENTS

The skill and grace of Nancy Jensen and Steve Lamont, together with generous support from the SDSC staff, made the production of this paper possible.

REFERENCES

1. Wilson, K. G.: "Grand Challenges to Computational Science," Cornell University, May 1987.

2. Peterson, V. L. and J. O. Arnold: "The Impact of Supercomputers on Experimentation: A View from a National Laboratory," ASEE Annual Conference Proceedings, 1985.

3. Hasslacher, B.: "Discrete Fluids," Los Alamos Science, No. 15, Special Issue, 1987, pp. 175-217.

4. Worlton, J.: "Toward a Science of Parallel Computation", *Computational Mechanics - Advances and Trends,* AMD Vol. 75, ASME, New York, 1987, pp. 23-35.

5. Johnson, G. M., "Requirements for Parallel Processing in Scientific Computation," Third International Conference on Supercomputing, Boston, Massachusetts, May 1988.

6. Weeks, C. L., "Concurrent Extensions to the Fortran Language," First World Congress on Computational Mechanics, Austin, Texas, September 1986.

7. Dongarra, J. J. and D. C. Sorenson, "SCHEDULE: An Environment for Developing Transportable Explicitly Parallel Codes in Fortran," Third SIAM Conference on Parallel Processing in Scientific Computing, Los Angeles, California, December 1987.

VERY LARGE-SCALE FINITE ELEMENT WAVE SIMULATIONS, INVERSE MODELING, AND "HARDWARE CODED" ALGORITHMS

Gregory L. Wojcik

Weidlinger Associates
4410 El Camino Real, Suite 110
Los Altos, California 94022

ABSTRACT

Our need to understand complex wave phenomena in fields ranging from geophysics to optics encourages the use of large-scale direct and inverse numerical simulations. Due to the extended size of most models, and their fine structure, these simulations often require very large grids with millions to tens of millions of nodes. They are only feasible on machines with resources comparable to that of the Cray 2 at a minimum. Explicit, time-domain finite element or finite difference wave propagation algorithms are well-suited to direct problems. Inverse problems are only just beginning to be attacked using these discrete algorithms in conjunction with least squares optimization techniques. The principal impediment to large-scale direct and inverse simulations is available computers. Resources on even the largest machines are simply inadequate for direct solutions in 3-D or realistic 2-D inverse modeling.

This paper illustrates these issues with some examples of direct and inverse modeling problems using finite elements. In particular, 3-D quarry blast simulations of multiple phased shots in a recent Air Force sponsored experiment are examined, along with 2-D simulations of crustal refraction experiments, and examples of inverse optical modeling using finite elements. The paper also explores the possibility of constructing specialized computers, so-called Wave Engines, that are hardware coded for these types of simulations. Specific issues include necessary conditions for feasibility and practicality, the "best" hardware, e.g., building block and/or RISC, to use, and where this approach fits in current technology.

INTRODUCTION

Many of the wave propagation problems encountered today in geophysics, acoustics, and electromagnetics are intractable by classical analysis methods based on separation of variables. The difficulty is caused by the small- and large-scale inhomogeneities necessary for adequate resolution of the physical domains, and also by the need to include nonlinear source and material effects. The only cost effective approach to solving such problems is a discrete numerical solution of the governing equations, typically using finite element or finite difference methods.

By virtue of their physical size and the high spatial/temporal frequency resolution required, these problems generate some of the largest numerical models currently solved in "production mode" on supercomputers. There are typically two classes encountered. The most common is of course the direct problem, for which model parameters and initial-boundary conditions are specified and the equations solved in either the time or frequency domain. Less common is the inverse problem, where a set of observations in conjunction with initial-boundary conditions are used to infer model parameters. Inversion effectively combines the direct problem with optimization methods in a very tricky iterative algorithm. It is not surprising that the inverse problem is of ultimate interest in applied technology.

We examine some of the practical issues encountered in direct and inverse hyperbolic problem solving in the context of large-scale finite element discretizations. The direct problem examples are based on some of our recent time-domain simulations of 2-D and 3-D seismic experiments performed by the United States Geological Survey, Air Force Geophysics Laboratory, and others. These experiments include quarry blast signature characterizations for nuclear test detection and discrimination, and long-baseline refraction surveys for crustal studies. The inversion examples are based on frequency domain simulations of submicron integrated circuit (IC) feature "imaging" from optical backscatter measurements. The integrated circuit size scale is at least nine orders of magnitude smaller than that of the seismic examples, but the simulation and inversion principles are identical.

Following these examples, the paper addresses practical modeling limitations of the current generation of supercomputers and an alternative hardware solution based on custom configurations of conventional chip sets implementing the discrete algorithms directly, rather than in software. The point is that the next generation of vector/parallel machines will be needed to begin solving large-scale 3-D hyperbolic problems effectively. Inversion in 2-D also awaits these machines because optimization requires many direct solves and Jacobian evaluations in an iterative loop. However, rather than wait for these machines, it may be possible to use today's chip technology and microcomputer expertise in a cost-effective alternative to Cray-?'s and CM-?'s, using hardware coded algorithms. We will discuss the conditions necessary for this alternative to be practical, and estimate performance and cost for a time-domain finite element wave engine.

3-D QUARRY BLAST SIMULATIONS

One aspect of our ongoing research program in wave propagation is the simulation of seismic experiments conducted under the auspices of the Air Force Geophysics Laboratory. The exercise addressed here is the San Vel Quarry experiment in Littleton, Massachusetts, performed during the summer of 1987 by Stump, et al. (1987). Our simulations support continuing field work and interpretation of these and future experiments. The calculations described here were performed by D. K. Vaughan and the author at Weidlinger Associates, Los Altos using the finite element code, FLEX, e.g., see Vaughan (1983), developed for a wide variety of transient phenomena in linear and nonlinear mechanics.

Problem

The experiment measured ground acceleration and/or velocity at ranges of hundreds of meters to tens of kilometers from a series of quarry blasts. The blasts were conventional quarrying explosions intended to fracture and rubbleize rock off one of the quarry pit's faces. The pit itself was approximately 400x800 feet at the time of the experiments. The typical blast configuration consisted of three staggered rows of 48-72 explosive-filled bore holes, each approximately 58-60 feet deep, along a portion of the pit's edge. The holes were detonated in succession along the edge, so-called ripple firing, with the detonation sequence (delay time) chosen to maximize rock fracturing while minimizing ground motions felt by nearby homes and businesses. Although seismic data were collected at various ranges, the set of principal interest here are accelerograms from instruments scattered around the pit within a circle approximately 1600 feet in diameter. The three-component acceleration records at the sites are given in Stump, et al. (1987). Three blasts, each on a different section of the pit perimeter, were recorded.

One purpose of the experiment was to examine effects of ripple firing on ground motion spectra. This type of detonation produces spectral scalloping that may be useful in discriminating small nuclear explosions from quarry blasts. It was noted, in both these and other experiments, that the actual delay times deviated significantly from those planned—apparently caused by repeatability problems with commercial blasting caps. The question is what effect do these random delay time errors and local site effects have on ground motion, and spectral scalloping in particular? Since this is difficult to answer in the field, our research explores the problem by means of numerical modeling. The objective is to simulate the detonation sequence in a discrete model of the quarry, generate a set of synthetic seismograms at the scaled instrument locations, and compare synthetic and measured seismograms. The calculations are expected to yield insight into effects of shot timing errors and local topography and geology on observed ground motions.

Solution

To perform the simulations, a three-dimensional, explicit finite element model of the quarry was built and executed on a Cray 2 at the Air Force Weapons Laboratory. The 240 x 240 x 74 element model (= 4.26 million elements) represented the quarry pit and surrounding area (2160 x 2160 x 656 feet) including all accelerometer sites. It was gridded to propagate 100 Hz shear (S-)waves without significant dispersion, since this was the upper bound on instrument frequency response. Figure 1 is a cartoon of the model showing instrument locations and shot detail. The model required significant new code development in order to accommodate the large number of small-diameter, explosive filled bore holes. Since the typical hole was much smaller than a free-field element, a subgridding capability was developed to grade the element size down from the free-field dimension, through an intermediate zone, to approximately twice the hole diameter.

Because the quarry blast is highly nonlinear near the explosive array—due to high-pressure, fracturing, and the resulting large strains and displacements—conventional linear source modeling techniques could not be used. Instead an "energy pill" source was implemented, where the explosive cylinder and its immediate neighborhood was replaced after detonation by a pressurized, outwardly moving region with the same total energy (kinetic plus potential) as the explosive products and included rock. Nonlinear rock fracturing was mimicked by tension cutoff in the element's material model. A nonlinear cap model of the inelastic constitutive behavior of rock was available but not used due to the prevalence of fracturing rather than cyclic nonlinear processes. Although the detailed fracture phenomenology is not simulated by this model, it was deemed adequate for calculating the seismic pulses radiated by the explosion, including shielding by fractured rock around neighboring boreholes. A two-dimensional example of the model soon after detonation of three sheets (i.e., two-dimensional boreholes) of explosive is shown in Fig. 2. This and other 2-D models were used extensively for development of the source and subgrid coding implemented finally in 3-D.

The principal difficulty encountered with the model is late time numerical noise contamination. This is due to the calculation's relatively long duration, about one second of simulated time, and the high frequency nature of the borehole detonations and rock fracturing. Waves in the model reverberate within the model many times over this period, and since the radiation boundary condition (absorbing boundary) on the outer sides of the model are not perfect, trapped energy eventually grows to a significant level. This difficulty was reduced to the point that reasonable results could be obtained by moving the bottom boundary deeper and introducing a small amount of viscous damping. The damping does not affect the seismic signals significantly, but it does reduce the ringing substantially. This experience clearly indicates the need for better time-domain radiation boundary conditions and time-domain material damping.

An example of results are shown in Fig. 3 from a full simulation of the Shot 3 test with 72 borehole explosions detonated according to the original delay time specifications. The model requires about 33 hours of CPU time to simulate one second of model response, and uses 90 million words of memory. Figure 3 shows synthetic vertical velocity seismograms on a circle surrounding the quarry, indicating the azimuthal dependence of ground motion on local topography. Polarization analysis shows that these seismograms are dominated by Rayleigh surface waves, both direct and scattering induced. Comparing synthetic vertical velocity seismograms with integrated accelerograms at the instrument sites shows much more

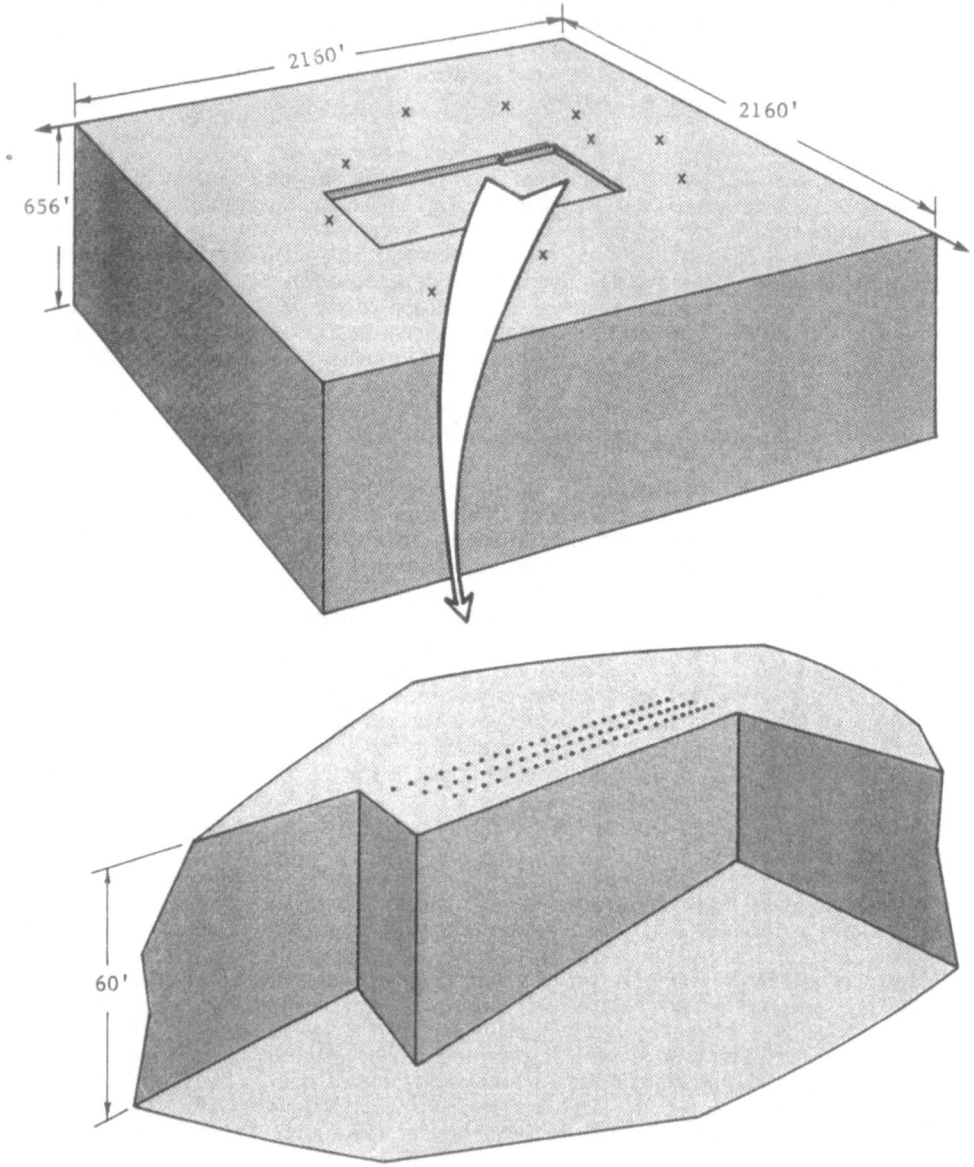

Figure 1. Three-dimensional perspective of the finite element quarry model showing instrument locations and explosive array detail. This is discretized into 240x240x74 elements.

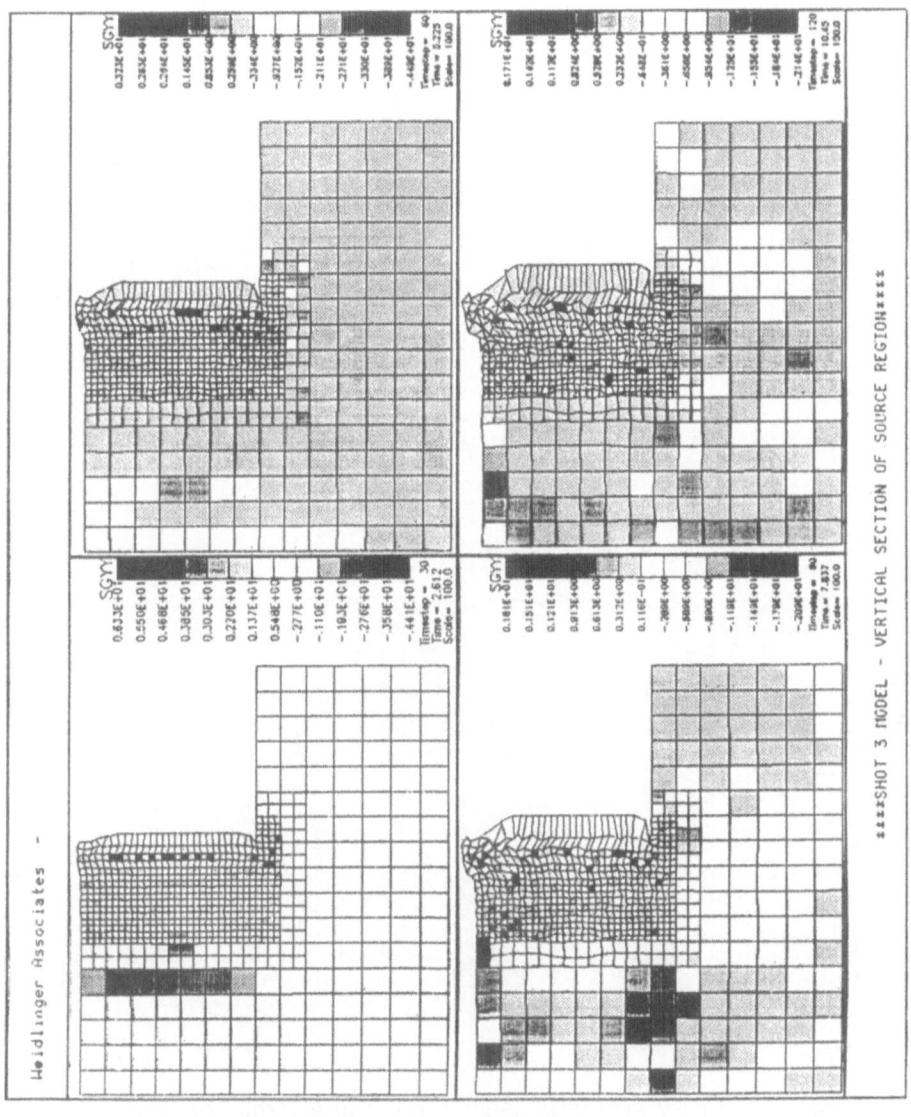

Figure 2. Snapshots of the explosive-filled region (2-D version) after detonation, illustrating element grading around the boreholes and the tension-cutoff model.

289

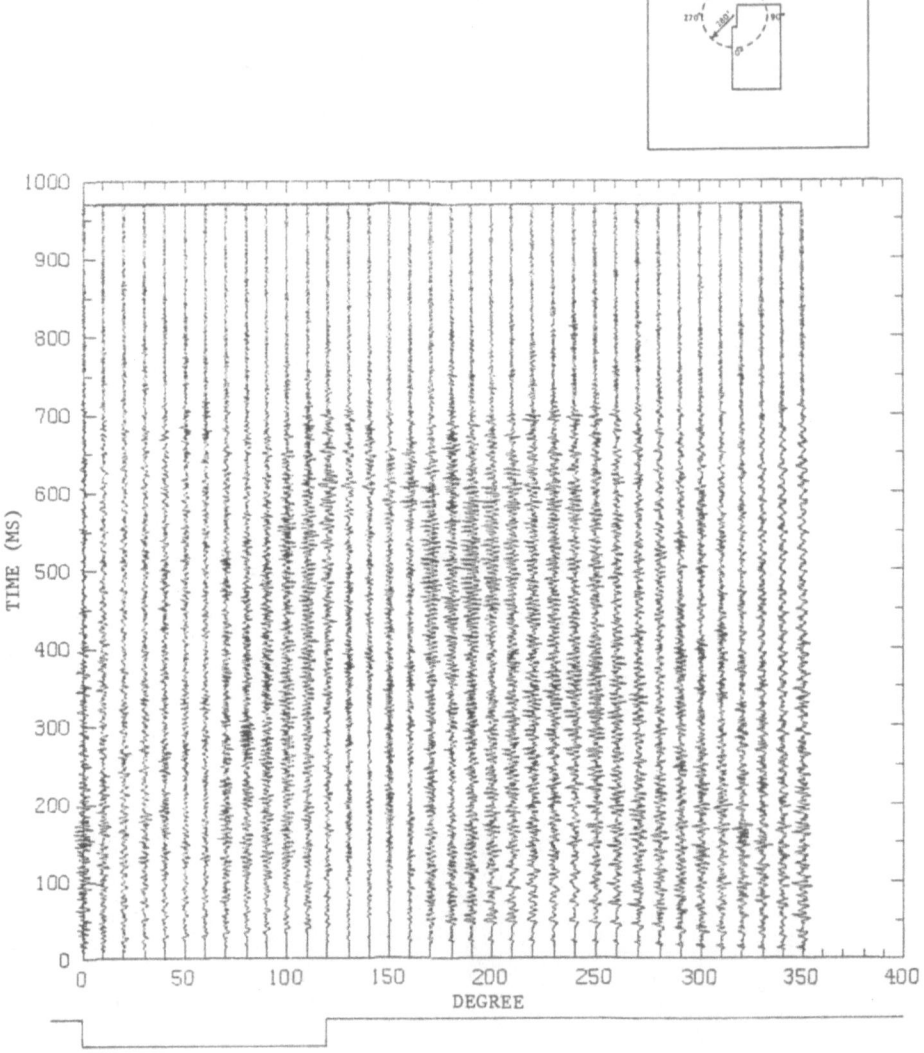

Figure 3. Vertical velocity synthetic seismograms at a radius of 280 feet from the center of the Shot 3 simulation. These show azimuthal variations of the Rayleigh waves dominating the seismic radiation on the surface.

coherent, high frequency motion in the synthetics. This is an artifact of the calculation's very severe, high frequency source environment and inadequate damping, as well as the assumption of perfect shot delays. Our conclusion from this comparison is that better phenomenological modeling is needed in the source region, which may require a source subgrid with a few million elements alone. In addition, random shot delay errors must be introduced

2-D REFRACTION SIMULATIONS

The previous example was inherently 3-D, however, many seismic propagation problems can be adequately modeled in 2-D. A typical example is the classical refraction experiment for investigating subsurface geologic structure. We consider the recent PACE experiment in Arizona, conducted by the United States Geological Survey. Harley Benz at USGS, Menlo Park, California is a principal contributor to this modeling effort, while D. K. Vaughan at Weidlinger Associates performed the calculations using FLEX.

Problem

Refraction experiments constitute a "simple" inverse problem. They are typically based on the assumption of plane, layered geology, and involve vertical velocity seismometers spaced along a surface line, recording reflected and refracted waves from near-surface explosions at the line's ends or center. In most cases the actual geology under a refraction line includes large and small scale structural perturbations that generally complicate the interpretation of data. Near-vertical and wider angle reflections are then used to model the gross inhomogeneity, with refractions controlling the distinct horizontal interfaces. Depending on the length of the line, e.g., hundreds of meters or kilometers, and size of the explosion, e.g., pounds or tons, these refraction experiments can probe from the near-surface to the upper mantle.

In 1987 the USGS conducted the Pacific Arizona Crustal Experiment (PACE) in west-central Arizona on a line 255 kilometers long, oriented southwest-northeast. The purpose of this and future experiments is to deduce crustal structure across the transition between the southwest Basin and Range Province and the Colorado Plateau. Since the Basin and Range is a highly extended crustal structure and the Colorado Plateau is a rather enigmatic block of thick crust, the interface between the two is of basic scientific interest. The seismograms collected by USGS constitute one of the best sets of crustal refraction data yet obtained in the United States, and the structural cross-sections deduced from this data provide an excellent basis for interpretation using large-scale numerical modeling.

The effects of smaller scale inhomogeneity and general scattering in the crust are poorly understood. It is likely that the usual interpretations based on refraction and ray tracing inadvertently lump these effects in with intrinsic attenuation. To better quantify the impact of small scale inhomogeneity, we have modeled the PACE experiment using time-domain finite elements on the Cray 2. Our purpose is twofold: first is to verify the gross geologic model deduced from the data by conventional means; and second is to examine the effects of random perturbations in the lower crust and see if real data supports the existence of random structure.

Solution

The PACE model consists of 1111 x 4444 = 4.9 million elements covering a 50 x 200 kilometer cross section under the refraction line. General wave speed functions are prescribed over the model. The cross-section is shown in the top illustration of Fig. 4. The 45 meter elements are sized to support at least 8 Hz S-waves near the surface where the seismic wave speeds are lowest and the wavelengths shortest. Note that using low order shape functions, ten elements per wavelength are generally the minimum needed to support a propagating wave with less than two percent error in phase velocity. The source was approximately 2000 pounds of ANFO detonated in a borehole at the southwest end of the line. Source region details were ignored in this model since propagation effects are of more

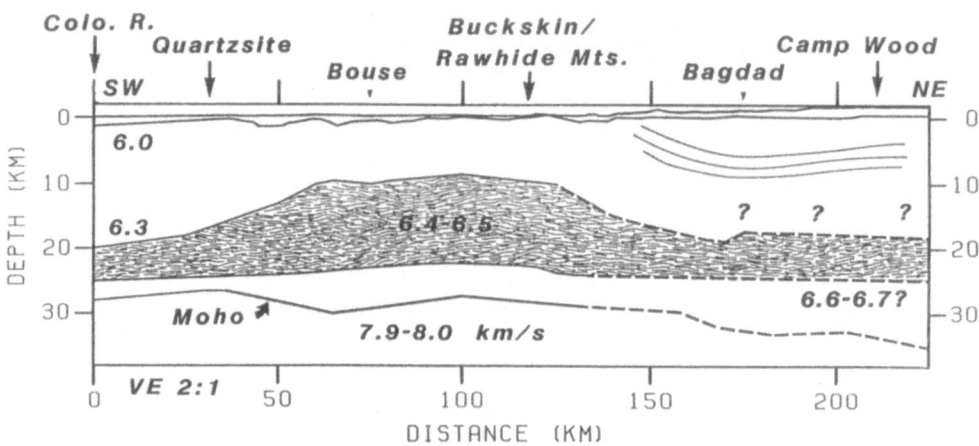

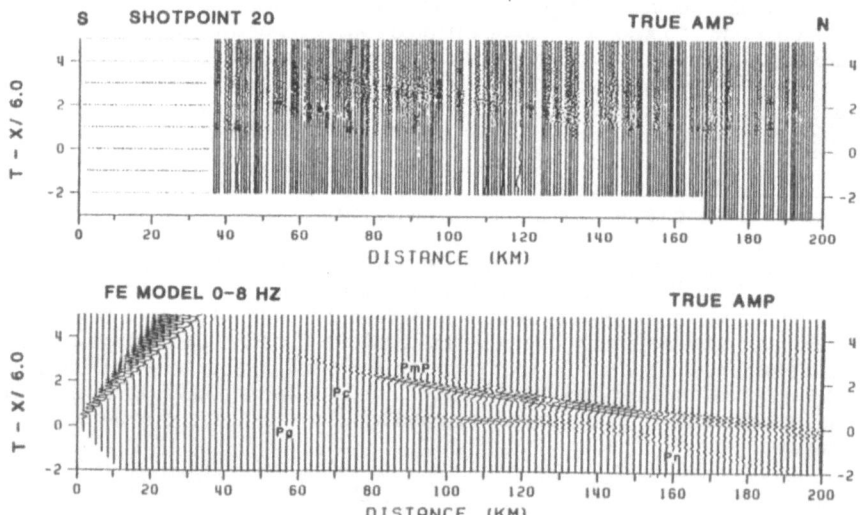

Figure 4. The PACE geologic model (top), vertical velocity data recorded across the refraction line (middle), and a 4.9 million element simulation of the experiment (bottom).

interest, hence, the source was simulated by simply pressurizing a 3x3 cross of elements near the surface.

The code development necessary for these simulations was minimal compared to that for the 3-D quarry simulations. The principal new development by Benz was a random wave speed routine based on either Gaussian or self-similar distributions. These distributions are applied to the model's lower crust. As a basis for comparison to the resulting inhomogeneous finite element solutions, we use a so-called reflectivity code on "equivalent" plane, layered half-space models of the PACE crust.

The finite element calculation requires about eleven hours of Cray 2 single CPU time to propagate body waves across the model. PACE data and results of an early calculation are shown in the lower two illustrations in Figs. 4. The calculation in Fig. 4 did not include a scattering model in the lower crust. Subsequent calculations have examined effects of various isotropic and anisotropic models, and indicate significant deviations in the arrivals beyond 140 km corresponding to travel paths through the lower crust. Comparisons with data are difficult in view of the limited resolution in the far-field, but nonetheless, correlations are being attempted.

2-D INVERSE OPTICAL SCATTERING

Another aspect of our research in wave propagation concerns electromagnetic waves at the optical wavelengths (.25 to .7 micrometers) typically used for microlithography and defect inspection in the integrated circuit (IC) industry. This involves solving Maxwell's equations instead of the elasticity equations, using very similar time- or frequency-domain algorithms tailored to the curl equations relating electric and magnetic fields in dielectrics. Much of this research necessarily concerns the inverse problem. One application is rapid remote sensing and measurement techniques for submicron features and calibration patterns on silicon wafers. This is very important for on-line process monitoring of future microlithography manufacturing processes, as a means to increase chip yield. This work was conducted in large part by J. Mould, Jr. of Weidlinger Associates, Los Altos, e.g., see Wojcik et al. (1988).

Problem

The canonical problem consists of a silicon chip upon which a two-dimensional photoresist line feature has been printed with some unknown, non-ideal cross-section. The feature is illuminated by a focused laser beam with wavelength comparable to the feature dimension. Light is "scattered" by the mechanism of diffraction, where the backscatter pattern depends on illumination angle, wavelength, cross-section, refractive index, and substrate. The idea is to measure the intensity and phase of the backscatter and use these data in an inversion algorithm to recover the feature's true shape. For the feature sizes currently used in microlithography, conventional optical microscopy merely yields an undecipherable diffraction pattern, while electron microscopy is too invasive and time consuming to be effective on the production line. Thus, so-called numerical microscopy may offer a viable alternative.

Given the measured complex backscatter at a set of discrete observation points in the near- or far-field, in principle, the inversion algorithm is straightforward. First, the scattering feature's geometry is estimated and parametrized using vertex coordinates if polygonal or polynomial coefficients otherwise. This estimate is not difficult since we know that the actual shape is reasonably close to the design shape given the controlled environment of a modern production line. Second, synthetic backscatter is calculated at the observation points using a finite element model of the estimated domain. In addition, the Jacobian must be calculated at the observation points, i.e., the derivative of field variables with respect to shape parameters. Note that partial wave solvers like ray tracing are inadequate for this step because the field is diffraction dominated. Third, field residuals (measurements minus synthetics) are found at the observation points and combined with the Jacobians in a least squares algorithm to give an incremental change in the feature shape parameters that minimizes the residual. We now return to the second step and iterate until a suitable convergence criterion is satisfied.

Although simple in principle, this inversion algorithm has formidable difficulties in practice. These include the existence of local minima in parameter space distinct from the one sought, and the large amount of numerical calculation required for synthetic field and Jacobian calculations.

<u>Solution</u>

Implementation of this inversion algorithm depends on the choice of discrete numerical wave solver, residual minimization technique, and linear system solver. The obvious wave solver is a frequency domain finite element algorithm, e.g., a Helmholtz solver in two dimensions, since this choice permits conforming material interfaces and analytical expressions for Jacobians. For minimization, the Gauss-Newton method appears adequate for our purpose, with enhancements added as needed, e.g., the Marquardt-Levenberg method and linear constraints. In order to solve the finite element system of equations (sparse, banded, complex, non-Hermitian), either direct or iterative solvers can be used. In two-dimensions it makes little difference, but in three-dimensions only iterative solvers are practical, e.g., the conjugate gradient method.

Application of the inversion algorithm to a small test problem is illustrated in Figs. 5-7. Figure 5 shows the method of parametrization for polygonal cross-sections and Fig. 6 illustrates a simple finite element grid. Only the vertices of the cross-section are free parameters, with all other nodes on a side of the polygon slaved to them. This approach reduces the number of inversion parameters with no loss in modeling generality. Figure 7 shows an example inversion for a somewhat deformed feature shape, demonstrating shape reconstruction in the least squares sense.

Although five iterations are shown in Fig. 7, effective convergence is obtained after the second. Each iteration requires one solve of the finite element system and one Jacobian solve for each shape parameter. There are six parameters for the case shown because the base is constrained to lie on the plane. If a direct method is used then the Jacobian evaluation is just a system backsolve since the factored matrix is available. In contrast, for an iterative method the Jacobian evaluation requires a full solve for each parameter. Because iterative methods allow control of solution accuracy, and Jacobians need not be very precise, the disadvantage of multiple full solves is ameliorated somewhat by allowing reduced accuracy (fewer iterations). A number of direct and iterative system solvers have been used successfully. It is clear that solvers must be capable of optimal performance on supercomputer architectures, since the wrong algorithm can absolutely destroy supercomputer performance due to memory and/or vectorization conflicts.

We have performed numerical inversion experiments with a variety of dielectric media, background noise types, and illumination angles. All results indicate that this is a practical approach for reconstructing images from diffraction data. One important observation is that electromagnetic intensity and phase information are both necessary for a successful inversion of transparent dielectric features, whereas metallic features can be inverted from far-field intensity only. Another observation is that even for two-dimensional problems inversion is a very computation-intensive process, particularly as the features and surrounding media increase in complexity. For example, curved feature cross-sections with many parameters, sitting on multi-layered dielectric media. It appears that on-line instrumentation for inspection will require supercomputer-level numerical performance in order for this approach to be feasible.

WAVE ENGINES: A HARDWARE SOLUTION

The previous sections illustrate forward modeling of some large-scale 2-D and 3-D geophysics problems and inverse modeling of a smaller-scale 2-D optics problem, all using the finite element algorithm. The optics problem is effectively large-scale by virtue of the

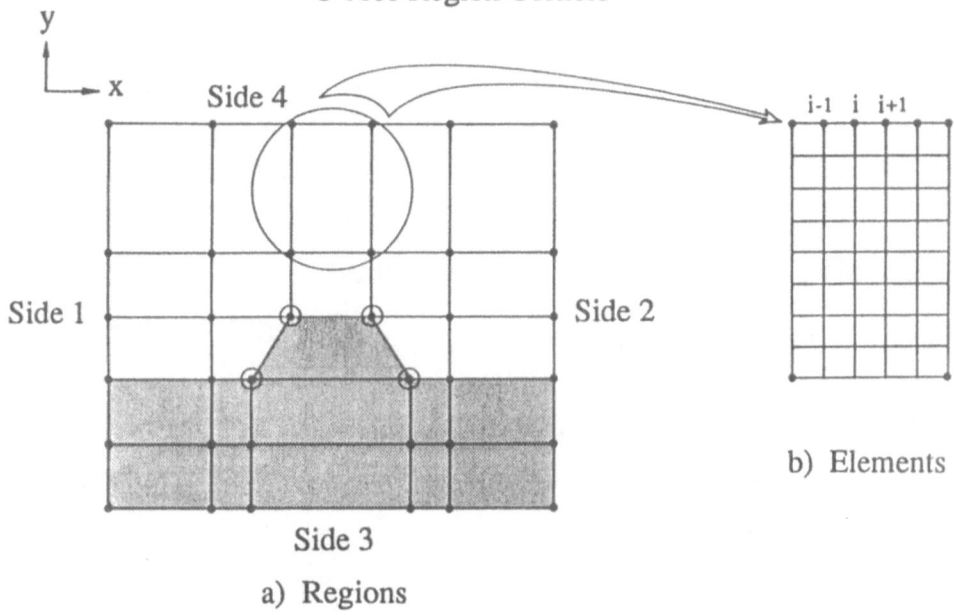

y

x

Side 4

Side 1

Side 2

Side 3

i-1 i i+1

b) Elements

a) Regions

Figure 5. Illustration of regions and elements in the finite element discretization used for inversion.

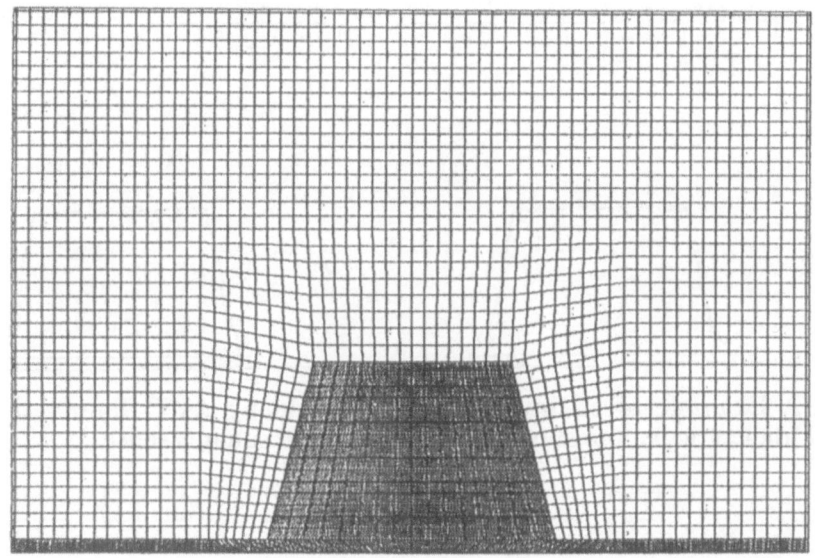

Figure 6. A typical finite element grid of a 2-D line model on a substrate.

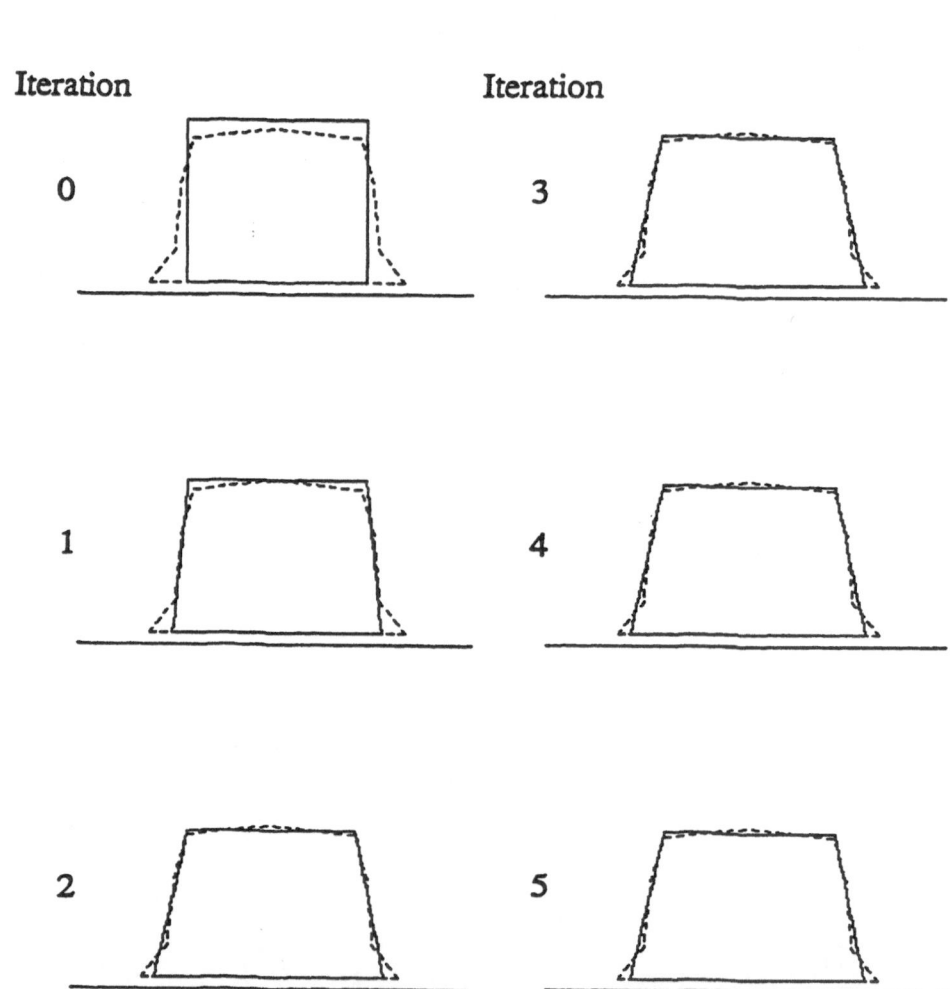

Figure 7. Example of an inversion sequence using ±45° illumination on a "rough" dielectric feature showing the fit in a least squares sense.

iterations and Jacobians required for an effective inversion. The principal limitation of these wave propagation examples is their need for supercomputer resources.

In practical applications, the geophysical analyst would often like to run 10 or 20 large-scale forward simulations for model parameter studies, while the IC process engineer would like very fast inversion simulations (seconds to minutes) for monitoring a production environment. Currently, the option is to spend hundreds of thousands of dollars over many weeks doing forward problems, and dedicating a supercomputer to process monitoring for inversion problems. Neither solution is practical.

Therefore, this section examines an alternative in the form of an application-specific computer designed around a finite element algorithm for the wave-type equations solved in the above examples. The means to achieve this is the integration of algorithms and ICs, given the current proliferation of building-block VLSI devices, microcomputer memory, and design expertise. Specialized machines (chips or boards) are fairly common in the military signal processing community, e.g., for FFTs and direct array processing, however, this approach has not been exploited extensively for more general engineering and scientific simulations requiring the solution of partial differential equations. The research reported below on an application-specific machine involves ourselves at Weidlinger Associates and T. Willis, president of Accent Research in Houston, Texas.

Background

The concept described here is a special-purpose, high-performance machine—the so-called Wave Engine—designed at the hardware level, i.e., hardware coded, by implementing the kernel finite element subroutine directly using building block chips. The goal is to achieve ten times the performance of a Cray 2 at 1/100 the hardware cost, albeit on a single application. This solution is warranted when hundreds of hours of Cray time are routinely devoted to a single, mature application algorithm, e.g., wave propagation, or in situations requiring fast, compact (portable), or relatively inexpensive hardware, e.g., for use in the field or in production systems where today's equipment is impractical by reason of size, cost, or performance.

One critical requirement for this approach to be practical is that the application must be solved by a mature and invariant algorithm. An invariant algorithm is one for which the fundamental computational kernel does not change from problem to problem, i.e., all problems eventually map to the same representation of the governing equations, e.g., explicit finite elements. Thus, a hardware implementation of this kernel is sufficient for a given class of applications. The algorithm must also be relatively simple and well-structured, which effectively means vectorizeable. Another technical requirement is of course that hardware and expertise be commercially available to implement the kernel.

Clearly, this approach is applicable to a variety of other simulation problems where a specialized, compact, and extremely fast processing capability is required and the economic tradeoffs warrant it. Examples include sonar and radar scattering in conjunction with inversion, seismic modeling and inversion for oil exploration, as well as some of the much more involved problems encountered in fluid flow simulations. In all cases there must be a well-understood, invariant, high-performance algorithm that is used extensively on general-purpose supercomputers and would benefit from the cost performance and turnaround of a specialized hardware engine. An example of a similar but more ambitious application-specific machine is given by Nosenchuck, et al (1986) in the context of the Navier-Stokes equations and computational fluid dynamics (CFD). Our approach differs by addressing simpler equations and algorithms.

Algorithm

The numerical calculation of wave fields can be performed in either the time- or frequency-domain using explicit or implicit integration schemes. Explicit time-domain solvers are often preferable when modeling transient phenomena, or when significant nonlinearities are present. Implicit frequency domain solvers, e.g., for the reduced wave equation (Helmholtz equation), are typically better suited to steady-state phenomena, or where global Jacobians are needed, i.e., inverse problems. In practice it is neither necessary nor wise to limit consideration to only one approach. However, for this example only one type of solver is considered. Since algorithm simplicity is a key feature for a robust hardware implementation, the explicit time-domain approach is the target algorithm.

The explicit algorithm uses a central difference approximation for time derivatives and a finite element approximation for spatial derivatives. These approximations result in a second order accurate discrete representation of the governing equations in terms of a sequence of adds and multiplies for each element in the model. For many applications, 32-bit arithmetic (single precision) is adequate. Using a convenient programming language, inner (element) and outer (model) processing loops are defined to evaluate this sequence of operations over the propagation environment. The discrete system is integrated forward in time (step-by-step) using a modified leapfrog scheme. Since the system is hyperbolic, integration is stable provided the timestep is smaller than the minimum transit time of a wave across the smallest grid element. Note that the data structures, simple boundary and interface processing, and flexibility of modeling tend to favor finite element over finite difference spatial discretization.

The basic algorithm reduces to a simple computational kernel consisting of a small number of program statements executed for each element during one timestep (an element cycle or element-timestep). This kernel is very stable, i.e., invariant with respect to a wide variety of problems governed by the wave equation. For example, our production finite element propagation code, with a history of thousands of Cray CP hours on many different structural and continuum problems, has never required problem-related modification of the fundamental 2-D and 3-D processing kernels.

Wave Engine

The implementation is conceptually quite simple. The kernel subroutine, written in a high-level language like FORTRAN, is dissected into fundamental arithmetic loops and memory operations, optimized for pipelined execution., and rewritten as a tree or flow diagram explicitly showing the relationship between data elements and arithmetic operations. An example of such a dissection for a minimal algorithm is illustrated in Fig. 8. Each arithmetic operation is replaced by an individual floating point unit (FPU), and all units are chained and interconnected according to the algorithm flow, with suitable delay elements for memory access. Further analysis of the algorithm provides the detailed timing sequence for a controller that generates signals to read-write memory and clock data through the vector pipeline.

The performance goal of this pipelined processor is execution of one element-timestep per clock cycle. Using building block chips with a 33 megahertz clock yields 33 million element-timesteps per second. Since 18 floating point operations—hence 18 FPUs—are required to evolve one element one timestep for the 2-D algorithm considered here, this translates to nearly 600 megaflops, or about five times the algorithm's performance on a single Cray 2 CPU. Note that performance is proportional to the number of floating point operations. For example, a 3-D scalar wave solver requires about 60 operations per element-timestep—hence 60 FPUs—but yields fifteen times the Cray 2's performance. Of course, higher performance can also be achieved by using faster chips, although attendant memory and manufacturing costs increase substantially.

The vector processor is built using floating point arithmetic units, vector registers, and control logic. Available FPUs are exemplified by Texas Instrument's SN74ACT8847 and Advanced Micro Devices' AM29325, which combine high speed, three bus architecture, and programming flexibility in a single chip about two inches on a side. These FPUs are one element in a complete building block family of 32-bit VLSI single-chip devices that readily allow custom design of high-performance processor systems. Although relatively new, this building block technology is evolving in terms of speed, flexibility, power, and integration, thereby providing an excellent basis for application-specific computers of the type described here.

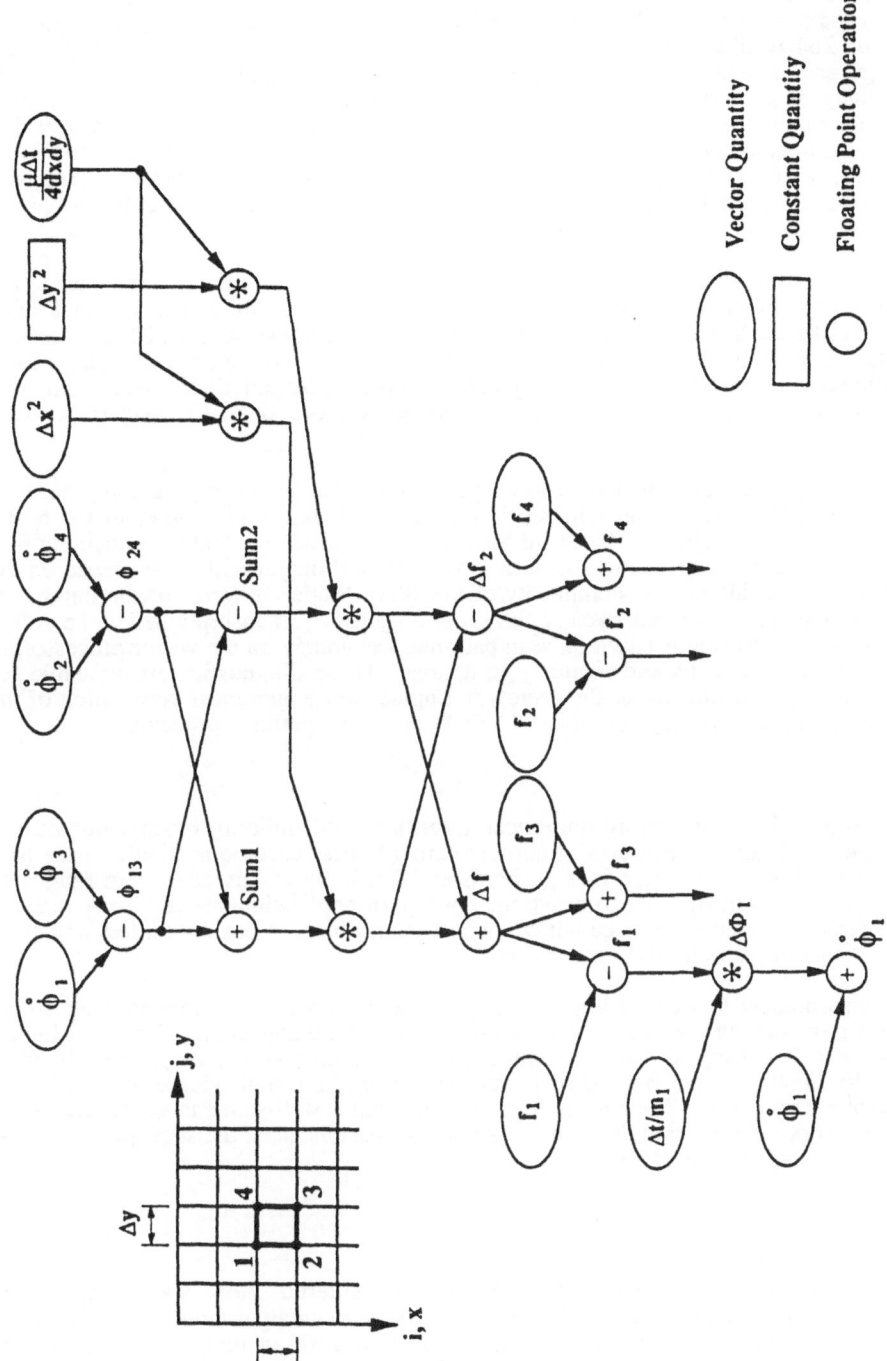

Figure 8. Dissection of the simplest explicit finite element algorithm for the scalar, 2-D wave equation.

Given current technology, design and construction of the Wave Engine's vector processor does not present any major difficulties. The real problem is memory bandwidth, determined by the product of memory references and element processing rate. The 2-D example considered here involves eight references times 33 million element-timesteps per second, or 264 million words per second. Since this pushes affordable single-path technology, an alternative involving multiple memory partitions and data paths must be used. Consequently, it turns out that memory issues constitute most of the design effort. It is essential for optimal performance that the Wave Engine's memory hold all pertinent data arrays, with no auxiliary results saved. This requires a sizable memory for most applications, say 64 to 128 megawords, however, because conventional personal computer dynamic random access memory (DRAM) can be used, the total memory cost is reasonable, particularly considering overall performance.

A high-speed, user programmable, scalar RISC processor was originally included in the design for executing boundary conditions and post-processing output. However, since a fast workstation like the Sun 4 will most likely host the Wave Engine, its own RISC processor can be used in place of a dedicated scalar unit. Since boundary conditions may change from problem to problem, and they typically require three orders of magnitude less work compared to element processing, it is natural to relegate boundary processing to a concurrent scalar processor.

The prototype machine will be wirewrapped rather than incurring the setup costs of printed circuit (PC) boards, and will reside on five VME bus cards, either in the host's chassis or a stand-alone chassis. One card holds the vector processor and the remaining four contain the principal vector memories. Real-time graphical output will be displayed on the host machine. Considering the complexity of the Wave Engine in terms of the number of FPUs and signal changes at each cycle, a full set of diagnostics are incorporated in the design using numerical verifications based on scan path-like technology on the vector processor, as well as byte parity checks and memory compares. These diagnostics are valuable for debugging during construction of the prototype Engine, while numerical verification of the vector processor prior to every use will probably be standard operating procedure.

The Future

This approach of integrating numerical algorithms and building block chips can be optimized as application specific integration circuits (ASICs) become available in the near future. The ultimate integration of algorithms and ICs is by means of custom fabricated VLSI chips. However, in order to prove the concept of application-specific computers for partial differential equations, it is certainly prudent and practical to begin with less ambitious chip-level rather than circuit-level integration.

It is also prudent to start by implementing a mature, robust, time-domain wave solver rather than more complicated partial differential equations and algorithms. The next logical step appears to be a frequency-domain wave solver based on a conjugate gradient iterative algorithm for the linear system of normal equations arising from finite element discretization of the Helmholtz equation. The conjugate gradient method is well-suited to vectorization and preliminary work indicates that the resulting Engine is feasible using the same processor and memory technology described above.

CONCLUSIONS

In terms of the partial differential equations considered, these wave propagation applications are quite mature and uninteresting, at least from an algorithms viewpoint and in comparison to modern CFD problems. In conjunction with an inversion strategy they become a bit more interesting, however, the issue is still not algorithms, but rather computers and resources. Production codes always seem to require more performance and memory than current computers can deliver.

The examples discussed here illustrate some practical limits of simulations on current Cray 2 level machines. Massively parallel machines like the CM 2 provide an alternative, although performance of production algorithms on this machine are disappointing. Presumably, more powerful nodes, as on the new Intel "hypercube-type" machines, will help. However, there always appears to be architectural constraints imposed by the need for generality that ultimately limit cost-performance.

Therefore, if the analyst has a mature, vectorized algorithm and needs hundreds of hours of simulation time, costing say $150,000 to $1,000,000 or more, then it seems logical that special purpose machines provide a practical alternative, particularly given the available building block chip families, application specific integrated circuits, cheap PC memory, and the future promise of custom chips. It is likely that a market niche will develop between workstations and supercomputers to fill this need.

ACKNOWLEDGMENTS

The work reported here was sponsored by the Air Force Geophysics Laboratory under Contract F19628-88-C-0067 (seismic models), by the National Science Foundation under SBIR Grant ISI-8760089 (optical inversion), and by Weidlinger Associates' internal funds (Wave Engines).

REFERENCES

Stump, B.W., S.K. Reamer, R.E. Reinke, J.A. Leverette, and R. Goerke (1987). *SAN VEL QUARRY EXPERIMENT: Near-Source Data*, Dept. of Geological Sciences, Southern Methodist University, Dallas, TX 75275.

Vaughan, D.K. (1983). *FLEX Users Guide*, UG8298, Weidlinger Associates, Los Altos.

Wojcik, G.W., J. Mould, Jr, V. Pereyra, and S. Wright (1988). *Optical Diffraction and Inverse Scattering from Features on Silicon Wafers*, Weidlinger Associates Phase I Report under NSF SBIR Grant ISI-8760089.

Nosenchuck, D.M., M.G. Littman, and W. Flannery (1986). *Two-Dimensional Nonsteady Viscous Flow Simulation on the Navier-Stokes Computer MiniNode*, Journal of Scientific Computing, Vol. 1, No. 1.

INDEX

Premutation matrix, 80
Poisson equation, 50, 170
Poisson kernels, 103
Polynomial approximation, 51
Preconditioning, 1, 6, 11, 29, 53, 73, 76, 98, 186, 252
Pressure analogy, 128
PRIME, 153
Productivity, 49
Profile storage, 73, 80
Pseudo-compressibility, 209

Q-R algorithm, 8, 12
Quadrature, 50

Radiation boundary conditions, 138
Reduction algorithms, 125
Refraction simulation, 291
Reorthogonalization, 79, 85
Reusable intrinsic sample point concept, 240
Reynolds number, 15, 49
Richardson's extrapolation, 167
Roundoff error, 44, 49

SCS-49, 69
Scaling transformation, 6
Schur complement, 252
Seismogram, 290
Self-adjoint, 51
Sequent, 69
Shared memory, 272
S/MD, 272
Single precision, 69, 158
Singularity, 96
Skyline, 4, 175
Solid state storage device (SSD), 81, 90, 117, 190
Sparsity, 35, 50, 98, 229, 252
Spectral method, 49, 50, 168
Speedup, 273
Static condensation, 5, 6, 229
STELLAR, 69
Stokes flow, 50

Structural/acoustics, 137
Structural/impedance, 139
Subdomain, 2, 252
Submatrix, 4, 5, 68
Subspace augmentation, 40
Subspace iteration, 35, 38, 42, 125, 126
Subspace repetition, 42
Substructure, 35, 37, 50, 56, 125, 229
Sum-factorization methods, 52
Switching network, 273
Synchronization overhead, 274
Synchronization penalty, 258

Task overhead, 274
Taylor's vortex array, 167
Tensor products, 51
Time-stepping, 24, 27, 74, 174
Topological embedding, 49
TRANAIR, 95
Trapezoidal rule, 173
Triangularization, 5, 12
Turbulence model, 213

Unique element method, 187
User interface, 67

Variational principle, 99
Variational scheme, 50
Vax, 153
Vector iteration, 37
Vectorization, 19, 74, 121, 187, 272
Vector parallel processor, 50
Vector processor, 54, 69, 272
Vibration, 36
Virtual parallel processor, 54
Viscous flow, 50
Vorticity, 169

Wake, 100
Wave engine, 285
Wavelength, 97
Wave number, 96
Wave simulation, 285